T0237627

Examples and Problems in Advanced Calculus: Real-Valued Functions

Bijan Davvaz

Examples and Problems in Advanced Calculus: Real-Valued Functions

 Springer

Bijan Davvaz
Department of Mathematics
Yazd University
Yazd, Iran

ISBN 978-981-15-9571-4   ISBN 978-981-15-9569-1   (eBook)
https://doi.org/10.1007/978-981-15-9569-1

This Springer imprint is published by the registered company Springer Nature Singapore Pte Ltd.
The registered company address is: 152 Beach Road, #21-01/04 Gateway East, Singapore 189721,
Singapore

Preface

This book is a result of the experience of teaching general mathematics (calculus) from 1990 to the present time at Yazd University, Iran. I had the idea of writing this book from many years ago but began working on it seriously about three years ago. The book is intended to serve as a textbook of calculus exercises with solutions for calculus courses that are usually taken by first-year students. Indeed, it can be used by all students of mathematics, statistics, computer science, engineering and basic sciences. A total of 501 problems have been selected from 20,000 problems in different books. In choosing problems, three major criteria have been considered, to be challenging, interesting and educational. Therefore, most issues have technical and educational aspects. Over the years, while teaching general mathematics, I have encountered many challenging exercises which have been solved with much time. Most of these exercises have been collected over the last thirty years from different references. Although references that have been used are listed at the end of the book, I do not know the sources and the main creators of most exercises and problems. Therefore, it is my duty to thank everyone who has designed these issues for the first time. The book is based on problems and—regardless of the assumption that readers have prior knowledge of classical contents from other textbooks—the beginning of each chapter includes necessary definitions, concepts and theorems. Moreover, as the reader will soon see, there are many exercises at the end of each chapter. They are divided into two categories: easier and harder. The purpose of these exercises is to allow students to test their assimilation of the material, challenge their mathematical ingenuity, and be a means of developing mathematical insight, intuition and techniques.

The book is composed of seven chapters. Chapter 1 contains problems about sets, numbers and functions. Problems of limits and continuity are presented in Chap. 2. Chapter 3 is devoted to the derivative where we present various beautiful problems related to Rolle's theorem and the mean value theorem. In Chap. 4, we consider optimization problems which often deal with the question: "what is the largest or smallest given some constraints?" in some manner that a function representing a problem can take. Chapters 5 and 6 are devoted to the integral and its applications. Indeed, the study of integration begins in Chap. 5, which contains problems related to the definition and properties of integration and problems related to the methods of integration and improper integrals. In Chap. 6, we investigate how to compute the area, length of a curve, volume of a solid and area of a surface of revolution. Problems related to sequences and series are presented in the last chapter. A study of this chapter should give the reader a thorough understanding of miscellaneous methods about convergence of sequences and series.

Finally, I hope readers enjoy this book and find this issue informative and helpful.

Yazd, Iran   Bijan Davvaz

Contents

About the Author

Bijan Davvaz is Professor at the Department of Mathematics, Yazd University, Iran. He earned his Ph.D. in Mathematics with a thesis entitled "Topics in Algebraic Hyperstructures" from Tarbiat Modarres University, Iran, and completed his M.Sc. in Mathematics from the University of Tehran. Apart from his role of Professor, he also has served as Head of the Department of Mathematics (1998–2002), Chairman of the Faculty of Science (2004–2006) and Vice-President for Research (2006–2008) at Yazd University, Iran. His areas of interest include algebra, algebraic hyperstructures, rough sets and fuzzy logic. A member of editorial boards for 25 mathematical journals, Prof. Davvaz has authored 5 books and over 550 research papers, especially on algebra, fuzzy logic, algebraic hyperstructures and their applications.

Chapter 1
Sets, Numbers and Functions

1.1 Basic Concepts and Theorems

Sets and Set Operations

If S is a set (whose elements may be numbers or any other objects) and x is an *element* of S, then we write $x \in S$. If it so happens that x is not an element of S, then we write $x \notin S$. For a property p and an element x of a set S, we write $p(x)$ to indicate that x has the property p. A set may be defined by a property. The notation $A = \{x \in S : p(x)\}$ indicates that the set A consists of all elements x of S having the property p. We can consider the notion of subset and set operations (union, intersection, complement and difference) of sets as usual, together the fundamental rules governing these operations. Indeed, suppose that A and B are two sets.

(1) A is called a *subset* of B, and we write $A \subseteq B$ if every element of A is an element of B.
(2) The *union* of A and B, denoted by $A \cup B$, is the set of all elements x such that x belongs to at least one of the two sets A or B.
(3) The *intersection* of A and B, denoted by $A \cap B$, is the set of all elements x which belongs to both A and B.
(4) The *complement* of A refers to elements of the universal set not in A and denoted by A^c.
(5) The *difference set* of A and B, denoted by $A - B$, is the set of all elements x which belong to A and not in B.

We can extend the notion of union and intersection to arbitrary collections of sets.

The *empty set* is the set having no elements which will be denoted by the symbol $\emptyset$. A set which contains only finitely many elements is called a *finite set*. A set is *infinite* if it is not finite.

For any three sets A, B and C we have

© The Author(s), under exclusive license to Springer Nature Singapore Pte Ltd. 2020
B. Davvaz, *Examples and Problems in Advanced Calculus: Real-Valued Functions*,
https://doi.org/10.1007/978-981-15-9569-1_1

(1) $A \cup B = B \cup A$,
(2) $A \cap B = B \cap A$,
(3) $A \cup (B \cup C) = (A \cup B) \cup C$,
(4) $A \cap (B \cap C) = (A \cap B) \cap C$,

(5) $A \cup (B \cap C) = (A \cup B) \cap (A \cup C)$,
(6) $A \cap (B \cup C) = (A \cap B) \cup (A \cap C)$,
(7) $(A \cup B)^c = A^c \cap B^c$,
(8) $(A \cap B)^c = A^c \cup B^c$.

The equalities (7) and (8) are called *De Morgan's laws*.

Let A and B be two sets. The set of all *ordered pair* (a, b) with $a \in A$ and $b \in B$ is called the *Cartesian product* of A and B and is denoted by $A \times B$. In symbols

$$A \times B = \{(a, b) : a \in A, \ b \in B\}.$$

For the ordered pair (a, b), a is called the *first coordinate* and b is called the *second coordinate*.

For any three sets A, B and C we have

(1) $A \times (B \cap C) = (A \times B) \cap (A \times C)$,
(2) $A \times (B \cup C) = (A \times B) \cup (A \times C)$,
(3) $A \times (B - C) = (A \times B) - (A \times C)$.

Numbers

The simplest numbers are the positive numbers, 1, 2, 3 and so on, used for counting. These are called *natural numbers*. The set of natural numbers is denoted by $\mathbb{N}$. The *whole numbers* are the natural numbers together with 0. The *integer set* includes whole numbers and negative whole numbers. Integers can be positive, negative or zero. The set of integers is denoted by $\mathbb{Z}$. A rational number is any number that can be expressed as the quotient or fraction $\frac{a}{b}$ of two integers, a numerator a and a non-zero denominator b. The set of all rational numbers is denoted by $\mathbb{Q}$. Numbers that are not rational are called *irrational numbers*. A *real number* is a number that can be found on the number line. These are the numbers that we normally use and apply in real-world applications. The set of all real numbers is denoted by $\mathbb{R}$. Figure 1.1 shows a diagram of the number sets. Four bounded intervals of real line may be used:

$$[a, b] = \{x : a \leq x \leq b\},$$
$$(a, b] = \{x : a < x \leq b\},$$
$$[a, b) = \{x : a \leq x < b\},$$
$$(a, b) = \{x : a < x < b\}.$$

Similarly, unbounded intervals can be defined. Suppose that a real number a and a positive number $\delta > 0$ are given. Then, the interval $(a - \delta, a + \delta)$ is called a *neighborhood* of a. The inequality $0 < |x - a| < \delta$ defines the same neighborhood with missing the midpoint a. This set of numbers is called a *deleted neighborhood* of a.

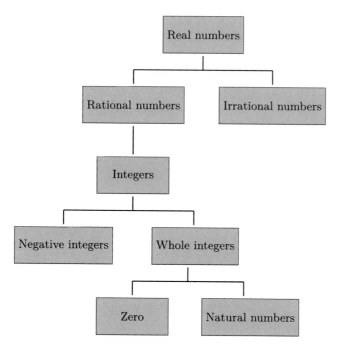

Fig. 1.1 Diagram of number sets

Now, we mention two important properties of real numbers.

(1) If $x \in \mathbb{R}$, $y \in \mathbb{R}$ and $x > 0$, then there exists a positive integer n such that $nx > y$.
(2) If $x \in \mathbb{R}$, $y \in \mathbb{R}$ and $x < y$, then there exists $r \in \mathbb{Q}$ such that $x < r < y$.

Part (1) is usually referred to as the *archimedean property* of $\mathbb{R}$. Part (b) may be stated by saying that $\mathbb{Q}$ is *dense* in $\mathbb{R}$, that is, between any two real numbers is a rational one.

Mathematical Induction

One of the methods of proof that is very useful in providing the validity of a statement $P(n)$ involving the natural number n is mathematical induction.

If $P(n)$ is a statement involving the natural number n such that

(1) $P(1)$ is true,
(2) $P(k)$ implies $P(k + 1)$ for any natural number k,

then $P(n)$ is true for all natural number n.

Fig. 1.2 Rectangular
coordinates of point P

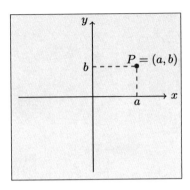

Coordinates in the Plane

Let (a, b) be an ordered pair of real numbers. Plotting the first number a as a point of
the x-axis and the second number b as a point of the y-axis, we erect a perpendicular
to the x-axis at a and a perpendicular to the y-axis at b. These perpendiculars intersect
in a point P, which we regard as representing the ordered pair (a, b) (see Fig. 1.2).
The point P is said to have *rectangular coordinates* a and b. The distance between
two points $P = (a_1, b_1)$ and $Q = (a_2, b_2)$ in the plane is given by formula

$$\overline{PQ} = \sqrt{(a_2 - a_1)^2 + (b_2 - b_1)^2}.$$

Let L be a non-vertical straight line in the plane and $P = (a_1, b_1)$ and $Q = (a_2, b_2)$ be
two distinct points of L. Then, the *slope* of the line L is equal to the ratio $m = \dfrac{b_2 - b_1}{a_2 - a_1}$.
The *inclination angle* of the line L is the smallest angle θ between the positive x-axis
and L. It is not difficult to see that $m = \tan \theta$. Two oblique lines are perpendicular
if and only if the product of their slopes is equal to -1. An equation of the line with
slope m going through a given point $P = (a, b)$ is $y - b = m(x - a)$.

The *distance d* between the point (x_0, y_0) and the line L with equation $ax + by +
c = 0$ is given by

$$d = \frac{|ax_0 + by_0 + c|}{\sqrt{a^2 + b^2}}.$$

The *conic sections* all have second degree equations of the form

$$Ax^2 + Bxy + Cy^2 + Dx + Ey + F = 0.$$

Well-known cases are:

(1) *Circle equation:* $(x - x_0)^2 + (y - y_0)^2 = a^2$.
(2) *Ellipse equation:* $\dfrac{(x - x_0)^2}{a^2} + \dfrac{(y - y_0)^2}{b^2} = 1$.

(3) *Hyperbola equation:* $\dfrac{(x - x_0)^2}{a^2} - \dfrac{(y - y_0)^2}{b^2} = 1.$

(4) *Parabola equation:* $(y - y_0)^2 = a(x - x_0)$ or $(y - y_0) = b(x - x_0)^2.$

Functions

Let A and B be two sets. A *function* f from A into B is a subset of $A \times B$ (and hence is a set of ordered pair (a, b)) with the property that each $a \in A$ belongs to precisely one pair (a, b). In other words no two distinct ordered pairs have the same first coordinate. Instead of $(x, y) \in f$ we usually write $y = f(x)$. Then y is called the *image of x under f*. The set A is called the *domain* of f. The *range* of f is the set

$$\{b \in B \ : \ b = f(a) \text{ for some } a \in A\}.$$

That is, the range of f is the subset of B consisting of all images of elements of A. If f is a function from A into B, we write $f : A \to B$.

For instance, the set $f = \{(x, x^2) \ : \ x \in \mathbb{R}\}$ is the function usually described by the equation

$$f(x) = x^2, \ (x \in \mathbb{R}).$$

The domain of f is the real line, and the range of f is $[0, \infty)$.

It must be emphasized that an equation such as $f(x) = x^4 - 1$ does not define a function until the domain is explicitly specified. Thus, the statements

$$f(x) = x^4 - 1, \ (2 \le x \le 5)$$

and

$$f(x) = x^4 - 1, \ (1 \le x \le 2)$$

define different functions according to our definition.

Note that in the definition of function neither A nor B needs to be a set of numbers. A *real function* is a function whose domain and range are subsets of $\mathbb{R}$. If f is a real function, then the *graph* of f is the set of all pairs (x, y) in $\mathbb{R}^2$ for which (x, y) is an ordered pair in f.

A *constant function* is a function whose range consists of a single element. A classic example is the *absolute value function* , that is,

$$f(x) = |x| = \begin{cases} x & \text{if } x \ge 0 \\ -x & \text{if } x < 0. \end{cases}$$

Triangle inequality: The inequality

$$|a + b| \le |a| + |b|$$

holds, for all $a, b \in \mathbb{R}$.

The symbol $[x]$ is defined as the greatest integer less than or equal to x, that is,

$$[x] = n \text{ if } n \leq x < n+1,$$

where n is an integer. This function is called the *greatest integer function*. We can define some operations on functions. New functions are formed from given functions by adding, subtracting, multiplying and dividing function values. These functions are known as the *sum*, *difference*, *product* and *quotient* of the original functions. More concisely, let f and g be two functions and D the largest set on which f and g are both defined. Then, we define

$$(f+g)(x) = f(x) + g(x),$$
$$(f-g)(x) = f(x) - g(x),$$
$$(fg)(x) = f(x) \cdot g(x),$$
$$\left(\frac{f}{g}\right)(x) = \frac{f(x)}{g(x)} \quad (g(x) \neq 0),$$

for all $x \in D$.

Let $f : A \to B$ and $g : B \to C$ be two functions. Then, we define a new function $g \circ f$, the *composition* of f and g by $g \circ f : A \to C$ whose value at x is given by

$$(g \circ f)(x) = g(f(x)),$$

and the domain of $g \circ f$ is the set of all elements x in the domain of f such that $f(x)$ is in the domain of g; see Fig. 1.3.

Shifting the graph of a function: Given a function

$$y = f(x),$$

with graph G, the graph of the function

$$y = f(x - c)$$

Fig. 1.3 Composition of two functions

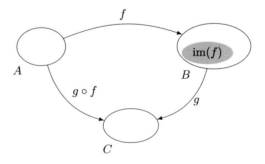

is the result of subjecting G to a *horizontal shift* of $|c|$ units, to the right if $c > 0$ and to the left if $c < 0$. Similarly, the graph of the function

$$y = f(x) + c$$

is the result of subjecting G to a *vertical shift* of $|c|$ units, upward if $c > 0$ and downward if $c < 0$.

A function $f : X \to Y$ is said to be

(1) *(strictly) increasing* if $x_1 < x_2$ implies $f(x_1) < f(x_2)$,
(2) *(strictly) decreasing* if $x_1 < x_2$ implies $f(x_1) > f(x_2)$,
(3) *one to one* if $f(x_1) = f(x_2)$ implies $x_1 = x_2$,
(4) *onto* provided that $y \in Y$, then there exists at least $x \in X$ such that $y = f(x)$,
(5) *bijective* if it is one to one and onto. A bijective is also called a *one-to-one correspondence*.

Two sets are said to have the *same cardinality* if there exists a one-to-one correspondence between them. An infinite set is said to be *countably infinite (enumerable)* if it has the same cardinality as the set of positive integers. An infinite set is *uncountably infinite (non-denumerable)* if it is infinite but not countably infinite.

Let f be a one-to-one function. The *inverse* of f, denoted by f^{-1}, is the unique function with the domain equal to the range of f that satisfies

$$f\left(f^{-1}(y)\right) = y, \text{ for all } y \text{ in the range of } f,$$
$$f^{-1}\left(f(x)\right) = x, \text{ for all } x \text{ in the domain of } f.$$

So, f^{-1} is the set of ordered pairs (y, x) defined by $x = f^{-1}(y)$ if and only if $y = f(x)$. The *elementary functions* include:

(1) *Powers of x*: x, x^2, x^3, etc.
(2) *Roots of x*: $\sqrt{x}, \sqrt[3]{x}$, etc.
(3) *Trigonometric functions*: $\sin x$, $\cos x$, $\tan x$, $\cot x$, $\sec x$ and $\csc x$.
(4) *Inverse trigonometric functions*: $\sin^{-1} x$, $\cos^{-1} x$, $\tan^{-1} x$, $\cot^{-1} x$, $\sec^{-1} x$ and $\csc^{-1} x$.
(5) *Logarithms and exponential functions* : $\ln x$, $\log_a x$, e^x and a^x.
(6) *Hyperbolic functions*: $\sinh x$, $\cosh x$, $\tanh x$, $\coth x$, $\operatorname{sech} x$ and $\operatorname{csch} x$.
(7) *Inverse hyperbolic functions*: $\sinh^{-1} x$, $\cosh^{-1} x$, $\tanh^{-1} x$, $\coth^{-1} x$, $\operatorname{sech}^{-1} x$ and $\operatorname{csch}^{-1} x$.
(8) All functions obtained by replacing x with any of the previous functions.
(9) All functions obtained by adding, subtracting, multiplying or dividing any of the previous functions.

There exist many identities between elementary functions. In the following we present some of them.

(1) $\sin(a + b) = \sin a \cos b + \cos a \sin b$,
(2) $\sin(a - b) = \sin a \cos b - \cos a \sin b$,

(3) $\cos(a + b) = \cos a \cos b - \sin a \sin b,$

(4) $\cos(a - b) = \cos a \cos b + \sin a \sin b,$

(5) $\sin a + \sin b = 2 \sin \left(\dfrac{a + b}{2} \right) \cos \left(\dfrac{a - b}{2} \right),$

(6) $\sin a - \sin b = 2 \sin \left(\dfrac{a - b}{2} \right) \cos \left(\dfrac{a + b}{2} \right),$

(7) $\cos a + \cos b = 2 \cos \left(\dfrac{a + b}{2} \right) \cos \left(\dfrac{a - b}{2} \right),$

(8) $\cos a - \cos b = -2 \sin \left(\dfrac{a + b}{2} \right) \sin \left(\dfrac{a - b}{2} \right),$

(9) $\sinh a = \dfrac{e^x - e^{-x}}{2},$

(10) $\cosh a = \dfrac{e^x + e^{-x}}{2},$

(11) $\log_a xy = \log_a x + \log_a y,$

(12) $\log_a \dfrac{x}{y} = \log_a x - \log_a y,$

(13) $\log_a x^r = r \log_a x,$

(14) $\log_a x = \dfrac{\log_b x}{\log_b a},$

(15) $\log_a x = \dfrac{\ln x}{\ln a},$

(16) $\log_e x = \ln x.$

A function f is said to be *bounded* if there exists a real number M such that $|f(x)| \leq M$ for all x. A function f is an *even function* if its domain contains the point $-x$ whenever it contains x, and if $f(-x) = f(x)$ for all x in the domain of f. A function f is an *odd function* if its domain contains the point $-x$ whenever it contains x, and if $f(-x) = -f(x)$ for all x in the domain of f. A function f is said to be *periodic* with period T if the domain of f contains $x + T$ whenever it contains x, and if $f(x + T) = f(x)$, for all x.

Let $f : X \to Y$ be a function and let $A \subseteq X$ and $B \subseteq Y$. The *image* of A under f, which we denote by $f(A)$, is $f(A) = \{f(x) : x \in A\}$. The *inverse image* of B under f, which we denote by $f^{-1}(B)$, is $f^{-1}(B) = \{x \in X : f(x) \in B\}$.

If the reader is interested to see the proof of theorems that are presented in this chapter, we refer him/her to [1–16].

1.2 Problems

1. *Given any numbers $a_0, a_1, \ldots, a_n$ and $b_0, b_1, \ldots, b_n$, show that*

$$\sum_{i=1}^{n} (a_i - a_{i-1})b_{i-1} = (a_n b_n - a_0 b_0) - \sum_{i=1}^{n} a_i (b_i - b_{i-1}).$$

This result is known as the formula for summation by parts.

Solution. We have

$$\sum_{i=1}^{n}(a_i - a_{i-1})b_{i-1}$$

$$= (a_1 - a_0)b_0 + (a_2 - a_1)b_1 + \cdots + (a_n - a_{n-1})b_{n-1}$$
$$= a_1b_0 - a_0b_0 + a_2b_1 - a_1b_1 + \cdots + a_nb_{n-1} - a_{n-1}b_{n-1}$$
$$= a_1b_0 - a_0b_0 + a_2b_1 - a_1b_1 + \cdots + a_nb_{n-1} - a_{n-1}b_{n-1} + (a_nb_n - a_nb_n)$$
$$= (a_nb_n - a_0b_0) - (a_1b_1 - a_1b_0) - (a_2b_2 - a_2b_1) - \cdots - (a_nb_n - a_nb_{n-1})$$
$$= (a_nb_n - a_0b_0) - \Big((a_1b_1 - a_1b_0) + (a_2b_2 - a_2b_1) + \cdots + (a_nb_n - a_nb_{n-1})\Big)$$
$$= (a_nb_n - a_0b_0) - \sum_{i=1}^{n} a_i(b_i - b_{i-1}).$$

2. *Find the set of solution of the inequality*

$$|x|^3 + 5x^2 - 2|x| - 24 < 0.$$

Solution. We observe that $|x| = 2$ is a root of the equation $|x|^3 + 5x^2 - 2|x| - 24 = 0$. Hence, the inequality can be written as follows:

$$(|x| - 2)(|x|^2 + 7|x| + 12) < 0,$$

and so we conclude that

$$(|x| - 2)(|x| + 3)(|x| + 4) < 0.$$

Since $|x| + 3$ and $|x| + 4$ are positive, it follows that $|x| - 2 < 0$, or equivalently $|x| < 2$. Hence, the solution set is the open interval $(-2, 2)$.

3. *Let a and b be two real numbers. Show that*

$$\frac{|a + b|}{1 + |a + b|} \le \frac{|a|}{1 + |a|} + \frac{|b|}{1 + |b|}.$$

Solution. We consider the following cases:

(1) $|a + b| \le |a|$,
(2) $|a + b| \le |b|$,
(3) $|a| \le |a + b|$ and $|b| \le |a + b|$.

If $|a + b| \le |a|$, then $|a + b| + |a| \cdot |a + b| \le |a| + |a| \cdot |a + b|$. This implies that $|a + b| \cdot (1 + |a|) \le |a| \cdot (1 + |a + b|)$. Consequently, we get

$$\frac{|a + b|}{1 + |a + b|} \le \frac{|a|}{1 + |a|} \le \frac{|a|}{1 + |a|} + \frac{|b|}{1 + |b|}.$$

If $|a + b| \leq |b|$, the proof is similar to the previous case.

Now, let $|a| \leq |a + b|$ and $|b| \leq |a + b|$. Since $|a + b| \leq |a| + |b|$, it follows that

$$\frac{|a + b|}{1 + |a + b|} \leq \frac{|a|}{1 + |a + b|} + \frac{|b|}{1 + |a + b|}$$
$$\leq \frac{|a|}{1 + |a|} + \frac{|b|}{1 + |b|}.$$

4. *Let $p, q, r, s \in \mathbb{R}$, $s > 0$, $q > 0$ and $\dfrac{p}{q} < \dfrac{r}{s}$. Show that*

$$\frac{p}{q} < \frac{p + r}{q + s} < \frac{r}{s}.$$

Solution. Since $sq > 0$, it follows that

$$sq\frac{p}{q} < sq\frac{r}{s},$$

which implies that $sp < qr$ or $qr - sp > 0$. Since $s, q > 0$, it follows that $s(q + s) > 0$ and $q(q + s) > 0$. So,

$$\frac{qr - sp}{s(q + s)} > 0 \text{ and } \frac{qr - sp}{q(q + s)} > 0.$$

Therefore, we have

$$\frac{p}{q} < \frac{p}{q} + \frac{qr - sp}{q(q + s)} = \frac{p(q + s) + qr - sp}{q(q + s)} = \frac{p + r}{q + s} \tag{1.1}$$

and

$$\frac{r}{s} > \frac{r}{s} - \frac{qr - sp}{s(q + s)} = \frac{r(q + s) - qr + sp}{s(q + s)} = \frac{p + r}{q + s}. \tag{1.2}$$

Now, from (1.1) and (1.2), the proof completes.

5. *If $n > 1$ is an integer, prove that*

$$0 < \sqrt[n]{x} - 1 < \frac{x - 1}{n}, \quad for\ all\ x > 1.$$

Solution. Since $1 < x$, it follows that $1 < \sqrt[n]{x}$. So,

$$0 < \sqrt[n]{x} - 1. \tag{1.3}$$

On the other hand, for each two real numbers a and b, we know that

$$a^n - b^n = (a - b)(a^{n-1} + a^{n-2}b + \cdots + b^{n-1}). \tag{1.4}$$

In (1.4), if we substitute $\sqrt[n]{x}$ and 1 instead of a and b, then we get

$$x - 1 = (\sqrt[n]{x} - 1)\left(x^{\frac{n-1}{n}} + x^{\frac{n-2}{n}} + \cdots + x^{\frac{1}{n}} + 1\right).$$

Since $x > 1$, it follows that

$$x - 1 > (\sqrt[n]{x} - 1)\underbrace{(1 + 1 + \cdots + 1)}_{n \text{ times}} = (\sqrt[n]{x} - 1)n.$$

Therefore,

$$\sqrt[n]{x} - 1 < \frac{x - 1}{n}. \tag{1.5}$$

By (1.3) and (1.5), the proof completes.

6. *For every natural number n, prove that*

$$\sum_{k=1}^{n} k^2 = \frac{n(n + 1)(2n + 1)}{6}. \tag{1.6}$$

Solution. We prove the validity of (1.6) by mathematical induction. First, the equality is clearly true for $n = 1$. To complete the proof, assume (1.6) holds for n and we prove this equality for $n + 1$. We have

$$\begin{aligned}
\sum_{k=1}^{n+1} k^2 &= \sum_{k=1}^{n} k^2 + (n + 1)^2 \\
&= \frac{n(n + 1)(2n + 1)}{6} + (n + 1)^2 \\
&= \frac{(n + 1)(n + 2)(2(n + 1) + 1)}{6}.
\end{aligned}$$

Thus, by mathematical induction, (1.6) is true for all natural numbers.

7. *Prove that*

$$\prod_{k=2}^{n}\left(1 - \frac{1}{k^2}\right) = \frac{n + 1}{2n}.$$

Solution. We apply mathematical induction. The equality is true for $n = 2$. Suppose that the equality is true for $n = m$. Then, we have

$$\prod_{k=2}^{m}\left(1 - \frac{1}{k^2}\right) = \frac{m + 1}{2m}.$$

Consequently, for $n = m + 1$, we obtain

$$\prod_{k=2}^{m+1} \left(1 - \frac{1}{k^2}\right) = \prod_{k=2}^{m} \left(1 - \frac{1}{k^2}\right) \cdot \left(1 - \frac{1}{(m+1)^2}\right)$$
$$= \frac{m+1}{2m} \cdot \left(1 - \frac{1}{(m+1)^2}\right)$$
$$= \frac{(m+1)\left((m+1)^2 - 1\right)}{2m(m+1)^2}$$
$$= \frac{m^2 + 2m}{2m(m+1)}$$
$$= \frac{m+2}{2(m+1)}.$$

8. *Let a and b be two real numbers and n be a natural number. Prove that*

$$(a+b)^n = \binom{n}{0}a^n + \binom{n}{1}a^{n-1}b + \cdots + \binom{n}{n-1}ab^{n-1} + \binom{n}{n}b^n, \qquad (1.7)$$

where $\binom{n}{m} = \dfrac{n!}{m!(n-m)!}$, *for* $0 \le m \le n$.

Solution. For the proof, we use mathematical induction. Obviously, (1.7) is true for $n = 1$. Suppose that (1.7) is true for $n = k$, i.e.,

$$(a+b)^k = \binom{k}{0}a^k + \binom{k}{1}a^{k-1}b + \cdots + \binom{k}{k-1}ab^{k-1} + \binom{k}{k}b^k. \qquad (1.8)$$

We multiply both sides of (1.8) by $a + b$. Since

$$\binom{k+1}{m} = \binom{k}{m-1} + \binom{k}{m},$$

we get

$$(a+b)^{k+1}$$
$$= a^{k+1} + \left(\binom{k}{0} + \binom{k}{1}\right)a^k b + \cdots + \left(\binom{k}{k-1} + \binom{k}{k}\right)ab^k + b^{k+1}$$
$$= a^{k+1} + \binom{k+1}{1}a^k b + \cdots + \binom{k+1}{k}ab^k + b^{k+1}.$$

Consequently, (1.7) is true for $n = k + 1$.

9. *Let* $m = 2^n$ *and* $a_1, a_2, \ldots, a_m$ *be m positive real numbers. Show that*

$$\sqrt[m]{a_1 a_2 \ldots a_m} \leq \frac{a_1 + a_2 + \cdots + a_m}{m} \tag{1.9}$$

and the equality holds if and only if $a_1 = \ldots = a_m$.

Solution. We do the proof by mathematical induction. Let $n = 1$, i.e., $m = 2$. We have

$$(a_1 - a_2)^2 \geq 0 \Leftrightarrow a_1 a_2 \leq \frac{a_1^2 + a_2^2}{2} \Leftrightarrow a_1 a_2 \leq \frac{a_1^2 + a_2^2}{4} + \frac{a_1 a_2}{2}$$
$$\Leftrightarrow a_1 a_2 \leq \frac{(a_1 + a_2)^2}{4} \Leftrightarrow \sqrt{a_1 a_2} \leq \frac{a_1 + a_2}{2}.$$

Moreover,

$$\sqrt{a_1 a_2} = \frac{a_1 + a_2}{2} \Leftrightarrow (a_1 - a_2)^2 = 0 \Leftrightarrow a_1 = a_2.$$

So, we are done with the initial step. Now, we consider the inductive step. Suppose that the inequality (1.9) holds for $n = k$, i.e., for $m = 2^k$. We prove the inequality for $n = k + 1$. Hence, $2^{k+1} = 2m$. Indeed, we must prove (1.9) for $2m$. Consider

$$b_1 = \frac{a_1 + a_2}{2}, \quad b_2 = \frac{a_3 + a_4}{2}, \ldots, \quad b_m = \frac{a_{2m-1} + a_{2m}}{2}.$$

Based on our assumption, $b_1, \ldots, b_m$ must satisfy (1.9). Thus,

$$\left(\frac{a_1 + a_2}{2} \cdot \ldots \cdot \frac{a_{2m-1} + a_{2m}}{2} \right)^{\frac{1}{m}} \leq \frac{a_1 + a_2 + \cdots + a_{2m-1} + a_{2m}}{2m}, \tag{1.10}$$

Consequently,

$$\left(\sqrt{a_1 a_2} \cdot \ldots \cdot \sqrt{a_{2m-1} a_{2m}} \right)^{\frac{1}{m}} \leq \frac{a_1 + a_2 + \cdots + a_{2m-1} + a_{2m}}{2m}, \tag{1.11}$$

which implies that

$$(a_1 a_2 \ldots a_{2m-1} a_{2m})^{\frac{1}{2m}} \leq \frac{a_1 + a_2 + \cdots + a_{2m-1} + a_{2m}}{2m}. \tag{1.12}$$

Now, if the equality holds in (1.12), then the equality holds in (1.11) and (1.10), and so by induction assumption we get

$$\frac{a_1 + a_2}{2} = \frac{a_3 + a_4}{2} = \ldots = \frac{a_{2m-1} + a_{2m}}{2}$$

or

$$a_1 + a_2 = a_3 + a_4 = \ldots = a_{2m-1} + a_{2m}. \tag{1.13}$$

Since the equalities in (1.13) hold for every permutation of $a_1, a_2, \ldots, a_{2m}$, it follows that $a_1 = a_2 = \ldots = a_{2m-1} = a_{2m}$.

10. *Show that in Problem (9), the inequality (1.9) between the geometric mean and the arithmetic mean holds even m is not a power of 2.*

Solution. Clearly, there is $n \in \mathbb{N}$ such that $2^{n-1} < m < 2^n = k$. Suppose that

$$b_j = \begin{cases} a_j & \text{for } j = 1, \ldots, m \\ b = \dfrac{a_1 + a_2 + \cdots + a_m}{m} & \text{for } j = m+1, \ldots, k. \end{cases}$$

Now, we apply Problem (9) to numbers $b_1, b_2, \ldots, b_k$. Then,

$$(b_1 \cdot \ldots \cdot b_k)^{\frac{1}{k}} \leq \frac{b_1 + \cdots + b_k}{k},$$

which implies that

$$(b_1 \cdot \ldots \cdot b_m)^{\frac{1}{k}} \cdot b^{\frac{k-m}{k}} \leq \frac{b_1 + \cdots + b_m + (k-m)b}{k}.$$

Consequently, we get

$$(b_1 \cdot \ldots \cdot b_m)^{\frac{1}{k}} \cdot b^{\frac{k-m}{k}} \leq b \text{ or } (b_1 \cdot \ldots \cdot b_m)^{\frac{1}{k}} \leq b^{\frac{m}{k}}.$$

Finally, we obtain

$$(b_1 \cdot \ldots \cdot b_m)^{\frac{1}{m}} \leq \frac{b_1 + \cdots + b_m}{m},$$

the inequality of arithmetic and geometric means.

11. *For each $n \in \mathbb{N}$, prove that*

$$n! \leq \left(\frac{n+1}{2}\right)^n.$$

Solution. According to the inequality of arithmetic and geometric means, Problem (10), we have

$$\sqrt[n]{n!} = \sqrt[n]{1 \cdot 2 \ldots n} \leq \frac{1 + 2 + \cdots + n}{n} = \frac{n+1}{2}.$$

Now, let us raise sides of the power of n, and we get the result.

12. *Let a and b denote two real numbers. Prove that if $0 \leq a - b < \epsilon$ for every positive real number ϵ, then $a = b$.*

Solution. If $a > b$, then $a - b > 0$. Take $\epsilon = \dfrac{a-b}{2}$, then $a - b < \dfrac{a-b}{2}$, a contradiction.

13. *Show that $\sqrt{12}$ is not a rational number.*

Solution. Suppose that $\sqrt{12} = \dfrac{p}{q}$, where p and q are positive integers with no common factors. Then, $12 = \dfrac{p^2}{q^2}$ or $p^2 = 12q^2 = 3(2q)^2$, which implies that p is a multiple of 3. Let $p = 3k$, for some integer k. So, $p^2 = 9k^2 = 3(2q)^2$. Consequently, $3k^2 = (2q)^2$. This implies that q is a multiple of 3. Thus, 3 is a common factor of p and q, and this is a contradiction.

14. *Let $n \in \mathbb{N}$. If n is not the square of a natural number, prove that $\sqrt{n}$ is irrational.*

Solution. Suppose that $\sqrt{n}$ is a rational number and

$$\sqrt{n} = \frac{a}{b},$$

where $a, b \in \mathbb{N}$ and b is the smallest positive integer denominator for which this is true. Then, we have

$$n = \frac{a^2}{b^2} \text{ or } a^2 = nb^2.$$

Since $n > 1$, it follows that $b^2 < nb^2 = a^2$. Consequently, $0 < b < a$. Now, if we divide a by b, then we obtain quotient q and remainder r such that

$$a = qb + r \text{ and } 0 \leq r < b.$$

If $r = 0$, then $a = qb$. Hence, $\dfrac{a}{b} = q$, and so $n = q^2$. This is a contradiction.
If $r \neq 0$, then $0 < r = a - qb < b$. Now, we have

$$nb^2 = a^2 \Rightarrow nb^2 - qab = a^2 - qab$$
$$\Rightarrow b(nb - qa) = a(a - qb).$$

Hence, we obtain

$$\sqrt{n} = \frac{a}{b} = \frac{nb - qa}{a - qb},$$

and this is a contradiction with minimality of b.

15. *If $n \in \mathbb{N}$, prove that $\sqrt{n(n+1)}$ is irrational.*

Solution. By Problem (14), it is enough to show that $n(n+1)$ cannot be square of a natural number.

Suppose that $n(n+1) = k^2$, for some $k \in \mathbb{N}$. Then, $k^2 - n^2 = n$ or $(k-n)(k+n) = n$. But $n + k$ is a positive integer greater than n and so $k - n$ is a positive integer. Now, we have $n = (k-n)(k+n) \geq k + n > n$. This is a contradiction.

16. *If $n \in \mathbb{N}$ and $n \neq 1$, prove that*

(1) $\sqrt{n^2 - 1}$ is irrational;
(2) $\sqrt{n - 1} + \sqrt{n + 1}$ is irrational.

Solution. (1) According to Problem (14), we show that $n^2 - 1$ cannot be square. Let $n^2 - 1 = k^2$, for some $k \in \mathbb{N}$. Then, $n^2 - k^2 = (n - k)(n + k) = 1$, a contradiction.

(2) Suppose that $\sqrt{n - 1} + \sqrt{n + 1} = r$ is rational. Then, we get $n - 1 + n + 1 + 2\sqrt{n^2 - 1} = r^2$, which implies that

$$\sqrt{n^2 - 1} = \frac{r^2 - 2n}{2}.$$

Since $\dfrac{r^2 - 2n}{2} \in \mathbb{Q}$, it follows that $\sqrt{n^2 - 1} \in \mathbb{Q}$. This is a contradiction with part (1).

17. *If a and b are any two real numbers, where $a < b$, prove that there exists an irrational number c such that $a < c < b$.*

Solution. Since $\dfrac{a}{\sqrt{2}} < \dfrac{b}{\sqrt{2}}$, by the rational density theorem there is $r \in \mathbb{Q}$ such that $\dfrac{a}{\sqrt{2}} < r < \dfrac{b}{\sqrt{2}}$. So, $a < r\sqrt{2} < b$. We know that $\sqrt{2}$ is an irrational number. Hence, $r\sqrt{2}$ is an irrational number. Now, it is enough to take $c = r\sqrt{2}$.

18. *Let a denote a positive real number. Prove that there exists a natural number n such that $\dfrac{1}{10^n} < a$.*

Solution. Since $a > 0$, it follows that there exists $n \in \mathbb{N}$ such that $\dfrac{1}{n} < a$. Clearly, $10^n > n$, so $\dfrac{1}{10^n} < \dfrac{1}{n}$. Thus, we have $\dfrac{1}{10^n} < a$.

19. *Prove that for any two real numbers a and b, where $a < b$, there exists an integer k and a natural number n such that $a < \dfrac{k}{10^n} < b$.*

Solution. We have $b - a > 0$. By Problem (18), there is $n \in \mathbb{N}$ such that $\dfrac{1}{10^n} < b - a$. On the other hand, there exists $k \in \mathbb{Z}$ such that $k \leq 10^n b < k + 1$, which implies that $\dfrac{k}{10^n} \leq b < \dfrac{k + 1}{10^n}$. Hence, $0 \leq b - \dfrac{k}{10^n} < \dfrac{1}{10^n}$. So, we conclude that $0 \leq b - \dfrac{k}{10^n} < b - a$. This implies that $a < \dfrac{k}{10^n} \leq b$. If $b = \dfrac{k}{10^n}$, it is enough one replace b by another real number b' such that $a < b' < b$.

20. *If a and b are two real numbers and $a < b$, show that*

$$(a, b) = \{x \in \mathbb{R} : \exists \lambda, \ 0 < \lambda < 1, \ x = \lambda a + (1 - \lambda)b\}.$$

Solution. Clearly, for each $0 < \lambda < 1$ we have $a < \lambda a + (1 - \lambda)b < b$. Now, let $x \in (a, b)$ be arbitrary. Then, $0 < x - a < b - a$, and so $0 < \dfrac{x - a}{b - a} < 1$. If $\alpha = \dfrac{x - a}{b - a}$, then $x - a = \alpha(b - a)$ or $x = a + \alpha b - \alpha a$. Hence, we have $x = (1 - \alpha)a + \alpha b$. Now, take $\lambda = 1 - \alpha$.

21. *If $A_n = (-\dfrac{1}{n}, \dfrac{1}{n})$ for all $n \in \mathbb{N}$, show that*

$$\bigcap_{n=1}^{\infty} A_n = \{0\}.$$

Solution. Obviously, $0 \in \bigcap_{n=1}^{\infty} A_n$. So, let $x \neq 0$ and $x \in \bigcap_{n=1}^{\infty} A_n$.

If $x > 0$, then there exists $n_0 \in \mathbb{N}$ such that $\dfrac{1}{n_0} < x$. This implies that $x \notin A_{n_0}$, a contradiction.

If $x < 0$, then there exists $m_0 \in \mathbb{N}$ such that $\dfrac{1}{m_0} < -x$. So, $x < -\dfrac{1}{m_0}$. This implies that $x \notin A_{m_0}$, a contradiction.

22. *If $A_n = [\dfrac{1}{n}, 1 - \dfrac{1}{n}]$ for all $n \geq 3$, show that*

$$\bigcup_{n=3}^{\infty} A_n = (0, 1).$$

Solution. Clearly, $\bigcup_{n=3}^{\infty} A_n \subseteq (0, 1)$. For the converse inclusion, let $x \in (0, 1)$ be arbitrary. Then, there is $n \in \mathbb{N}$ such that $\dfrac{1}{n} < x$. Since $1 - x > 0$, there is $m \in \mathbb{N}$ such that $\dfrac{1}{m} < 1 - x$ or $x < 1 - \dfrac{1}{m}$.

If $m = n$, then $\dfrac{1}{n} < x < 1 - \dfrac{1}{n}$, and so $x \in A_n$.

If $m > n$, then we have $\dfrac{1}{m} < \dfrac{1}{n} < x$. Since $x < 1 - \dfrac{1}{m}$, it follows that $\dfrac{1}{m} < x < 1 - \dfrac{1}{m}$. Thus, $x \in A_m$.

If $m < n$, then $-\dfrac{1}{m} < -\dfrac{1}{n}$ or $1 - \dfrac{1}{m} < 1 - \dfrac{1}{n}$. Since $x < 1 - \dfrac{1}{m}$, it follows that $x < 1 - \dfrac{1}{n}$. So, we have $\dfrac{1}{n} < x < 1 - \dfrac{1}{n}$. This implies that $x \in A_n$.

Therefore, for each $x \in (0, 1)$, there exists $n \in \mathbb{N}$ such that $x \in A_n$.

23. *Prove that*

$$\frac{1}{2} \cdot \frac{3}{4} \cdot \frac{5}{6} \cdots \frac{2n-1}{2n} < \frac{1}{\sqrt{2n}},$$

for all $n \in \mathbb{N}$.

Solution. Suppose that

$$f(n) = \frac{1}{2} \cdot \frac{3}{4} \cdot \frac{5}{6} \cdots \frac{2n-1}{2n},$$

$$g(n) = \frac{2}{3} \cdot \frac{4}{5} \cdot \frac{6}{7} \cdots \frac{2n}{2n+1}.$$

The, we have

$$f(n)^2 \le f(n) \cdot g(n) < \frac{1}{2n+1} < \frac{1}{2n}.$$

This implies that $f(n) < \dfrac{1}{\sqrt{2n}}$.

24. *If*

$$f(x) = \sqrt{1 - \sqrt{3 - \sqrt{2 - x}}}.$$

find D_f, the domain of f

Solution. We have

$$
\begin{aligned}
D_f &= \{x \ : \ 2 - x \ge 0, \ 3 - \sqrt{2-x} \ge 0 \text{ and } 1 - \sqrt{3 - \sqrt{2-x}} \ge 0\} \\
&= \{x \ : \ 2 \ge x, \ 3 \ge \sqrt{2-x} \text{ and } 1 \ge \sqrt{3 - \sqrt{2-x}}\} \\
&= \{x \ : \ 2 \ge x, \ 9 \ge 2 - x \text{ and } 1 \ge 3 - \sqrt{2-x}\} \\
&= \{x \ : \ 2 \ge x, \ x \ge -7 \text{ and } \sqrt{2-x} \ge 2\} \\
&= \{x \ : \ 2 \ge x, \ x \ge -7 \text{ and } 2 - x \ge 4\} \\
&= \{x \ : \ 2 \ge x, \ x \ge -7 \text{ and } -2 \ge x\} \\
&= [-7, -2].
\end{aligned}
$$

25. *Let f be a real function. Prove that if*

(1) $f(0) = 5$,

(2) $f(xy + 1) = f(x)f(y) - 2f(x) - y + 5$, for all $x, y \in \mathbb{R}$,

then $f(x) = \dfrac{10 + x}{2}$.

Solution. By (2), we have $f(yx + 1) = f(y)f(x) - 2f(y) - x + 5$. Thus, we conclude that

$$f(x)f(y) - 2f(x) - y + 5 = f(y)f(x) - 2f(y) - x + 5,$$

and hence $2f(x) + y = 2f(y) + x$, for all $x, y \in \mathbb{R}$. Afterward, if $y = 0$, then $2f(x) = 2f(0) + x$. Since $f(0) = 5$, it follows that $f(x) = \dfrac{10 + x}{2}$.

26. *If $x \in [0, 1]$, define*

$$f(x) = \begin{cases} x & if \ x \in \mathbb{Q} \\ 1 - x & if \ x \notin \mathbb{Q}. \end{cases}$$

Prove that

(1) $f(f(x)) = x$, *for all $x \in [0, 1]$,*
(2) $f(x) + f(1 - x) = 1$, *for all $x \in [0, 1]$,*
(3) $f(x + y) - f(x) - f(y)$ *is a rational number, for all $x, y \in [0, 1]$.*

Solution. (1) If $x \in \mathbb{Q}$, then $f(f(x)) = f(x) = x$; if $x \notin \mathbb{Q}$, then $1 - x \notin \mathbb{Q}$, which implies that $f(f(x)) = f(1 - x) = 1 - (1 - x) = x$.

(2) If $x \in \mathbb{Q}$, then $f(x) + f(1 - x) = x + 1 - x = 1$. If $x \notin \mathbb{Q}$, then $f(x) + f(1 - x) = 1 - x + 1 - (1 - x) = 1$.

(3) We consider three cases:

(i) If $x, y \in \mathbb{Q}$, then $f(x + y) - f(x) - f(y) = x + y - x - y = 0 \in \mathbb{Q}$.
(ii) If $x \in \mathbb{Q}$ and $y \notin \mathbb{Q}$, then $f(x + y) - f(x) - f(y) = 1 - (x + y) - x - (1 - y) = 1 - x - y - x - 1 + y = -2x \in \mathbb{Q}$.
(iii) Suppose that $x, y \notin \mathbb{Q}$. Then, $x + y$ can be rational or irrational. If $x + y$ is irrational, then $f(x + y) - f(x) - f(y) = 1 - (x + y) - 1 + x - 1 + y = -1 \in \mathbb{Q}$. If $x + y$ is rational, then $f(x + y) - f(x) - f(y) = x + y - 1 + x - 1 + y = 2(x + y - 1) \in \mathbb{Q}$.

27. *Is the function $f : \mathbb{N} \to \mathbb{R}$ defined by*

$$f(n) = \left(1 + \frac{1}{n}\right)^n$$

monotonic?

Solution. We show that f is increasing. Using the inequality of arithmetic and geometric means, Problem (9), we obtain

$$\left(1 + \frac{1}{n}\right)^n = 1 \cdot \underbrace{\left(1 + \frac{1}{n}\right) \cdots \left(1 + \frac{1}{n}\right)}_{n \text{ times}} \leq \left(\frac{1 + \left(1 + \frac{1}{n}\right) + \cdots + \left(1 + \frac{1}{n}\right)}{n + 1}\right)^{n+1}$$

$$= \left(\frac{1 + n\left(1 + \frac{1}{n}\right)}{n + 1}\right)^{n+1}$$

$$= \left(\frac{1}{n + 1} + \frac{n + 1}{n + 1}\right)^{n+1}$$

$$= \left(1 + \frac{1}{n + 1}\right)^{n+1}.$$

So, $f(n) \leq f(n + 1)$, for all $n \in \mathbb{N}$.

28. *Is the following function one to one on* $[-1, 1]$*?*

$$f(x) = \frac{x}{x^2 + 1}.$$

Solution. Suppose that $x_1 \neq x_2$ such that $f(x_1) = f(x_2)$. Then,

$$\frac{x_1}{x_1^2 + 1} = \frac{x_2}{x_2^2 + 1},$$

or equivalently, $x_1(x_2^2 + 1) = x_2(x_1^2 + 1)$. This implies that $x_1 - x_2 = (x_1 - x_2)x_1 x_2$. Since $x_1 - x_2 \neq 0$, it follows that $1 = x_1 x_2$. Since $x_1, x_2 \in [-1, 1]$, it follows that $x_1 = x_2 = 1$ or $x_1 = x_2 = -1$, and it is a contradiction.

29. *Let* f *be a real function and* A*,* B *be two subsets of domain of* f*. If* f *is one to one, show that* $f(A \cap B) = f(A) \cap f(B)$*.*

Solution. Since $A \cap B \subseteq A$ and $A \cap B \subseteq B$, we have $f(A \cap B) \subseteq f(A)$ and $f(A \cap B) \subseteq f(B)$. So, $f(A \cap B) \subseteq f(A) \cap f(B)$. For the converse inclusion, let $x \in f(A) \cap f(B)$. Then, $x \in f(A)$ and $x \in f(B)$. So, there exist $a \in A$ and $b \in B$ such that $x = f(a)$ and $x = f(b)$. Since f is one to one, it follows that $a = b$. Hence, $x \in f(A \cap B)$, and consequently, we have $f(A) \cap f(B) \subseteq f(A \cap B)$.

30. *If* $\{A_i\}_{i=1}^{\infty}$ *is a collection of denumerable sets, show that*

$$A = \bigcup_{i=1}^{\infty} A_i$$

is also denumerable.

Solution. Without loss of generality, we may assume that the sets A_i are pairwise disjoint. Since each A_i is denumerable, we can suppose that

$$A_i = \{a_{ij} : j = 1, 2, \ldots\}.$$

Now, we arrange the elements a_{ij} in an array:

$$
\begin{array}{ccccc}
a_{11} & a_{12} & a_{13} & a_{14} & \cdots \\
a_{21} & a_{22} & a_{23} & a_{24} & \cdots \\
a_{31} & a_{32} & a_{33} & a_{34} & \cdots \\
a_{41} & a_{42} & a_{43} & a_{44} & \cdots \\
\vdots & \vdots & \vdots & \vdots & \vdots
\end{array}
$$

Now, we define $f : \mathbb{N} \to A$ by

$$
\begin{aligned}
f(1) &= a_{11}, \\
f(2) &= a_{21}, \\
f(3) &= a_{12}, \\
f(4) &= a_{31}, \\
f(5) &= a_{22}, \\
f(6) &= a_{13}.
\end{aligned}
$$

Clearly, f is one to one and onto.

31. *Find a one-to-one and onto function*

(1) from $[0, 1]$ to $[-1, 1]$;
(2) from $(0, 1)$ to $(-1, 1)$.

Solution. (1) We consider the points $(0, -1)$ and $(1, 1)$ in the plane $\mathbb{R}^2$. The equation of the line going through the points $(0, -1)$ and $(1, 1)$ is $y = 2x - 1$. Now, it is easy to see that the function $f : [0, 1] \to [-1, 1]$ defined by $f(x) = 2x - 1$ is a one-to-one and onto function.

(2) Similar to part (1), the function $f : (0, 1) \to (-1, 1)$ given by $f(x) = 2x - 1$ is a one-to-one and onto function.

32. *Find a one-to-one and onto function*

(1) from $(-1, 1)$ to $\mathbb{R}$;
(2) from $(0, 1)$ to $\mathbb{R}$.

Solution. (1) It is easy to see that the function $g : (-1, 1) \to \mathbb{R}$ given by $g(x) = \tan(\frac{\pi x}{2})$ is a one-to-one and onto function.

(2) By the previous part, and part (2) of Problem (31), the functions $f : [0, 1] \to [-1, 1]$ defined by $f(x) = 2x - 1$ and $g : (-1, 1) \to \mathbb{R}$ defined by $g(x) = \tan(\frac{\pi x}{2})$ are one to one and onto. So, $g \circ f : (0, 1) \to \mathbb{R}$ is a one-to-one and onto function.

33. *Prove that the open unit interval $(0, 1)$ of real numbers is a non-denumerable set.*

Solution. Suppose that the set $(0, 1)$ is denumerable. Then, there is a one-to-one and onto function $f : \mathbb{N} \to (0, 1)$. We consider each number $x \in (0, 1)$ as a decimal expansion in the form $0.x_1x_2x_3\ldots$, which $x_n \in \{0, 1, \ldots, 9\}$, for all $n \in \mathbb{N}$. So, we can list all elements of $(0, 1)$ as follows:

$$f(1) = 0.a_{11}a_{12}a_{13}\ldots$$
$$f(2) = 0.a_{21}a_{22}a_{23}\ldots$$
$$\vdots$$
$$f(n) = 0.a_{n1}a_{n2}a_{n3}\ldots$$
$$\vdots$$

Now, we construct a number $y \in (0, 1)$ which cannot appear in the above listing of $f(n)$'s. Suppose that $y = 0.y_1y_2y_3\ldots$ is defined by

$$y_n = \begin{cases} 3 & \text{if } a_{nn} \neq 3 \\ 1 & \text{if } a_{nn} = 3. \end{cases}$$

for all $n \in \mathbb{N}$. Now, we have

$$y_1 \neq a_{11} \Rightarrow y \neq f(1),$$
$$y_2 \neq a_{22} \Rightarrow y \neq f(2),$$
$$\vdots$$
$$y_n \neq a_{nn} \Rightarrow y \neq f(n),$$
$$\vdots$$

This contradiction implies that our assumption that $(0, 1)$ is denumerable is wrong and the set $(0, 1)$ is non-denumerable.

34. *Show that the set $\mathbb{R}$ of all real numbers is non-denumerable.*

Solution. By Problem (33), $(0, 1)$ is non-denumerable, and by the second part of Problem (32), there is a one to one corresponding between $(0, 1)$ and $\mathbb{R}$. Therefore, its equipotent set $\mathbb{R}$ must be non-denumerable.

35. *Show that*

(1) the set $\mathbb{Q}$ of all rational numbers is denumerable;
(2) the set $\mathbb{R} - \mathbb{Q}$ of all irrational numbers is non-denumerable.

Solution. (1) We represent each rational number uniquely as $\dfrac{a}{b}$, where $a \in \mathbb{Z}, b \in \mathbb{N}$ and the greatest common divisor of a and b is 1. Let $\mathbb{Q}^+$ be the set of all positive rational numbers and $\mathbb{Q}^-$ be the set of all negative rational numbers. Then, $\mathbb{Q} = \mathbb{Q}^+ \cup \{0\} \cup \mathbb{Q}^-$. So, it is enough to show that $\mathbb{Q}^+$ is denumerable. We define $f : \mathbb{Q}^+ \to \mathbb{N} \times \mathbb{N}$ by $f(\dfrac{a}{b}) = (a, b)$. It is easy to see that f is one to one. Note that $f(\mathbb{Q}^+)$ is an infinite subset of the denumerable set $\mathbb{N} \times \mathbb{N}$. Consequently, $f(\mathbb{Q}^+)$ is denumerable, and so $\mathbb{Q}^+$ is denumerable.

(2) Suppose that $\mathbb{R} - \mathbb{Q}$ is denumerable. Since $\mathbb{Q}$ is denumerable, it follows that $\mathbb{Q} \cup (\mathbb{R} - \mathbb{Q}) = \mathbb{R}$ is denumerable. This contradicts Problem (34). Consequently, the set of all irrational numbers is non-denumerable.

36. *Given a triangle with sides a, b, c and angles A, B, C opposite these sides, respectively. Verify the law of sines*

$$\frac{a}{\sin A} = \frac{b}{\sin B} = \frac{c}{\sin C}.$$

Solution. We consider two cases:

(1) Suppose that all angles are acute. We assume the triangle resembles Fig. 1.4.

If h is the length of the line segment perpendicular to AC and passing through vertex B, using basic trigonometry we have

$$\sin A = \frac{h}{c} \text{ and } \sin C = \frac{h}{a}.$$

Consequently, we obtain $a \sin C = c \sin A$. This implies that $\dfrac{a}{\sin A} = \dfrac{c}{\sin C}$. The rest of the proof is similar.

(2) Suppose that one angle is obtuse. Let the obtuse angle be C as shown in Fig. 1.5.

Fig. 1.4 All angles are acute

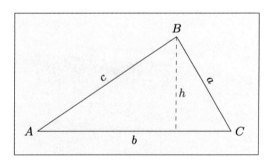

Fig. 1.5 One angle is obtuse

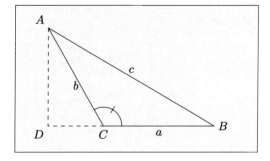

Extend the line segment BC and again let h be the length of the line segment perpendicular to BC passing through vertex A. Let D be the intersection of this line with BC. Again, using basic trigonometry, we have

$$\sin B = \frac{h}{c} \text{ and } \sin(2\pi - C) = \frac{h}{b}.$$

Since $\sin(2\pi - C) = \sin C$, it follows that $\sin C = \frac{h}{b}$. So, we have $c \sin B = b \sin C$. Therefore, we obtain $\frac{b}{\sin B} = \frac{c}{\sin C}$. The proof of other combinations is analogous.

37. *Expressing*

$$\frac{1}{2} + \cos x + \cos 2x + \cdots + \cos nx$$

as a quotient of two terms.

Solution. We apply the identity $\sin \alpha \cos \beta = \frac{1}{2}\sin(\beta + \alpha) - \frac{1}{2}\sin(\beta - \alpha)$. We obtain

$$\sum_{k=0}^{n} \sin \frac{x}{2} \cos kx = \frac{1}{2}\sum_{k=0}^{n}\left(\sin(k + \frac{1}{2})x - \sin(k - \frac{1}{2})x\right)$$

$$= \frac{1}{2}\sum_{k=0}^{n}\left(\sin(k + 1 - \frac{1}{2})x - \sin(k - \frac{1}{2})x\right).$$

Let $f(k) = \sin(k - \frac{1}{2})x$. It is easy to see that

$$\sum_{k=0}^{n}\left(f(k+1) - f(k)\right) = f(n+1) - f(0).$$

Consequently, we get

$$\sum_{k=0}^{n} \sin \frac{x}{2} \cos kx = \frac{1}{2}\left(\sin(n + \frac{1}{2})x + \sin \frac{x}{2}\right),$$

or

$$\sum_{k=0}^{n} \cos kx = \frac{1}{2\sin\frac{x}{2}}\left(\sin(n + \frac{1}{2})x + \sin \frac{x}{2}\right)$$

$$= \frac{\sin(n + \frac{1}{2})x}{2\sin\frac{x}{2}} + \frac{1}{2}.$$

This implies that

$$\frac{1}{2} + \cos x + \cos 2x + \cdots + \cos nx = \frac{\sin(n + \frac{1}{2})x}{2 \sin \frac{x}{2}}.$$

38. *Let* $\sin^2 \theta$, $\cos^2 \theta$ *and* $-\csc^2\theta$ *be all the roots of polynomial*

$$P(x) = a_0 + a_1 x + x^2 + x^3. \tag{1.14}$$

Find $P(\tan^2 \theta)$.

Solution. Since $\sin^2 \theta$, $\cos^2 \theta$ and $-\csc^2\theta$ are all the roots of the polynomial, it follows that
$$P(x) = (x - \sin^2 \theta)(x - \cos^2 \theta)(x + \csc^2 \theta).$$

Then, we get

$$P(x) = \cos^2 \theta + (\sin^2 \theta \cos^2 \theta - \csc^2 \theta)x + (\csc^2\theta - 1)x^2 + x^3. \tag{1.15}$$

From (1.14) and (1.15) we deduce that $\csc^2\theta - 1 = 1$, or equivalently $\csc^2\theta = \frac{1}{\sin^2 \theta} = 2$. This yields that

$$\sin^2 \theta = \cos^2 \theta = \frac{1}{2}.$$

Hence, we conclude that $P(x) = \left(x - \frac{1}{2}\right)^2 (x + 2)$. Therefore, we obtain $P(\tan^2 \theta) = P(1) = \frac{3}{4}$.

39. *Prove that the inverse of a real linear function is also linear and the two slopes are reciprocals of each other.*

Solution. Suppose that $f(x) = ax + b$. If $a = 0$, then the inverse does not exist. If $a \neq 0$, then the function is one to one. Let f^{-1} be the inverse of f. Then, we must have $f(f^{-1}(x)) = x$. Hence, we get $af^{-1}(x) + b = x$, which implies that $f^{-1}(x) = \frac{1}{a}x - \frac{b}{a}$; i.e., the inverse is a line with slope $\frac{1}{a}$.

40. *Verify the formulas*

(1) $\sin^{-1} x + \sin^{-1} y = \sin^{-1}\left(x\sqrt{1 - y^2} + y\sqrt{1 - x^2}\right)$,

(2) $\tan^{-1} x + \tan^{-1} y = \tan^{-1}\dfrac{x + y}{1 - xy}$,

where in (1) it is assumed that $|\sin^{-1} x + \sin^{-1} y| \leq \dfrac{\pi}{2}$ *and in (2) that* $|\tan^{-1} x + \tan^{-1} y| < \dfrac{\pi}{2}$ *and* $xy \neq 1$.

Solution. (1) Let $\alpha = \sin^{-1} x$ and $\beta = \sin^{-1} y$. Then, $\sin \alpha = x$, $\sin \beta = y$, $\cos \alpha = \sqrt{1 - \sin^2 \alpha} = \sqrt{1 - x^2}$ and $\cos \beta = \sqrt{1 - \sin^2 \beta} = \sqrt{1 - y^2}$. Now, we have

$$\sin(\alpha + \beta) = \sin \alpha \cos \beta + \sin \beta \cos \alpha$$
$$= x\sqrt{1 - y^2} + y\sqrt{1 - x^2}.$$

Thus, we observe that $\sin^{-1} x + \sin^{-1} y = \sin^{-1} \left(x\sqrt{1 - y^2} + y\sqrt{1 - x^2} \right)$.

(2) Let $\alpha = \tan^{-1} x$ and $\beta = \tan^{-1} y$. Then, $\tan \alpha = x$ and $\tan \beta = y$. So, we get

$$\tan(\alpha + \beta) = \frac{\tan \alpha + \tan \beta}{1 - \tan \alpha \tan \beta}$$
$$= \frac{x + y}{1 - xy}.$$

Therefore, we find that $\tan^{-1} x + \tan^{-1} y = \tan^{-1} \dfrac{x + y}{1 - xy}$.

41. *If*

$$f(x) = \frac{2}{3 \sin^{-1}(2x) - \pi},$$

find D_f, the domain of f.

Solution. Assume that $g(x) = \dfrac{2}{3x - \pi}$ and $h(x) = \sin^{-1}(2x)$. Then, f is the composition of g and h, i.e., $f = g \circ h$. So, we observe that

$$D_f = \left\{ x \ : \ 2x \in [-1, 1] \text{ and } \sin^{-1}(2x) \neq \frac{\pi}{3} \right\}$$
$$= \left\{ x \ : \ x \in [-\frac{1}{2}, \frac{1}{2}] \text{ and } 2x \neq \frac{\sqrt{3}}{2} \right\}$$
$$= [-\frac{1}{2}, \frac{\sqrt{3}}{4}) \cup (\frac{\sqrt{3}}{4}, \frac{1}{2}].$$

42. *Find some functions that are equal to their inverses.*

Solution. Each of the following functions is its own inverse, since $(f \circ f)(x) = x$, for all x in the domain of f.

(1) $f(x) = \left(a^{\frac{2}{3}} - x^{\frac{2}{3}} \right)^{\frac{3}{2}}$ $(0 \leq x \leq a)$.

(2) $f(x) = \dfrac{9x + 11}{13x - 9}$.

(3) $f(x) = \sqrt[3]{27 - x^3}$.

(4) $f(x) = \sqrt[4]{16 - x^4}$ $(0 \leq x \leq 2)$.

43. *Consider the function f defined by*

$$f(x) = \begin{cases} -3x^2 - 2 & if\ x < 0 \\ 0 & if\ x = 0 \\ 5x + 4 & if\ x > 0. \end{cases}$$

Show that f is invertible and write the explicit expression of its inverse.

Solution. Clearly, if $x_1 = 0$ and $x_1 \neq x_2$, then $f(x_1) \neq f(x_2)$. If $x_1, x_2 \in (-\infty, 0)$ and $x_1 \neq x_2$, then $f(x_1) - f(x_2) = -3x_1^2 - 2 + 3x_2^2 + 2 = 3(x_2 - x_1)(x_1 + x_2) \neq 0$, and so $f(x_1) \neq f(x_2)$. If $x_1, x_2 \in (0, \infty)$ and $x_1 \neq x_2$, then $5x_1 + 4 \neq 5x_2 + 4$, which implies that $f(x_1) \neq f(x_2)$. If $x_1 \in (-\infty, 0)$ and $x_2 \in (0, \infty)$, then $f(x_1) < -2 < 4 < f(x_2)$. Thus, f is a one-to-one function, and so f^{-1} exists. The range of f is $(-\infty, -2) \cup \{0\} \cup (4, \infty)$. If $y \in (-\infty, -2)$, then $x = \sqrt{-\dfrac{y+2}{3}}$ satisfies

$$-3x^2 - 2 = -3\left(-\frac{y+2}{3}\right) - 2 = y.$$

If $y \in (4, \infty)$, then $x = \dfrac{y-4}{5}$ satisfies

$$-5x + 4 = 5\left(\frac{y-4}{5}\right) + 4 = y.$$

Therefore, we have

$$f^{-1}(x) = \begin{cases} \sqrt{-\dfrac{y+2}{3}} & if\ x < -2 \\ 0 & if\ x = 0 \\ \dfrac{y-4}{5} & if\ x > 4. \end{cases}$$

44. *Let A and B be two non-empty subsets of $\mathbb{R}$ and $f : A \to B$ be a function. Prove that*

(1) If there exists a function $g : B \to A$ such that $g \circ f$ is the identity function on A, then f is one to one.

(2) If there exists a function $h : B \to A$ such that $f \circ h$ is the identity function on B, then f is onto.

(3) If f is one to one and onto, then the inverse of f is unique.

Solution. (1) Suppose that $x_1, x_2 \in A$ and $f(x_1) = f(x_2)$. Then, we have

$$x_1 = (g \circ f)(x_1) = g(f(x_1)) = g(f(x_2)) = (gof)(x_2) = x_2.$$

Consequently, f is one to one.

(2) For each $y \in B$, there is $x = h(y) \in A$ such that

$$f(x) = f(h(y)) = (f \circ h)(y) = y.$$

So, we conclude that f is onto.

(3) Let g_1 and g_2 be two inverses of f. Then, $f \circ g_2$ and $g_1 \circ f$ are identity functions. So, we have

$$g_1 = g_1 \circ (f \circ g_2) = (g_1 \circ f) \circ g_2 = g_2.$$

45. *Let A, B and C be non-empty subsets of $\mathbb{R}$ and $f : A \to B$ and $g : B \to C$ be one-to-one and onto functions. Prove that*

$$(g \circ f)^{-1} = f^{-1} \circ g^{-1}.$$

Solution. We have

$$(f^{-1} \circ g^{-1}) \circ (g \circ f) = f^{-1} \circ (g^{-1} \circ g) \circ f = f^{-1} \circ f = \text{identity function},$$
$$(g \circ f) \circ (f^{-1} \circ g^{-1}) = g \circ (f \circ f^{-1}) \circ g^{-1} = g \circ g^{-1} = \text{identity function}.$$

So, $f^{-1} \circ g^{-1}$ is the inverse of $g \circ f$. By part (3) of Problem (44), the inverse of a function is unique, so we conclude that $(g \circ f)^{-1} = f^{-1} \circ g^{-1}$.

46. *Prove that*

$$\log_a b \cdot \log_b c \cdot \log_c a = 1.$$

Solution. At the beginning we show that

$$\log_z x = \frac{\log_y x}{\log_y z}. \tag{1.16}$$

Assume that $\log_z x = t$, then $x = z^t$. This implies that $\log_y x = \log_y z^t$, or equivalently $\log_y x = t \log_y z$, and this completes the proof of (1.16). Now, by using (1.16) we get

$$\log_a b \cdot \log_b c \cdot \log_c a = \log_a b \cdot \frac{\log_a c}{\log_a b} \cdot \frac{\log_a a}{\log_a c} = 1.$$

47. *Consider the following function:*

$$f(x) = \frac{1}{\log_{\frac{1}{5}}(x - 3)} - 2.$$

(1) *Find the domain of f.*
(2) *Establish if f is invertible for $x > 4$ and if the answer is positive, find the explicit expression of the inverse.*

Solution. (1) We must have $x - 3 > 0$, or $x > 3$. Moreover,

$$\log_{\frac{1}{5}}(x - 3) \neq 0 \text{ if } x - 3 \neq 1.$$

Consequently, the domain of f is $(3, 4) \cup (4, \infty)$.

(2) For $x > 4$, f is increasing. So, the restriction of f to $(4, \infty)$ is invertible. It is easy to see that $f((4, \infty)) = (-\infty, -2)$. Now, if $y < -2$, then $y = f(x)$ if and only if

$$y = \frac{1}{\log_{\frac{1}{5}}(x-3)} - 2 \Leftrightarrow \frac{1}{\log_{\frac{1}{5}}(x-3)} = y + 2$$

$$\Leftrightarrow \log_{\frac{1}{5}}(x-3) = \frac{1}{y+2}$$

$$\Leftrightarrow x - 3 = \left(\frac{1}{5}\right)^{\frac{1}{y+2}}$$

$$\Leftrightarrow x = 3 + \left(\frac{1}{5}\right)^{\frac{1}{y+2}}.$$

So, we have

$$f^{-1}(x) = 3 + \left(\frac{1}{5}\right)^{\frac{1}{x+2}}.$$

48. *Show that*

$$\sum_{n=10}^{39} \ln\left(1 + \frac{1}{n}\right) = 2\ln 2.$$

Solution. Since $1 + \dfrac{1}{n} = \dfrac{n+1}{n}$, it follows that

$$\ln\left(1 + \frac{1}{n}\right) = \ln\left(\frac{n+1}{n}\right) = \ln(n+1) - \ln n.$$

So, we obtain

$$\sum_{n=10}^{39} \ln\left(1 + \frac{1}{n}\right) = \sum_{n=10}^{39}\left(\ln(n+1) - \ln n\right)$$
$$= \left(\ln 11 - \ln 10\right) + \left(\ln 12 - \ln 11\right) + \cdots + \left(\ln 40 - \ln 39\right)$$
$$= \ln 40 - \ln 10 = \ln(4 \cdot 10) - \ln 10 = \ln 4 + \ln 10 - \ln 10$$
$$= \ln 2^2 = 2\ln 2.$$

49. *Let $f(x) = \ln(\dfrac{1+x}{1-x})$, where $-1 < x < 1$. If a and b are given numbers in the domain of f, find all x such that*

$$f(x) = f(a) + f(b).$$

Solution. Indeed, x must satisfies the following condition:

$$\ln\left(\frac{1+x}{1-x}\right) = \ln\left(\frac{1+a}{1-a}\right) + \ln\left(\frac{1+b}{1-b}\right)$$
$$= \ln\left(\frac{(1+a)(1+b)}{(1-a)(1-b)}\right)$$
$$= \ln\left(\frac{1+a+b+ab}{1-a-b+ab}\right).$$

So, we must have $\dfrac{1+x}{1-x} = \dfrac{1+a+b+ab}{1-a-b+ab}$. Therefore, $x = \dfrac{a+b}{1+ab}$.

50. *Solve the equation*

$$5\cosh 2x + 3\sinh 2x = 5$$

giving the answer in terms of natural logarithm.

Solution. We can write

$$5\left(\frac{e^{2x}+e^{-2x}}{2}\right) + 3\left(\frac{e^{2x}-e^{-2x}}{2}\right) = 5 \Rightarrow 5e^{2x} + 5e^{-2x} + 3e^{2x} - 3e^{-2x} = 10$$
$$\Rightarrow 8e^{2x} + 2e^{-2x} = 10$$
$$\Rightarrow 4e^{2x} + e^{-2x} - 5 = 0$$
$$\Rightarrow 4e^{4x} + 1 - 5e^{2x} = 0$$
$$\Rightarrow \left(e^{2x} - 1\right)\left(4e^{2x} - 1\right)$$
$$\Rightarrow e^{2x} = 1 \text{ or } 4e^{2x} - 1 = 0$$
$$\Rightarrow x = 0 \text{ or } 2x = -\ln 4$$
$$\Rightarrow x = 0 \text{ or } x = -\ln 2.$$

51. *(1) Show that a function f is even if and only if its graph is symmetric with respect to the y-axis.*

(2) Show that a function f is odd if and only if its graph is symmetric with respect to the origin.

Solution. (1) Let f be even. Suppose that (x, y) is on the graph of f. Since (x, y) is on the graph of f, it follows that $f(x) = y$. Since f is even, it follows that $f(-x) = y$. So, we conclude that $(-x, y)$ is on the graph of f. For the converse, suppose that the graph of f is symmetric with respect to the y-axis. Suppose that $f(x)$ is defined and $y = f(x)$. Then, (x, y) is on the graph of f. Hence, by assumption $(-x, y)$ is on the graph of f, too. Thus, we have $f(-x) = y$, which implies that $f(-x) = f(x)$, and consequently f is even.

(2) Let f be odd. Assume that (x, y) be on the graph of f. Then, $f(x) = y$. Since f is odd, it follows that $f(-x) = -f(x) = -y$, and so $(-x, -y)$ is on the graph f. Now, we observe that (x, y) and $(-x, -y)$ are symmetric with respect to the origin. For the converse, suppose that the graph of f is symmetric with respect to the origin. Suppose that $f(x)$ is defined and $y = f(x)$. Then, (x, y) is on the graph of f. Hence, by assumption $(-x, -y)$ is on the graph of f, too. Thus, $f(-x) = -f(x)$. This implies that f is odd.

52. *Let f be any function whose domain contains* $-x$ *whenever it contains* x. *Show that f can be uniquely represented as the sum of an even function and an odd function.*

Solution. Note that for each x in the domain of f, we have

$$f(x) = \frac{f(x) + f(-x)}{2} + \frac{f(x) - f(-x)}{2}. \tag{1.17}$$

It is easy to see that the functions in the right side of (1.17) are even and odd, respectively. Now, let $f = f_e + f_o$, where f_e is an even function and f_o is an odd function. So, we have

$$f(x) = f_e(x) + f_o(x) \tag{1.18}$$

and $f_e(-x) = f_e(x)$ and $f_o(-x) = -f_o(x)$, for all x in the domain of f. Since $f(-x) = f_e(-x) + f_o(-x)$, it follows that

$$f(-x) = f_e(x) - f_o(x) \tag{1.19}$$

From (1.18) and (1.19), we obtain

$$f_e(x) = \frac{f(x) + f(-x)}{2} \text{ and } f_o(x) = \frac{f(x) - f(-x)}{2}.$$

53. *Let f be a non-zero function such that*

$$f(x + y) + f(x - y) = 2f(x)f(y).$$

Show that

(1) $f(0) = 1$;
(2) *f is an even function.*

Solution. (1) If $y = 0$, then $f(x) + f(x) = 2f(x)f(0)$, or equivalently $\left(f(0) - 1\right)f(x) = 0$. Since f is not zero function, it follows that $f(0) = 1$.
   (2) Let $x = 0$ and y be arbitrary. Then, we have $f(y) + f(-y) = 2f(0)f(y)$. Since $f(0) = 1$, it follows that $f(-y) = f(y)$. So, f is even.

54. *Let f be a real function such that*

$$f(x + 1) = \frac{1 + f(x)}{1 - f(x)}, \text{ for all } x \in \mathbb{R}.$$

Show that f is a periodic function.

Solution. We have

$$f(x+2) = \frac{1+f(x+1)}{1-f(x+1)} = \frac{1+\dfrac{1+f(x)}{1-f(x)}}{1-\dfrac{1+f(x)}{1-f(x)}} = -\frac{1}{f(x)},$$

and so

$$f(x+4) = -\frac{1}{f(x+2)} = -\frac{1}{-\dfrac{1}{f(x)}} = f(x).$$

This means that f is periodic with period 4.

55. *Let f be a real function with values in $[0, 1]$ such that*

$$f(x+a) = \frac{1}{2} + \sqrt{f(x) - f^2(x)}, \tag{1.20}$$

where a is a non-zero number. Show that f is a periodic function.

Solution. We have $f(x+a) - \dfrac{1}{2} = \sqrt{f(x) - f^2(x)}$ or $(f(x+a) - \dfrac{1}{2})^2 = f(x) - f^2(x)$. Hence, $f^2(x+a) - f(x+a) + \dfrac{1}{4} = f(x) - f^2(x)$. This implies that

$$\sqrt{f(x+a) - f^2(x+a)} = \sqrt{\left(f(x) - \frac{1}{2}\right)^2}$$

$$= \left| f(x) - \frac{1}{2} \right|$$

$$= f(x) - \frac{1}{2}, \quad \text{(since } f(x) \geq \frac{1}{2}\text{)}.$$

Now, we put $x + a$ instead of x in (1.20). Then, we obtain

$$f(x+2a) = \frac{1}{2} + \sqrt{f(x+a) - f^2(x+a)} = \frac{1}{2} + f(x) - \frac{1}{2} = f(x).$$

Therefore, f is a periodic function.

56. *Let $f(x) = x \sin x$. Show that f is unbounded on $\mathbb{R}$.*

Solution. Suppose that M is an arbitrary real number. We choose integers k_1 and k_2 such that

$$2\pi k_1 + \frac{\pi}{2} > M \text{ and } \frac{\pi}{2} - 2\pi k_2 < M.$$

Since $\sin(2\pi k_1 + \dfrac{\pi}{2}) = 1$ and $\sin(2\pi k_2 - \dfrac{\pi}{2}) = -1$, we obtain

$$f(2\pi k_1 + \frac{\pi}{2}) = 2\pi k_1 + \frac{\pi}{2} > M \text{ and } f(2\pi k_2 - \frac{\pi}{2}) = \frac{\pi}{2} - 2\pi k_2 < M.$$

Since M is arbitrary, it follows that f is unbounded on $\mathbb{R}$.

57. *Give an example of a function $f : \mathbb{R} \to \mathbb{R}$ which is unbounded in every open interval.*

Solution. We define $f : \mathbb{R} \to \mathbb{R}$ as follows: $f(x) = n$ if $x = \dfrac{m}{n} \in \mathbb{Q}$ is in the standard form; i.e., n is positive and the integers m and n have no common divisor other than 1; $f(x) = 0$ elsewhere. Suppose that a and b are two real numbers such that $a < b$. By the archimedean property of real numbers there exists $n_0 \in \mathbb{N}$ such that $n_0(b - a) > 1$. Then, for every $n \geq n_0$ we have $n(b - a) > 1$ or $na - nb > 1$. Hence, there exists $m \in \mathbb{Z}$ such that $na < m < nb$. Consequently, $a < \frac{m}{n} < b$. This shows that the image of the interval (a, b) under f includes all natural numbers $n \geq n_0$, and so it is unbounded.

58. *Let $f : \mathbb{R} \to \mathbb{R}$ be a one-to-one function such that*

$$f(x + y) = f(x) + f(y), \tag{1.21}$$

and

$$f(xy) = f(x)f(y). \tag{1.22}$$

Prove that $f(x) = x$, for all $x \in \mathbb{R}$.

Solution. Since $f(0) = f(0 + 0) = f(0) + f(0)$, it follows that $f(0) = 0$. So, we obtain $0 = f(0) = f(x + (-x)) = f(x) + f(-x)$. Thus, $f(-x) = -f(x)$. By Eq. (1.21), for each positive integer n we have

$$f(n) = f(\underbrace{1 + \cdots + 1}_{n \text{ times}}) = \underbrace{f(1) + \cdots + f(1)}_{n \text{ times}} = nf(1).$$

If m is a negative integer, then $f(m) = f(-(-m)) = -f(-m) = -(-m)f(1) = mf(1)$. Therefore, for every integer n we showed that $f(n) = nf(1)$.

Now, let $r = \dfrac{p}{q}$, where $p \in \mathbb{Z}$ and $q \in \mathbb{N}$. Then, we have

$$q\left(f(\frac{p}{q})\right) = \underbrace{f(\frac{p}{q}) + \cdots + f(\frac{p}{q})}_{q \text{ times}} = f(\underbrace{\frac{p}{q} + \cdots + \frac{p}{q}}_{q \text{ times}}) = f(p) = pf(1).$$

Thus, we conclude that $f(\dfrac{p}{q}) = \dfrac{p}{q}f(1)$. By (1.22), we observe that $f(1) = f(1 \cdot 1) = f(1) \cdot f(1)$, which implies that $f(1) = 1$ or 0. Since f is one to one, it follows that $f(1) \neq 0$ and so $f(1) = 1$. Therefore, $f(x) = x$, for all $x \in \mathbb{Q}$.

Now, suppose that there exists an irrational number a such that $f(a) \neq a$. Then, $f(a) < a$ or $f(a) > a$. If $f(a) < a$, then there exists $r \in \mathbb{Q}$ such that $f(a) < r < a$. Hence $a - r > 0$. So, there exists $b \in \mathbb{R}$ such that $b^2 = a - r$. Then, we observe that

$f(a - r) = f(b^2) = (f(b))^2 \geq 0$. Since f is one to one, it follows that $f(a - r) > 0$. This implies that $f(a) > f(r) = r$, a contradiction. Similarly, if $f(a) > a$, then we obtain a contradiction again.

Therefore, $f(x) = x$, for all $x \in \mathbb{R}$.

1.3   Exercises

Easier Exercises

1. If $A, B_1, B_2, \ldots, B_n$ are sets, prove that

$$A \cap (B_1 \cup B_2, \cup \ldots \cup B_n) = (A \cap B_1) \cup (A \cap B_2) \cup \ldots \cup (A \cap B_n).$$

2. If A and B are sets and there is a set C such that $A \cup C = B \cup C$ and $A \cap C = B \cap C$, prove that $A = B$.
3. Let A and B are sets. Prove that

$$(A \cap B) \cup C = A \cap (B \cup C) \Leftrightarrow C \subseteq A.$$

4. Let A and B are sets. Prove that

$$(A - B) \cup (B - A) = (A \cup B) - (A \cap B).$$

5. If a set A consists of n elements, prove that the set of all subsets of A contains exactly 2^n elements.
6. Let $(1 + \sqrt{2})^5 = a + b\sqrt{2}$, where a and b are positive integers. Determine the value of $a + b$.

Find all x satisfying the given inequality:

7. $|x| - |x - 1| < 1$,   9. $|x| + |x + 1| + |x + 2| < x^2 + x$,

8. $\left| \dfrac{x - 1}{x + 1} \right| < 1$,   10. $\left| \dfrac{1}{x} - 1 \right| < |x + 1|$.

11. Let a, b and c are real numbers. Show that $a \leq b \leq c$ or $c \leq b \leq a$ if and only if $|a - b| + |b - c| = |a - c|$.

Describe each of the following sets geometrically by sketching a graph on the Cartesian plane:

12. $\{(x, y) \in \mathbb{R} \times \mathbb{R} : x > y\}$,

13. $\{(x, y) \in \mathbb{R} \times \mathbb{R} : |x| = |y|\}$,

14. $\{(x, y) \in \mathbb{R} \times \mathbb{R} : |x + y| = 1\}$,

15. $\{(x, y) \in \mathbb{R} \times \mathbb{R} : |x| + |y| \leq 1\}$.

16. Is the sum or product of two irrational numbers always irrational?

17. Show that $\sqrt[3]{2}$ is not a rational number. Describe the graph of the given equation:

18. $x^2 + y^2 + x = 0$,

19. $x^2 + y^2 + 4x - 6y + 12 = 0$.

20. Find an equation of the circle with $(3, 2)$ and $(-1, 6)$ as the endpoint of a diameter.

21. Find an equation of the circle going through the points $(1, 1)$, $(2, 2)$ and $(3, 1)$.

22. Sketch the graph of

$$f(x) = \begin{cases} 1 - |x - 1| & \text{if } x \geq 0 \\ |x + 1| & \text{if } x < 0. \end{cases}$$

23. If

$$f(x) = \sqrt{\frac{(x - 3)(x + 2)}{x - 1}} \quad \text{and} \quad g(x) = \frac{x^2 - 16}{x - 7}\sqrt{x^2 - 9},$$

find D_f, $D_{f \pm g}$, D_{fg} and $D_{f/g}$. Find the domain and range of the following functions:

24. $f(x) = \dfrac{x^4 + x^3 - 9x^2 - 3x + 18}{x^2 + x - 6}$,

25. $f(x) = \dfrac{|x|}{[x]}$,

26. $f(x) = \dfrac{1}{\sqrt{1 - |\sin x|}}$,

27. $f(x) = \dfrac{x + 1}{\sqrt{x^4 - 6x^3 + 9x^2}}$,

28. $f(x) = \ln\left(\sqrt{x - 3} + \sqrt{7 - x}\right)$

29. $f(x) = \ln\left(\ln(1 - x^2)\right)$,

30. $e^{\left(f(x)\right)^2} = x$, where $f(x) \geq 0$,

31. $f(x) = \dfrac{\tan x}{2^x \log_3 x}$.

32. Suppose that $f(x + 1) - f(x) = 8x + 3$, where $f(x) = ax^2 + bx + 5$. Find a and b.

33. Show that the area of the triangle in Problem (36) is equal to

$$\frac{1}{2}bc \sin A = \frac{1}{2}ac \sin B = \frac{1}{2}ac \sin C.$$

34. What is the area of an equilateral triangle with side length l?

35. What is the area of an equilateral triangle with hypotenuse of length h?

36. Verify the following trigonometric identities:

(a) $\sin 3x = 3 \sin x - 4 \sin^3 x$;
(b) $\cos 3x = 4 \cos^3 x - 3 \cos x$;
(c) $\sin 4x = 8 \sin x \cos^3 x - 4 \sin x \cos x$;
(d) $\cos 4x = 8 \cos^4 x - 8 \cos^2 x + 1$.

For each of the following functions, determine if it is even, odd or neither.

37. $f(x) = \dfrac{x \sin x}{x^2 + 1}$,

38. $f(x) = \dfrac{\sin x}{x^3 + x}$,

39. $f(x) = \dfrac{\sin(\sqrt{x})}{x^3 + x}$,

40. $f(x) = \ln \left(x + \sqrt{1 + x^2} \right)$.

41. Determine the least periods of $f(x) = \sin(8x) + \cos(4x)$.
42. Determine the least periods of $f(x) = \sin(\cos x)$ and $g(x) = \cos(\sin x)$.
43. Prove that if f is defined by $f(x) = x - [x]$, then f is periodic. What is the smallest positive period of f?
44. Show that $|\sin nx| \le n|\sin x|$, for all $n \in \mathbb{N}$ and $x \in \mathbb{R}$.
45. Let X be a finite set with m elements and Y be a finite set with n elements. Prove that

   (a) If $m > n$, then there cannot be no one-to-one function from X to Y.
   (b) If $m \le n$, then there exist exactly $\dfrac{n!}{(n - m)!}$ one-to-one functions from X to Y.

46. Let $f : \mathbb{R} \to \mathbb{R}$ be a function given by

$$f(x) = \begin{cases} 2 & \text{if } x \text{ is rational} \\ -1 & \text{if } x \text{ is irrational.} \end{cases}$$

Find

   (a) $f(\{-1,\ 0,\ 1\})$, $f(\{\sqrt{3},\ \pi\})$ and $f(\{2,\ \ln 2,\ \frac{1}{5}\})$;
   (b) $f^{-1}(\{0,\ 1\})$, $f^{-1}(\{-2,\ 2\})$ and $f^{-1}(-3,\ 2,\ 6\})$.

47. Let $f(x) = \sin x$, for all $x \in \mathbb{R}$. Find $f([0,\ \frac{\pi}{2}])$, $f([\frac{\pi}{6},\ \frac{\pi}{2}]$ and $f([0,\ \frac{\pi}{6}])$.
48. Let $f(x) = \ln x$, for all $x \in (0,\ \infty)$. if $A = [0,\ 1]$ and $B = [1,\ 2]$, then determine $f^{-1}(A)$, $f^{-1}(B)$, $f^{-1}(A \cup B)$ and $f^{-1}(A \cap B)$.

Harder Exercises

49. The equations $|x|^2 - 3|x| + 2 = 0$ and $x^4 - ax^2 + 4 = 0$ have the same roots. Determine the value of a.

50. For all real numbers a, b and c, prove that

$$27(a^4 + b^4 + c^4) \geq (a + b + c)^4.$$

51. If a, b and c are positive real numbers and if $a + b + c = 1$, prove that

$$(1 - a)(1 - b)(1 - c) \geq 8abc.$$

Guess the following sets and then prove your guesses:

52. $\bigcup\limits_{n=1}^{\infty} (n, n + 1]$,

54. $\bigcap\limits_{n=1}^{\infty} [0, \frac{1}{n}]$,

53. $\bigcup\limits_{n=1}^{\infty} (-n, -n + 1]$,

55. $\bigcap\limits_{n=1}^{\infty} (0, \frac{1}{n})$.

56. For every natural number n, prove that

$$\sum_{k=1}^{n} k^3 = \frac{n^2(n + 1)^2}{4}.$$

57. Prove that

$$\frac{1}{1 \cdot 2} + \frac{1}{2 \cdot 3} + \frac{1}{3 \cdot 4} + \cdots + \frac{1}{n(n + 1)} = \frac{n}{n + 1}.$$

58. Prove that

$$(2n + 1) + (2n + 3) + (2n + 5) + \cdots + (4n - 1) = 3n^2,$$

for all $n \in \mathbb{N}$.

59. Prove that for any positive integer n,

$$\binom{n}{0}^2 + \binom{n}{1}^2 + \cdots + \binom{n}{n}^2 = \binom{2n}{n}.$$

60. Show that

$$\binom{n}{k} + \binom{n}{k - 1} = \binom{n + 1}{k},$$

for $k = 1, 2, \ldots, n$.

61. Prove that for any positive integers $k \leq m, n$,

$$\sum_{i=0}^{k} \binom{n}{i} + \binom{m}{k - i} = \binom{m + n}{k}.$$

62. Prove that for any positive integer n,

$$\sum_{k=0}^{n}(-1)^k \binom{n}{k} = 0 \text{ and } \sum_{k=0}^{n}\binom{n}{k} = 2^n.$$

63. Let $a_1, \ldots, a_n$ be real numbers and $b_1, \ldots, b_n$ be positive real numbers. Prove that

$$\min\left\{\frac{a_1}{b_1}, \ldots, \frac{a_n}{b_n}\right\} \le \frac{a_1 + \cdots + a_n}{b_1 + \cdots + b_n} \le \max\left\{\frac{a_1}{b_1}, \ldots, \frac{a_n}{b_n}\right\}.$$

64. If $1 < a < b$, determine which of the following numbers is bigger than another number,

$$\left(\log_a b\right)^{\log_b a} \text{ and } \left(\log_b a\right)^{\log_a b}.$$

65. If m and n are two integers greater than 1, prove that

$$\sum_{k=1}^{n} \frac{(2k)^m}{n} > (n+1)^m.$$

66. If $S = a_1 + a_2 + \cdots + a_n$, prove that

$$\prod_{k=1}^{n}(1 + a_k) \le \sum_{k=0}^{n} \frac{S^k}{k!}.$$

67. (a) Suppose that $f(x) = a_0 + a_1 x + \cdots + a_n x^n$, where $a_0, a_1, \ldots, a_n$ are integers and $a_n \ne 0$. If $f(r) = 0$, where $r = \frac{p}{q}$, and p, q are integers having no common factors and $q \ne 0$, prove that q divides a_n and p divides a_0.
    (b) Use (a), to show that $\sqrt{7}$ cannot represent a rational number.
68. Show that $\left(2 + 5^{\frac{1}{3}}\right)^{\frac{1}{2}}$ does not represent a rational number.
69. Show that $\left(3 + \sqrt{2}\right)^{\frac{2}{3}}$ does not represent a rational number.
70. Show that $\log_2 3$ does not represent a rational number.
71. Prove that for each prime number p and for each real number $\epsilon > 0$, there exist rational numbers x and y such that

$$x^2 < p < y^2 < x^2 + \epsilon.$$

72. Prove that analytically that the three medians of any triangle meet in a point.
73. Prove that analytically that if the diagonals of a rectangle are perpendicular, then the rectangle is a square.
74. Prove that two lines $a_1 x + b_1 y + c_1 = 0$ and $a_2 x + b_2 y + c_2 = 0$ are parallel if and only if $a_1 b_2 - a_2 b_1 = 0$.
75. Let $f(x) = \dfrac{1}{x^3 + 3x^2 + 2x}$. Determine the smallest positive integer n such that

$$f(1) + f(2) + \cdots + f(n) > \frac{503}{2014}.$$

76. Let $f : \mathbb{R} \to \mathbb{R}$ be a function. If A and B are subsets of domain and range of f, respectively, prove that

   (a) $A \subseteq f^{-1}(f(A))$;
   (b) $f(f^{-1}(B)) \subseteq B$;
   (c) If f is one to one, then $A = f^{-1}(f(A))$;
   (d) If f is onto, then $B = f(f^{-1}(B))$.

77. Let $f : \mathbb{R} \to \mathbb{R}$ be a function and let A and B be subsets of range of f. Prove that

   (a) $f^{-1}(A \cup B) = f^{-1}(A) \cup f^{-1}(B)$;
   (b) $f^{-1}(A \cap B) = f^{-1}(A) \cap f^{-1}(B)$.

78. Let $f : \mathbb{R} \to \mathbb{R}$ be a function and let A and B be two subsets of domain of f.

   (a) Prove that $f(A \cup B) = f(A) \cup f(B)$;
   (b) Give an example which shows that the equality in Problem (29) is false if f is not one to one.

79. Let $f : \mathbb{R} \to \mathbb{R}$ be an onto function and let A and B be subsets of $\mathbb{R}$. Prove that $A = B$ if $f^{-1}(A) = f^{-1}(B)$. Give an example which shows that the assertion is false if f is not onto.

80. Let $f : \mathbb{R} \to \mathbb{R}$ be a function and let A be a subset of range of f. Prove that

$$f^{-1}(\mathbb{R} - A) = \mathbb{R} - f^{-1}(A).$$

81. Let $f : \mathbb{R} \to \mathbb{R}$ be a function and let A and B be subsets of in domain of f. Give an example showing that it is not true that

$$f(A - B) = f(A) - f(B).$$

Let a, b, c and d be real numbers such that $a < b$ and $c < d$. Prove that there exists a one-to-one correspondence between the following intervals:

82. $[a, b]$ and $[c, d]$,
83. $[a, b]$ and $(c, d]$,
84. $[a, b]$ and $[c, d)$,
85. $[a, b]$ and (c, d).
86. Let $S^1 = \{(x, y) \in \mathbb{R} \times \mathbb{R} : x^2 + y^2 = 1\}$. Prove that S^1 and $\mathbb{R}$ have the same cardinality and hence S^1 is non-denumerable.
87. Prove that the set of all polynomials

$$a_0 + a_1 x + \cdots + a_n x^n$$

with integer coefficients is denumerable.

88. An *algebraic number* is, by definition, any real root of an equation

$$a_0 + a_1 x + \cdots + a_n x^n = 0$$

with integer coefficients. Prove that the set of all algebraic numbers is denumerable.

89. Prove that there exist real numbers which are not algebraic.

90. Prove that the set of all finite subsets of a denumerable set is a denumerable set.

Chapter 2
Limits and Continuity

2.1 Basic Concepts and Theorems

The Limit of a Function

Let f be a function which is defined on a deleted neighborhood of a. We say that $f(x)$ approaches the *limit* L as x approaches to a, or that $f(x)$ has the limit L at a, if $f(x)$ gets closer and closer to L as x gets closer and closer to a. In other words, this definition states the function values $f(x)$ approach a limit L as x approaches a if the absolute value of the difference between $f(x)$ and L can be made as small as we please by taking sufficiently near a, but not equal to a. This is expressed by writing $f(x) \to L$ as $x \to a$, or

$$\lim_{x \to a} f(x) = L. \tag{2.1}$$

So, we can say that (2.1) is equivalent to the following statement: for every $\epsilon > 0$ there exists $\delta > 0$ such that

$$|f(x) - L| < \epsilon \text{ whenever } 0 < |x - a| < \delta.$$

If f has a limit at a, this limit is unique. This means that a function can not approach two different limits at the same time. If $f(x)$ has a limit L at a, then $f(x)$ is bounded in a deleted neighborhood of a. If $f(x)$ has a non-zero limit L at a, then there exists a deleted neighborhood of a in which $f(x)$ is non-zero and has the same sign as L.

Let f and g be two functions for which

$$\lim_{x \to a} f(x) = L \text{ and } \lim_{x \to a} g(x) = M.$$

© The Author(s), under exclusive license to Springer Nature Singapore Pte Ltd. 2020   41
B. Davvaz, *Examples and Problems in Advanced Calculus: Real-Valued Functions*,
https://doi.org/10.1007/978-981-15-9569-1_2

Then

(1) $\lim\limits_{x \to a} \big(f(x) \pm g(x)\big) = L \pm M$;

(2) $\lim\limits_{x \to a} f(x) \cdot g(x) = L \cdot M$;

(3) $\lim\limits_{x \to a} \dfrac{f(x)}{g(x)}) = \dfrac{L}{M}$, whenever $M \neq 0$.

Preservation of approach to zero: If $f(x)$ is bounded in a deleted neighborhood of a and $\lim\limits_{x \to a} g(x) = 0$, then $\lim\limits_{x \to a} f(x)g(x) = 0$.

Sandwich theorem for functions: Let f, g and h be three functions such that $f(x) \leq g(x) \leq h(x)$ in some deleted neighborhood of a. If $\lim\limits_{x \to a} f(x) = \lim\limits_{x \to a} h(x) = L$, then $\lim\limits_{x \to a} g(x) = L$.

To evaluate the limits of trigonometric functions, we may use the following limit:

$$\lim_{x \to 0} \frac{\sin x}{x} = 1.$$

One Sided Limits

Let f be a function which is defined at every number in open interval (a, c). Then, the limit of $f(x)$ as x approaches a from the right, is L, written $\lim\limits_{x \to a^+} f(x) = L$ if for every $\epsilon > 0$ there exists $\delta > 0$ such that $|f(x) - L| < \epsilon$ whenever $a < x < a + \delta$; this limit is called the *right hand limit*. Analogously, let f be a function which is defined at every number in open interval (d, a). Then, the limit of $f(x)$ as x approaches a from the left, is L, written $\lim\limits_{x \to a^-} f(x) = L$ if for every $\epsilon > 0$ there exists $\delta > 0$ such that $|f(x) - L| < \epsilon$ whenever $a - \delta < x < a$; this limit is called the *left hand limit*.

Condition for existence of a limit: The limit $\lim\limits_{x \to a} f(x)$ exists if and only if $\lim\limits_{x \to a^+} f(x)$ and $\lim\limits_{x \to a^-} f(x)$ both exist and are equal. In this case, we have

$$\lim_{x \to a} f(x) = \lim_{x \to a^+} f(x) = \lim_{x \to a^-} f(x).$$

Infinite Limits and Limits at Infinity

Let f be a function which is defined on a deleted neighborhood of a. We write $\lim\limits_{x \to a} f(x) = \infty$ if for every $M > 0$ there exists a corresponding $\delta > 0$ such that $f(x) > M$ whenever $0 < |x - a| < \delta$. Analogously, we write $\lim\limits_{x \to a} f(x) = -\infty$ if for every $M > 0$ there exists a corresponding $\delta > 0$ such that $f(x) < -M$ whenever $0 < |x - a| < \delta$. We can define *vertical asymptotes*, which occur when $f(x) \to$

$\pm\infty$. More precisely, the line $x = a$ is a vertical asymptote of the graph of a function f if either $\lim_{x\to a^+} f(x) = \pm\infty$ or $\lim_{x\to a^-} f(x) = \pm\infty$.

Let f be a function defined on an interval (a, ∞). Then, $\lim_{x\to\infty} f(x) = L$ if the values of $f(x)$ can be made arbitrarily close to L by taking x sufficiently large or equivalently if for every $\epsilon > 0$ there exists $N > 0$ such that $|f(x) - L| < \epsilon$ whenever $x > N$. Similarly, if f is defined on an interval $(-\infty, a)$, then we say $\lim_{x\to -\infty} f(x) = L$ if the values of $f(x)$ can be made arbitrarily close to L by taking x sufficiently large and negative or equivalently if for every $\epsilon > 0$ there exists $N > 0$ such that $|f(x) - L| < \epsilon$ whenever $x < -N$. We can define *horizontal asymptotes* by looking at functions with a finite limit as $x \to \pm\infty$. Indeed, the line $y = L$ is a horizontal asymptote of the graph of a function f if either $\lim_{x\to\infty} f(x) = L$ or $\lim_{x\to -\infty} f(x) = L$.

When a linear asymptote is not parallel to the x axis or y axis, it is called an *oblique asymptote* or *slant asymptote*. A function f is asymptote to the straight line $y = mx + b$ $(m \neq 0)$ if

$$\lim_{x\to\infty} \big(f(x) - (mx + b)\big) = 0 \quad \text{or} \quad \lim_{x\to -\infty} \big(f(x) - (mx + b)\big) = 0.$$

The value for m is computed first and is given by

$$m = \lim_{x\to a} \frac{f(x)}{x},$$

where a is either $-\infty$ or ∞ depending on the case being studied. If this limit does not exist, then there is no oblique asymptote in that direction. Having m then the value for b can be computed by

$$b = \lim_{x\to a} \big(f(x) - mx\big),$$

where a should be the same value used before. If this limit fails to exist, then there is no oblique asymptote in that direction.

Infinite limits at infinity can also be considered. There are formal definitions for each of the following.

$$\lim_{x\to\infty} f(x) = \infty, \quad \lim_{x\to -\infty} f(x) = \infty$$
$$\lim_{x\to\infty} f(x) = -\infty, \quad \lim_{x\to -\infty} f(x) = -\infty.$$

List of Indeterminate Forms

The following expressions are all called indeterminate forms

$$\frac{0}{0}, \frac{\infty}{\infty}, 0 \times \infty, \infty - \infty, 1^{\infty}, 0^{0} \text{ and } \infty^{0},$$

which are typically considered

Continuous Functions

A function f is said to be *continuous* at a if the following conditions are satisfied:

(1) $f(a)$ exists, i.e., $a \in D_f$,
(2) $\lim\limits_{x \to a} f(x)$ exists,
(3) $\lim\limits_{x \to a} f(x) = f(a)$.

If one or more of the above conditions fails to hold at a, then the function f is said to be *discontinuous* at a. If the functions f and g are continuous at a, then so are $f \pm g$, $f \cdot g$ and f/g provided that $g(a) \neq 0$ in the last case. A function f is said to be

(1) *continuous on an open interval* if it is continuous at every number in the open interval,
(2) *continuous from the right* at a if $\lim\limits_{x \to a^+} f(x) = f(a)$,
(3) *continuous from the left* at a if $\lim\limits_{x \to a^-} f(x) = f(a)$,

A function whose domain includes the closed interval $[a, b]$ is said to be *continuous on $[a, b]$* if it is continuous on (a, b), as well as continuous from the right at a and continuous from the left at b. All elementary functions are continuous in their domains. Given a function f with both right and left hand limits at a and these limits are unequal. Then, f is said to have a *jump discontinuity* at a.

Limit of a composite function: If $\lim\limits_{x \to a} f(x) = L$ and g is continuous at L, then

$$\lim\limits_{x \to a} g\big(f(x)\big) = g(L).$$

If f is continuous at a and g is continuous at $f(a)$, then $g \circ f$ is continuous at a.

Continuity of polynomials and rational functions: Any polynomial

$$P(x) = a_0 + a_1 x + a_2 x^2 + \cdots + a_n x^n, \quad (a_n \neq 0)$$

is continuous for all x. Any rational function

$$R(x) = \frac{a_0 + a_1 x + a_2 x^2 + \cdots + a_n x^n}{b_0 + b_1 x + b_2 x^2 + \cdots + b_m x^m}, \quad (a_n \neq 0, \ b_m \neq 0)$$

is continuous at every point of its domain of definition, that is, at every point where its denominator is non-zero.

Continuity of trigonometric functions: Each of the trigonometric functions $\sin x$, $\cos x$, $\tan x$, $\cot x$, $\sec x$ and $\csc x$ is continuous at every point of its domain of definition.

Continuity of elementary functions: All elementary functions are continuous over their entire domain.

Let f be a function defined on an interval I and let there is $p \in I$ such that $f(p) \leq f(x)$ for all $x \in I$. Then, $f(p) = m$ is called the (*absolute*) *minimum* of f on I. Similarly, if there is $q \in I$ such that $f(x) \leq f(q)$ for all $x \in I$, then $f(q) = M$ is called the (*absolute*) *maximum* of f on I.

Extreme value theorem: If f is a continuous function on $[a, b]$, then f is bounded on $[a, b]$ and f has both a maximum M and a minimum m on $[a, b]$, that is, there exist $p, q \in [a, b]$ such that $m = f(p) \leq f(x) \leq f(q) = M$, for all $x \in [a, b]$.

Sign-preserving property of continuous functions: Let f be a continuous function at a and suppose that $f(a) \neq 0$. Then there exists an interval $(a - \delta, a + \delta)$ in which f has the same sign at $f(c)$. Note that if there is one-sided continuity at a, then there is a corresponding one-sided interval $[a, a + \delta)$ or $(a - \delta, a]$ in which f has the same sign as $f(c)$.

Intermediate value theorem: If f is a continuous function on an interval I, then f has the intermediate value property: Whenever $a, b \in I$, $a < b$ and k lies between $f(a)$ and $f(b)$ (i.e., $f(a) < k < f(b)$ or $f(b) < k < f(a)$), there exists $c \in (a, b)$ such that $f(c) = k$.

Interval mapping theorem: If f is a continuous function on an interval I of any kind, then the image of I, $f(I) = \{f(x) : x \in I\}$ is also an interval.

If the reader is interested to see the proof of theorems that are presented in this chapter, we refer him/her to [1–16]. Mastery of the basic concepts in Chaps. 2, 3, 5 and 7 should make the analysis in such areas as mathematical analysis and analytic geometry more meaningful. These chapters can also save as a foundation for an in depth study of mathematical analysis and analytic geometry given in [2–4, 10, 11, 16].

2.2   Problems

59. *By using the definition of limit, show that* $\lim\limits_{x \to 0} \dfrac{x^3 + 2x}{x} = 2$.

Solution. Given any $\epsilon > 0$, we look for a $\delta > 0$ such that

$$\left| \frac{x^3 + 2x}{x} - 2 \right| = |x^2 + 2 - 2| = |x^2| = |x|^2 < \epsilon,$$

whenever $0 < |x| < \delta$. Since $|x|^2 < \epsilon$, it follows that $|x| < \sqrt{\epsilon}$. So, an appreciate choice of δ is $\delta = \sqrt{\epsilon}$.

60. *Prove that* $\lim\limits_{x \to 5} x^2 = 25$.

Solution. We must show that for any $\epsilon > 0$, there is $\delta > 0$ such that

$$|x^2 - 25| < \epsilon, \text{ whenever } 0 < |x - 5| < \delta,$$

or, equivalently,

$$|(x + 5)(x - 5)| < \epsilon, \text{ whenever } 0 < |x - 5| < \delta.$$

If one choose an initial radius $\delta_1 = 1$, then $|x - 5| < 1$. This implies that $4 < x < 6$ and so $|x + 5| = x + 5 < 11$. Hence, we get

$$|x^2 - 25| = |x + 5| \cdot |x - 5| < 11 \cdot |x - 5| < 11\delta$$

whenever $|x - 5| < \delta$ and $\delta \leq \dfrac{\epsilon}{11}$. Therefore, it is enough to consider $\delta = \min\{1, \dfrac{\epsilon}{11}\}$.

61. *Let f and g be two functions such that for all $0 < \epsilon < \dfrac{1}{2}$,*

$$|x - 2| < \frac{\epsilon}{10} \Rightarrow |f(x) + 7| < \epsilon \text{ and } |x - 2| < \frac{\epsilon^2}{3} \Rightarrow |g(x) - 5| < \epsilon.$$

Find $\delta > 0$ such that

$$|x - 2| < \delta \Rightarrow |f(x) + g(x) + 2| < \frac{1}{4}.$$

Solution. Assume that $\epsilon = \dfrac{1}{8}$. Then, we have

$$|x - 2| < \frac{1}{80} \Rightarrow |f(x) + 7| < \frac{1}{8} \text{ and } |x - 2| < \frac{1}{192} \Rightarrow |g(x) - 5| < \frac{1}{8}.$$

So, if we pick $\delta = \min\{\dfrac{1}{80}, \dfrac{1}{192}\} = \dfrac{1}{192}$, then we obtain

$$\begin{aligned} |f(x) + g(x) + 2| &= |(f(x) + 7) + (g(x) - 5)| \\ &\leq |f(x) + 7| + |g(x) - 5| \\ &< \frac{1}{8} + \frac{1}{8} = \frac{1}{4}, \end{aligned}$$

whenever $|x - 2| < \dfrac{1}{192}$.

62. *Prove that $\lim\limits_{x \to 0} \sin \dfrac{1}{x}$ does not exist.*

Solution. Suppose that $\lim\limits_{x \to 0} \sin \dfrac{1}{x} = L$. Then, choosing $\epsilon = \dfrac{1}{2}$, we can find a number $\delta > 0$ such that

$$\left|\sin\frac{1}{x} - L\right| < \frac{1}{2} \tag{2.2}$$

whenever $0 < |x| < \delta$. Let n be any positive integer such that

$$\frac{1}{(2n - \frac{1}{2})\pi} < \delta.$$

Then, both points

$$x_1 = \frac{1}{(2n + \frac{1}{2})\pi} \quad \text{and} \quad x_2 = \frac{1}{(2n - \frac{1}{2})\pi}$$

belong to the deleted neighborhood $0 < |x| < \delta$ and

$$\sin\frac{1}{x_1} = \sin(2n + \frac{1}{2})\pi = \sin\frac{\pi}{2} = 1,$$
$$\sin\frac{1}{x_2} = \sin(2n - \frac{1}{2})\pi = \sin\frac{-\pi}{2} = -1.$$

Now, from (2.2) we obtain

$$\left|\sin\frac{1}{x_1} - L\right| = |1 - L| < \frac{1}{2} \quad \text{and} \quad \left|\sin\frac{1}{x_2} - L\right| = |-1 - L| < \frac{1}{2}.$$

These two inequalities are incompatible, since the first implies $L > \frac{1}{2}$ and the second implies that $L < -\frac{1}{2}$. Thus the assumption that $\sin\frac{1}{x}$ has a limit at $x = 0$ leads to a contradiction.

63. *Prove that* $\lim\limits_{x\to 0}\cos\frac{1}{x}$ *does not exist.*

Solution. Suppose that $\lim\limits_{x\to 0}\cos\frac{1}{x} = L$. Let $\epsilon = \frac{1}{2}$. Then, there exists $\delta > 0$ such that

$$\left|\cos\frac{1}{x} - L\right| < \frac{1}{2} \tag{2.3}$$

whenever $0 < |x| < \delta$. Let n be any integer whose absolute value is so large that both points

$$x_1 = \frac{1}{(2n)\pi} \quad \text{and} \quad x_2 = \frac{1}{(2n + 1)\pi}$$

belong to $(-\delta, \delta)$. Then, we obtain

$$\cos \frac{1}{x_1} = \cos(2n)\pi = 1 \text{ and } \cos \frac{1}{x_2} = \cos(2n + 1)\pi = -1.$$

Hence, by (2.3) we have

$$\left| \cos \frac{1}{x_1} - L \right| = |1 - L| < \frac{1}{2} \text{ and } \left| \cos \frac{1}{x_2} - L \right| = |-1 - L| < \frac{1}{2}.$$

Therefore, we find that $L > \frac{1}{2}$ and $L < \frac{1}{2}$. This is a contradiction.

64. *Determine the following statements are true or false, justify the answer.*

(1) If $\lim\limits_{x \to a} f(x)$ exists and $\lim\limits_{x \to a} g(x)$ does not exist, then $\lim\limits_{x \to a} \left(f(x) + g(x) \right)$ does not exist.

(2) If neither $\lim\limits_{x \to a} f(x)$ nor $\lim\limits_{x \to a} g(x)$ exists, then $\lim\limits_{x \to a} \left(f(x) + g(x) \right)$ does not exist.

Solution. (1) This statement is true. If $\lim\limits_{x \to a} \left(f(x) + g(x) \right)$ exists, then

$$\lim_{x \to a} g(x) = \lim_{x \to a} \left(f(x) + g(x) - f(x) \right)$$
$$= \lim_{x \to a} \left(f(x) + g(x) \right) - \lim_{x \to a} f(x).$$

would also exist, and this is a contradiction.

(2) This statement is false. If we consider $f(x) = \sin \frac{1}{x}$ and $g(x) = -\sin \frac{1}{x}$, then neither $\lim\limits_{x \to 0} \sin \frac{1}{x}$ nor $\lim\limits_{x \to 0} -\sin \frac{1}{x}$ exist, but $\lim\limits_{x \to 0} \left(f(x) + g(x) \right) = \lim\limits_{x \to 0} \left(\sin \frac{1}{x} - \sin \frac{1}{x} \right) = 0$ exists.

65. *If r is any positive integer, prove that*

(1) $\lim\limits_{x \to \infty} \frac{1}{x^r} = 0$,

(2) $\lim\limits_{x \to -\infty} \frac{1}{x^r} = 0.$

Solution. (1) We must show that for any $\epsilon > 0$, there exists $N \in \mathbb{N}$ such that

$$\left| \frac{1}{x^r} - 0 \right| < \epsilon, \text{ whenever } x > N,$$

or, equivalently,

$$|x|^r > \frac{1}{\epsilon}, \text{ whenever } x > N,$$

or, equivalently,

$$|x| > (\frac{1}{\epsilon})^{\frac{1}{r}}, \text{ whenever } x > N,$$

Now, it is enough to take $N = (\frac{1}{\epsilon})^{\frac{1}{r}}$. So, if $N = (\frac{1}{\epsilon})^{\frac{1}{r}}$, then

$$\left| \frac{1}{x^r} - 0 \right| < \epsilon, \text{ whenever } x > N.$$

This proves (1).

(2) The proof of (2) is similar to the proof of (1).

66. *Let*

$$f(x) = \begin{cases} 1 \text{ if } x = \dfrac{1}{n}, \text{ where } n \text{ is a positive integer} \\ 0 \text{ otherwise.} \end{cases}$$

Show that $\lim_{x \to 0} f(x)$ *does not exist.*

Solution. Suppose that $\lim_{x \to 0} f(x) = L$. Then, for each $\epsilon > 0$, there exist $\delta > 0$ such that $|f(x) - L| < \epsilon$, whenever $0 < |x| < \delta$..

If $L \neq 0$, then take $\epsilon = \dfrac{|L|}{2}$. Hence, there exists $\delta > 0$ such that

$$|f(x) - L| < \frac{|L|}{2},$$

whenever $0 < |x| < \delta$. Now, if we consider $x_0 = -\dfrac{\delta}{2}$, then $0 < |x_0| < \delta$ but

$$|L| = |0 - L| = |f(x_0) - L| < \frac{|L|}{2},$$

and this is a contradiction.

If $L = 0$, then take $\epsilon = \dfrac{1}{2}$. Then, there exists $\delta > 0$ such that

$$|f(x)| < \frac{1}{2},$$

whenever $0 < |x| < \delta$. If n is a positive integer satisfying $n > \dfrac{1}{\delta}$, we consider $x_0 = \dfrac{1}{n}$. Then, $0 < |x_0| < \delta$ but

$$1 = |f(x_0)| < \frac{1}{2},$$

again, this is a contradiction.

Consequently, $\lim_{x \to 0} f(x)$ does not exist.

67. *Show that*

(1) $\sin x = 2^n \sin \dfrac{x}{2^n} \cos \dfrac{x}{2} \cos \dfrac{x}{4} \cos \dfrac{x}{8} \ldots \cos \dfrac{x}{2^n}$, *for all* $n \in \mathbb{N}$;

(2) $\dfrac{\sin x}{x} = \cos \dfrac{x}{2} \cos \dfrac{x}{4} \cos \dfrac{x}{8} \ldots$;

(3) $\dfrac{2}{\pi} = \dfrac{\sqrt{2}}{2} \cdot \dfrac{\sqrt{2 + \sqrt{2}}}{2} \cdot \dfrac{\sqrt{2 + \sqrt{2 + \sqrt{2}}}}{2} \cdots$

Solution. (1) We can write

$$
\begin{aligned}
\sin x &= 2 \sin \frac{x}{2} \cos \frac{x}{2} \\
&= 2(2 \sin \frac{x}{4} \cos \frac{x}{4}) \cos \frac{x}{2} \\
&= 2\left(2(2 \sin \frac{x}{8} \cos \frac{x}{8}) \cos \frac{x}{4}\right) \cos \frac{x}{2} \\
&\cdots \\
&= 2^n \sin \frac{x}{2^n} \cos \frac{x}{2} \cos \frac{x}{4} \cos \frac{x}{8} \ldots \cos \frac{x}{2^n}.
\end{aligned}
$$

(2) From part (1), we obtain

$$
\frac{\sin x}{x} \cdot \frac{\dfrac{x}{2^n}}{\sin \dfrac{x}{2^n}} = \cos \frac{x}{2} \cos \frac{x}{4} \cos \frac{x}{8} \ldots \cos \frac{x}{2^n},
$$

which implies that

$$
\lim_{n \to \infty} \left(\frac{\sin x}{x} \cdot \frac{\dfrac{x}{2^n}}{\sin \dfrac{x}{2^n}} \right) = \lim_{n \to \infty} \left(\cos \frac{x}{2} \cos \frac{x}{4} \cos \frac{x}{8} \ldots \cos \frac{x}{2^n} \right).
$$

This completes the proof.

(3) By consideration the formula $\cos x = \sqrt{\dfrac{1}{2}(1 + \cos 2x)}$, it is enough to take $x = \dfrac{\pi}{2}$ in part (2).

68. *A regular polygon is a polygon whose sides are all the same length and whose interior angles all have the same measure.*

(1) Show that P_n, the perimeter of a regular polygon with n sides inscribed in a circle with radius r is

$$
P_n = 2nr \sin \frac{\pi}{n}.
$$

(2) Show that $\lim_{n \to \infty} P_n$ is equal to the environment of its peripheral circle, that is, $2\pi r$.

Fig. 2.1 A regular polygon

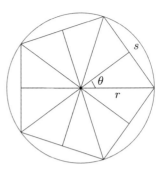

Fig. 2.2 A regular polygon when $n \to \infty$

Solution. (1) Let s be the length of any side of polygon and $\theta = \dfrac{\pi}{n}$ (see Fig. 2.1). Then, we have

$$\sin \frac{\pi}{n} = \frac{\frac{s}{2}}{r}.$$

This implies that $s = 2r \sin \dfrac{\pi}{n}$, and so $P_n = 2rn \sin \dfrac{\pi}{n}$.

(2) We have

$$\lim_{n \to \infty} P_n = \lim_{n \to \infty} 2rn \sin \frac{\pi}{n} = \lim_{n \to \infty} 2\pi r \frac{\sin \frac{\pi}{n}}{\frac{\pi}{n}} = 2\pi r.$$

Note that $2\pi r$ is the perimeter of the circle as we expected (see Fig. 2.2).

69. *Find* $\lim\limits_{x \to 0} \dfrac{2x \sin x}{\sec x - 1}$ *if it exists.*

Solution. We have

$$\lim_{x \to 0} \frac{2x \sin x}{\sec x - 1} = \lim_{x \to 0} \frac{2x \sin x}{\dfrac{1}{\cos x} - 1} = \lim_{x \to 0} \frac{2x \sin x \cos x}{1 - \cos x} = \lim_{x \to 0} \frac{2x \sin x \cos x}{2 \sin^2 \dfrac{x}{2}}$$

$$= \lim_{x \to 0} \frac{x \sin 2x}{2 \sin^2 \dfrac{x}{2}} = \lim_{x \to 0} \frac{\sin 2x}{\sin \dfrac{x}{2}} \cdot \frac{\dfrac{x}{2}}{\sin \dfrac{x}{2}} = 4.$$

70. Find $\lim\limits_{x \to \frac{\pi}{4}} \dfrac{\tan 2x}{\cot(\dfrac{\pi}{4} - x)}$ *if it exists.*

Solution. By using the formula $\tan(\alpha - \beta) = \dfrac{\tan \alpha - \tan \beta}{1 + \tan \alpha \tan \beta}$ we obtain

$$\lim_{x \to \frac{\pi}{4}} \frac{\tan 2x}{\cot(\dfrac{\pi}{4} - x)} = \lim_{x \to \frac{\pi}{4}} \frac{1 - \tan x}{1 + \tan x} \cdot \frac{\sin 2x}{\cos 2x}$$

$$= \lim_{x \to \frac{\pi}{4}} \frac{\cos x - \sin x}{\cos x + \sin x} \cdot \frac{2 \sin x \cos x}{(\cos x - \sin x)(\cos x + \sin x)}$$

$$= \lim_{x \to \frac{\pi}{4}} \frac{2 \sin x \cos x}{(\cos x + \sin x)^2}$$

$$= \frac{1}{2}.$$

71. Find $\lim\limits_{x \to \infty} \left(\sin \sqrt{x + 1} - \sin \sqrt{x} \right)$ *if it exists.*

Solution. We have

$$\sin \sqrt{x + 1} - \sin \sqrt{x} = 2 \cos \left(\frac{\sqrt{x + 1} + \sqrt{x}}{2} \right) \cdot \sin \left(\frac{\sqrt{x + 1} - \sqrt{x}}{2} \right).$$

We set $f(x) = 2 \cos \left(\dfrac{\sqrt{x + 1} + \sqrt{x}}{2} \right)$ and $g(x) = \sin \left(\dfrac{\sqrt{x + 1} - \sqrt{x}}{2} \right)$. Then, $|f(x)| \le 2$ and

$$\lim_{x \to \infty} g(x) = \lim_{x \to \infty} \sin \left(\frac{\sqrt{x + 1} - \sqrt{x}}{2} \cdot \frac{\sqrt{x + 1} + \sqrt{x}}{\sqrt{x + 1} + \sqrt{x}} \right) = \lim_{x \to \infty} \sin \left(\frac{1}{2(\sqrt{x + 1} + \sqrt{x})} \right) = 0.$$

Therefore, $\lim\limits_{x \to \infty} \left(\sin \sqrt{x + 1} - \sin \sqrt{x} \right) = \lim\limits_{x \to \infty} f(x)g(x) = 0.$

72. *Assuming that the following limit*

$$\lim_{x \to 0} \frac{x - \sin x}{x^3}$$

exists, evaluate it.

Solution. Suppose that $L = \lim\limits_{x \to 0} \dfrac{x - \sin x}{x^3}$. Let $x = 3y$. Then, we get

$$\lim_{y \to 0} \frac{3y - \sin 3y}{27 y^3}.$$

Using the formula $\sin 3y = 3 \sin y - 4 \sin^3 y$ we find that

$$
\begin{aligned}
L &= \lim_{y \to 0} \frac{3y - 3 \sin y + 4 \sin^3 y}{27 y^3} \\
&= \lim_{y \to 0} \left(\frac{y - \sin y}{9 y^3} + \frac{4 \sin^3 y}{27 y^3} \right) \\
&= \frac{1}{9} \lim_{y \to 0} \frac{y - \sin y}{y^3} + \frac{4}{27} \lim_{y \to 0} \left(\frac{\sin y}{y} \right)^3 \\
&= \frac{1}{9} L + \frac{4}{27}.
\end{aligned}
$$

Hence, we obtain $9L = L + \dfrac{4}{3}$, and so $L = \dfrac{1}{6}$.

73. *Find* $L = \lim\limits_{x \to 0} (\cot^2 x - \dfrac{1}{x^2})$.

Solution. We set $x = 2y$. Then, $L = \lim\limits_{y \to 0} (\cot^2 2y - \dfrac{1}{4 y^2})$. By using the formula $\cot 2y = \dfrac{\cot y - \tan y}{2}$ we obtain

$$
\begin{aligned}
L &= \lim_{y \to 0} \left(\frac{\cot^2 y + \tan^2 y - 2}{4} - \frac{1}{4 y^2} \right) \\
&= \lim_{y \to 0} \left(\frac{1}{4} (\cot^2 y - \frac{1}{y^2}) + \frac{\tan^2 y - 2}{4} \right) \\
&= \frac{1}{4} L - \frac{1}{2}.
\end{aligned}
$$

Therefore, $L = \dfrac{-2}{3}$.

74. *Find* $\lim\limits_{x \to 0} \sin x \left[\dfrac{1}{x} \right]$ *if it exists.*

Solution. Clearly, we have

$$\frac{1}{x} - 1 < \left[\frac{1}{x} \right] \le \frac{1}{x}. \tag{2.4}$$

If $0 < x < \dfrac{\pi}{2}$, then $\sin x > 0$. So, if we multiply (2.4) by $\sin x$, we obtain

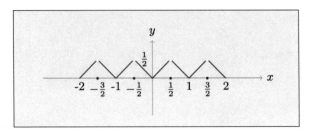

Fig. 2.3 Graph of the function f in Problem 75

$$\frac{\sin x}{x} - \sin x < \sin x \left[\frac{1}{x}\right] \le \frac{\sin x}{x}. \tag{2.5}$$

Since the limits of the left and right sides of (2.5) are equal, it follows that $\lim\limits_{x \to 0^+} \sin x \left[\frac{1}{x}\right] = 1$. If $-\frac{\pi}{2} < x < 0$, then $\sin x < 0$. In this case, in a similar way we obtain $\lim\limits_{x \to 0^-} \sin x \left[\frac{1}{x}\right] = 1$. Since the left and right hand limits are equal, we conclude that the limit exists and it is equal to 1 as $x \to 0$.

75. *Let f be defined on $\mathbb{R}$ by setting*

$$f(x) = \begin{cases} x & if\ 0 \le x < \dfrac{1}{2} \\ 0 & if\ x = \dfrac{1}{2} \\ 1 - x & if\ \dfrac{1}{2} < x \le 1 \end{cases}$$

and $f(x + n) = f(x)$, where n is any integer. Determine the points of discontinuity of f.

Solution. The function f graphed in Fig. 2.3.

Examining the graph of f, we find that f is discontinuous at $\dots, -\dfrac{3}{2}, -\dfrac{1}{2}, \dfrac{1}{2}, \dfrac{3}{2}, \dots$

76. *Suppose that $f : (0, \infty) \to \mathbb{R}$ is defined by*

$$f(x) = \frac{1}{x} - \left[\frac{1}{x}\right].$$

(1) Show that f is discontinuous at $x = \dfrac{1}{n}$, for all $n \in \mathbb{N}$.

(2) Show that $\lim\limits_{x \to 0^+} \left(\dfrac{1}{x} - \left[\dfrac{1}{x}\right]\right)$ does not exist.

Solution. (1) For each $n \in \mathbb{N}$, we have $f(\frac{1}{n}) = 0$. Now, if $\frac{1}{n} < x < \frac{1}{n-1}$, then $\left[\frac{1}{x}\right] = n - 1$ and $f(x) = \frac{1}{x} - (n-1)$. So, we obtain

$$\lim_{x \to \frac{1}{n}^+} f(x) = 1 \neq 0 = f(\frac{1}{n}).$$

Consequently, f is not continuous at $x = \frac{1}{n}$.

(2) Suppose that $\lim\limits_{x \to 0^+} (\frac{1}{x} - \left[\frac{1}{x}\right]) = L$. If $\epsilon = \frac{1}{4}$, then we can find $\delta > 0$ such that

$$\left|\frac{1}{x} - \left[\frac{1}{x}\right] - L\right| < \frac{1}{4}, \tag{2.6}$$

whenever $0 < x < \delta$. Let n be any large positive integer such that $\frac{1}{n} < \delta$. Then both

$$x_1 = \frac{1}{n} \text{ and } x_2 = \frac{1}{n + \dfrac{1}{2}}.$$

belong to the interval $(0, \delta)$. Then, we get $f(x_1) = 0$ and $f(x_2) = \frac{1}{2}$. Now, from (2.6), we have

$$|0 - L| < \frac{1}{4} \text{ and } \left|\frac{1}{2} - L\right| < \frac{1}{4}.$$

Thus, we get $-\frac{1}{4} < L < \frac{1}{4}$ and $\frac{1}{4} < L < \frac{3}{4}$. This is a contradiction.

77. *Give an example of a function* $f : [0, 1] \to \mathbb{R}$ *such that*

$$A = \{a \in [0, 1] : \lim_{x \to a} f(x) \text{ does not exist}\}$$

is countably infinite.

Solution. We define $f : [0, 1] \to \mathbb{R}$ as follows:

$$f(x) = \begin{cases} \dfrac{1}{n} \text{ if } x \in \left(\dfrac{1}{n+1}, \dfrac{1}{n}\right], & n \in \mathbb{N} \\ 0 \text{ if } x = 0. \end{cases}$$

Then, $A = \{\frac{1}{n} : n \in \mathbb{N}\}$ which is countably infinite.

78. *Suppose that* $f : \mathbb{R} \to \mathbb{R}$ *is a function such that* $f \circ f$ *is continuous. Determine whether* f *is continuous.*

Solution. f is not continuous in general. Consider the following function:

$$f(x) = \begin{cases} 0 \text{ if } x < -1 \\ 1 \text{ if } x \geq -1. \end{cases}$$

Clearly, $(f \circ f)(x) = 1$, for all $x \in \mathbb{R}$, but f is discontinuous.

79. *Find the largest value of k for which the function defined by $f(x) = [x^2 - 2]$ is continuous on the interval $[3, 3 + k)$.*

Solution. If $3 \leq x$, then $9 \leq x^2$ or $7 \leq x^2 - 2$. So, by considering $3 \leq x < \sqrt{10}$, we obtain $7 \leq x^2 - 2 < 8$ and $f(x) = 7$ is a continuous function. Consequently, $k = \sqrt{10} - 3$.

80. *Let f be a function defined on $\mathbb{R}$ by setting*

$$f(x) = \sqrt{x - [x]}, \quad for \; all \; x \in \mathbb{R}.$$

Show that f is discontinuous at the points $x = n$, where n is any integer, and is continuous at all other points.

Solution. If $x = n$ for some $n \in \mathbb{Z}$, then $f(x) = 0$.

If $x \neq n$ for all $n \in \mathbb{Z}$, then there is $m \in \mathbb{Z}$ such that $m < x < m + 1$, and so $f(x) = \sqrt{x - m}$. Consequently, we obtain

$$f(x) = \begin{cases} \sqrt{x - m} \text{ if } m < x < m + 1, \text{ for some } m \in \mathbb{Z} \\ 0 \qquad \text{ if } x \in \mathbb{Z}. \end{cases}$$

If $n \in \mathbb{Z}$, then $f(n) = 0$, and the left and right limits are not equal. Indeed, we have

$$\lim_{x \to n^+} f(x) = \lim_{x \to n^+} \sqrt{x - n} = 0 \text{ and } \lim_{x \to n^-} f(x) = \lim_{x \to n^-} \sqrt{x - (n - 1)} = 1.$$

Thus, for each $n \in \mathbb{Z}$, f is discontinuous at n. Obviously, f is continuous at all other points.

81. *Let f be a function defined on $\mathbb{R}$ by setting*

$$f(x) = \lim_{n \to \infty} \frac{x^{2n} - 1}{x^{2n} + 1}.$$

Examine f for continuity at all points $x \in \mathbb{R}$.

Solution. We consider the following cases:

(1) If $x = 1$ or $x = -1$, then $f(x) = 0$.
(2) If $-1 < x < 1$, then $f(x) = -1$.

(3) If $x > 1$ or $x < -1$, then $f(x) = \lim\limits_{n \to \infty} \dfrac{1 - \dfrac{1}{x^{2n}}}{1 + \dfrac{1}{x^{2n}}} = 1.$

Therefore, f is discontinuous at $x = 1$ and $x = -1$.

82. *If f is continuous at a and $f(a) > 0$, show that there exist positive numbers $\epsilon, \delta > 0$ with $|x - a| < \delta$ implying $f(x) > \epsilon$.*

Solution. Suppose that $\epsilon = \dfrac{f(a)}{2}$. Since f is continuous at a, there is $\delta > 0$ so that if $0 < |x - a| < \delta$ we have $|f(x) - f(a)| < \epsilon$, which implies that

$$f(x) > f(a) - \epsilon = f(a) - \frac{f(a)}{2} = \frac{f(a)}{2} = \epsilon.$$

83. *If f is a continuous function such that $f(x) = f(x^3)$ for all $x \in \mathbb{R}$, show that f is constant.*

Solution. Let $x \in \mathbb{R}$. Then, $x^{\frac{1}{3}} \in \mathbb{R}$, and $f(x) = f(x^{\frac{1}{3}})$. Thus, we conclude that

$$f(x) = f(x^{\frac{1}{3}}) = f(x^{\frac{1}{3^2}}) = \ldots = f(x^{\frac{1}{3^n}}) = \ldots.$$

Since f is continuous, it follows that

$$f(x) = \lim_{n \to \infty} f\left(x^{\frac{1}{3^n}}\right) = f\left(\lim_{n \to \infty} x^{\frac{1}{3^n}}\right) = f(1).$$

84. *How many continuous functions f are there which satisfy the equation*

$$(f(x))^2 = x^2, \quad for\ all\ x \in \mathbb{R}. \tag{2.7}$$

Solution. Suppose that f is a continuous function that satisfies (2.7)). Clearly, $f(0) = 0$.

If $I = (0, \infty)$, then f is continuous on I and $f(x) \neq 0$, for all $x \in I$. So,

$$f(x) = x, \ \text{for all } x \in I \tag{2.8}$$

or

$$f(x) = -x, \ \text{for all } x \in I. \tag{2.9}$$

If $J = (-\infty, 0)$, then f is continuous on J and $f(x) \neq 0$, for all $x \in J$. Thus,

$$f(x) = x, \ \text{for all } x \in J \tag{2.10}$$

or

$$f(x) = -x, \text{ for all } x \in J. \tag{2.11}$$

If we combine (2.8), (2.9), (2.10) and (2.11), then the following cases occur, $f(x) = x$ or $f(x) = -x$ or $f(x) = |x|$ or $f(x) = -|x|$.

85. *Let* $f : \mathbb{R} \to \mathbb{R}$ *be a continuous function such that the equation*

$$(f(x))^2 = x^2, \tag{2.12}$$

does not have solution. Show that f *is neither onto nor one to one.*

Solution. Since $(f(x))^2 = x^2$ does not have solution, it follows that there is not any x such that $f(x) = x$ or $f(x) = -x$. The graphs of $y = x$ and $y = -x$ divided the plane $\mathbb{R}^2$ into four sections. Since f is continuous, the graph of f does not have any intersection with graphs of $y = x$ or $y = -x$, and the graph of f is inside one of the four sections, thanks to continuity of f. Assume that the graph of f is in the above of the graphs $y = x$ and $y = -x$. In otherwise, we can substitute f with $-f$. Therefore, $f(x) \geq |x|$, for all $x \in \mathbb{R}$, and consequently, f is not onto.

Let $a = f(0)$ Then, $f(a) > a$ and $f(-a) > a$. Then, we have $f(0) < f(a)$ and $f(0) < f(-a)$. It is possible to find a number λ such that $f(0) < \lambda < f(a)$ and $f(0) < \lambda < f(-a)$. So, by the intermediate value theorem there exist $x_1 \in (0, a)$ and $x_2 \in (-a, 0)$ such that $f(x_1) = f(x_2) = \lambda$. Therefore, f is not one to one.

86. *Let* f *be a real continuous function. If* f *is one to one, show that* f *is either strictly decreasing or strictly increasing.*

Solution. Assume that f neither strictly decreasing nor strictly increasing. Then, there exist $a, b, c \in \mathbb{R}$ such that $a < b < c$, $f(a) > f(b)$ and $f(b) < f(c)$. We consider $\lambda \in \mathbb{R}$ such that $f(b) < \lambda < \min\{f(a), f(c)\}$. Then, by intermediate value theorem, there exist $x_1 \in (a, b)$ and $x_2 \in (b, c)$ such that $f(x_1) = \lambda$ and $f(x_2) = \lambda$. Since f is one to one, it follows that $x_1 = x_2$, and this is a contradiction.

87. *Let* $f : \mathbb{R} \to [0, \infty)$ *be a bijective function. Show that* f *is not continuous on* $\mathbb{R}$.

Solution. Assume that f is a continuous function. Then, by Problem (86), f is either strictly decreasing or strictly increasing. Since f is onto, it follows that there is $c \in \mathbb{R}$ such that $f(c) = 0$.

If f is strictly decreasing, then for every $c < x$ we get $f(x) < f(c)$, which is a contradiction.

If f is strictly increasing, then for every $x < c$ we have $f(x) < f(c)$, which is again a contradiction.

88. *Suppose that* f *is a real function defined on* $\mathbb{R}$ *which satisfies*

$$\lim_{h \to 0} \left(f(x + h) - f(x - h) \right) = 0,$$

for every $x \in \mathbb{R}$. *Does this imply that* f *is continuous?*

Solution. No, because if we consider the following function:

$$f(x) = \begin{cases} x & \text{if } x \neq 0 \\ 1 & \text{if } x = 0, \end{cases}$$

then f is discontinuous at 0, while $\lim\limits_{h \to 0} f(x + h) = \lim\limits_{h \to 0} f(x - h) = 0$ or $\lim\limits_{h \to 0} (f(x + h) - f(x - h)) = 0$.

89. *Let f and g be two real continuous functions. If $f(x) = g(x)$, for all $x \in \mathbb{Q}$, show that $f(x) = g(x)$, for all $x \in \mathbb{R}$.*

Solution. Suppose that a is an irrational number. Then, there exists a sequence $\{a_n\}$ of rational numbers such that $a_n \to a$ as $n \to \infty$. Since f and g are continuous, it follows that

$$g(a) = \lim_{n \to \infty} g(a_n) = \lim_{n \to \infty} f(a_n) = f(a).$$

Therefore, we conclude that $f(a) = g(a)$, for all $a \in \mathbb{R}$.

90. *Let f and g be two real continuous functions such that for any two arbitrary numbers $\alpha < \beta$, there exists $c \in (\alpha, \beta)$ with $f(c) = g(c)$. Prove that $f(x) = g(x)$, for all $x \in \mathbb{R}$.*

Solution. Let $x \in \mathbb{R}$ be fixed and $n \in \mathbb{N}$ be arbitrary. By assumption, we can find $c_n \in \mathbb{R}$ such that $x < c_n < x + \dfrac{1}{n}$ and $f(c_n) = g(c_n)$. This implies that $\lim\limits_{n \to \infty} c_n = x$ and $\lim\limits_{n \to \infty} f(c_n) = \lim\limits_{n \to \infty} g(c_n)$. Since f and g are continuous, it follows that

$$f\left(\lim_{n \to \infty} c_n \right) = g\left(\lim_{n \to \infty} c_n \right).$$

Consequently, we get $f(x) = g(x)$.

91. *Let f and g be two periodic functions such that $\lim\limits_{x \to \infty} (f(x) - g(x)) = 0$. Prove that $f(x) = g(x)$, for all $x \in \mathbb{R}$.*

Solution. Suppose that the positive numbers p and q are the period of f and g, respectively. We have

$$\lim_{n \to \infty} (f - g)(x + np) = \lim_{n \to \infty} (f - g)(x + nq) = \lim_{n \to \infty} (f - g)(x + np + nq).$$

On the other hand, for every $x \in \mathbb{R}$ and every $n \in \mathbb{N}$, we have

$$(f - g)(x) = (f - g)(x + np) + (f - g)(x + nq) - (f - g)(x + np + nq).$$

This implies that

$$(f - g)(x) = \lim_{n \to \infty} \Big((f - g)(x + np) + (f - g)(x + nq) - (f - g)(x + np + nq) \Big)$$
$$= 0.$$

Therefore, we conclude that $f(x) = g(x)$.

92. *Let*

$$f(x) = \begin{cases} x & if \ x \in \mathbb{Q} \\ 1 - x & if \ x \notin \mathbb{Q}. \end{cases}$$

Show that f is continuous only at $x = \dfrac{1}{2}$.

Solution. Suppose that f is continuous at x. Since rational and irrational numbers are dense in real numbers, it follows that there exist a sequence $\{a_n\}$ of rational numbers and a sequence $\{b_n\}$ of irrational numbers such that $a_n \to x$ and $b_n \to x$ as $n \to \infty$. Since f is continuous, it follows that

$$f(x) = f\Big(\lim_{n \to \infty} a_n \Big) = \lim_{n \to \infty} f(a_n) = \lim_{n \to \infty} a_n = x \qquad (2.13)$$

and

$$f(x) = f\Big(\lim_{n \to \infty} b_n \Big) = \lim_{n \to \infty} f(b_n) = \lim_{n \to \infty} (1 - b_n) = 1 - x. \qquad (2.14)$$

From (2.13) and (2.14), we get $x = \dfrac{1}{2}$.

93. *Notice that every rational x can be written in the form $x = \dfrac{n}{m}$, where $m > 0$, and m and n are integers without any common divisors. When $x = 0$, we take $n = 1$. Consider the function f defined on $\mathbb{R}$ by*

$$f(x) = \begin{cases} 0 & if \ x \ is \ irrational \\ \dfrac{1}{n} & if \ x = \dfrac{m}{n}. \end{cases}$$

Prove that f is continuous at x if and only if x is irrational.

Solution. Let f be continuous at $a \in \mathbb{Q}$. Then, there exists a sequence of irrational numbers $\{a_n\}$ such that $a_n \to a$ as $n \to \infty$. Since f is continuous at a, it follows that

$$\lim_{n \to \infty} f(a_n) = f(\lim_{n \to \infty} a_n) = f(a).$$

So, we obtain $f(a) = 0$, a contradiction. Thus, f is discontinuous at every rational number.

Now, let b be an irrational number. We show that $\lim_{x \to b} f(x) = f(b) = 0$. Let $\epsilon > 0$ be given. Then, there exists $n \in \mathbb{N}$ such that $\dfrac{1}{n} < \epsilon$. Pick $\delta > 0$ such that all rational numbers $x = \dfrac{r}{s}$ in interval $|x - b| < \delta$ have the denominator greater than n, i.e., $s > n$. So, if $x \in \mathbb{R}$ is an irrational number such that $|x - b| < \delta$, then $|f(x) - f(b)| = 0 < \epsilon$. But if $x = \dfrac{r}{s}$ is a rational number, then

$$|f(x) - f(b)| = |f(x)| = \left|f\left(\frac{r}{s}\right)\right| = \frac{1}{s} < \frac{1}{n} < \epsilon.$$

Therefore, f is continuous at b.

94. *Show that the function*

$$f(x) = \begin{cases} (-1)^r & if \ \dfrac{1}{r+1} \le x < \dfrac{1}{r}, \ r = 1, 2, 3, \ldots \\ 0 & if \ x = 0 \\ 1 & if \ x = 1 \end{cases}$$

is discontinuous at $0, 1, \dfrac{1}{2}, \ldots, \dfrac{1}{r}, \ldots$.

Solution. We have

$$\lim_{x \to 0^+} f(x) = \lim_{r \to \infty} (-1)^r, \ \text{and this limit not exist.}$$

Hence, we conclude that f is discontinuous at $x = 0$. Similarly, we have

$$f(1) = 1 \text{ and } \lim_{x \to 1^-} f(x) = -1.$$

So, f is discontinuous at $x = 1$. For $r = 2k$ or $r = 2k - 1$ $(k = 1, 2, \ldots)$, we have

$$f\left(\frac{1}{2k}\right) = (-1)^{2k-1} = -1, \quad \lim_{x \to \frac{1}{2k}^+} f(x) = -1 \text{ and } \lim_{x \to \frac{1}{2k}^-} f(x) = 1;$$

$$f\left(\frac{1}{2k-1}\right) = (-1)^{2k-2} = 1, \quad \lim_{x \to \frac{1}{2k-1}^+} f(x) = 1 \text{ and } \lim_{x \to \frac{1}{2k-1}^-} f(x) = -1.$$

Therefore, f is discontinuous at $x = \dfrac{1}{r}$, for $r = 1, 2, \ldots$.

95. *Let f be a function defined on $[0, 1]$ by setting*

$$f(x) = \begin{cases} \dfrac{1}{2^n} & if \ \dfrac{1}{2^{n+1}} < x \le \dfrac{1}{2^n}, \ for \ n = 0, 1, 2, \ldots \\ 0 & if \ x = 0. \end{cases}$$

Show that f is continuous except at points $x = \dfrac{1}{2}, \dfrac{1}{2^2}, \ldots, \dfrac{1}{2^n}, \ldots$. Describe the nature of discontinuity at each of these points.

Solution. Since $f(0) = \lim\limits_{x\to 0^+} f(x) = 0$, it follows that f is continuous at $x = 0$.

Similarly, since $f(1) = \lim\limits_{x\to 1^-} f(x) = 1$, it follows that f is continuous at $x = 1$.

Now, let n be an arbitrary non-negative integer. Then, $f\left(\dfrac{1}{2^n}\right) = \dfrac{1}{2^n}$, and

$$\lim_{x\to \frac{1}{2^n}^+} f(x) = \frac{1}{2^{n-1}} \quad \text{and} \quad \lim_{x\to \frac{1}{2^n}^-} f(x) = \frac{1}{2^n}.$$

So, we conclude that f is discontinuous at $x = \dfrac{1}{2^n}$, for all $n \in \mathbb{N}$. Since the right-hand and left-hand limits exist but are not equal, a jump discontinuity occurs. Clearly, f is continuous at all other points.

96. *Let $a_m \neq 0$ and $f(x) = a_0 + a_1 x + \cdots + a_m x^m$. Prove that*

$$\lim_{n\to\infty} \frac{f(n+1)}{f(n)} = 1.$$

Solution. We can write

$$
\begin{aligned}
\frac{f(n+1)}{f(n)} &= \frac{a_0 + a_1(n+1) + \cdots + a_m(n+1)^m}{a_0 + a_1 n + \cdots + a_m n^m} \\
&= \frac{a_m n^m \left(\dfrac{a_0}{a_m n^m} + \dfrac{a_1(n+1)}{a_m n^m} + \cdots + \dfrac{a_{m-1}(n+1)^{m-1}}{a_m n^m} + (\dfrac{n+1}{n})^m \right)}{a_m n^m \left(\dfrac{a_0}{a_m n^m} + \dfrac{a_1}{a_m n^{m-1}} + \cdots + \dfrac{a_{m-1}}{a_m n} + 1 \right)} \\
&= \frac{g(n) + (\dfrac{n+1}{n})^m}{h(n) + 1},
\end{aligned}
$$

where $g(n) \to 0$ and $h(n) \to 0$ as $n \to \infty$. Since $\lim\limits_{n\to\infty} (\dfrac{n+1}{n})^m = 1$, it follows that

$\lim\limits_{n\to\infty} \dfrac{f(n+1)}{f(n)} = 1$.

97. *Let f and g be two periodic functions with period p and q, respectively, such that*

$$\lim_{x\to 0} \frac{f(x)}{x} = a \quad \text{and} \quad \lim_{x\to 0} \frac{g(x)}{x} = b \neq 0,$$

where a and b are real numbers. Prove that

$$\lim_{n\to\infty} \frac{f\left((2+\sqrt{3})^n p\right)}{g\left((4+\sqrt{11})^n q\right)} = 0.$$

Solution. We can write

$$\frac{f\left((2+\sqrt{3})^n p\right)}{g\left((4+\sqrt{11})^n q\right)} = \frac{f\left((2+\sqrt{3})^n p + (2-\sqrt{3})^n p - (2-\sqrt{3})^n p\right)}{g\left((4+\sqrt{11})^n q + (4-\sqrt{11})^n q - (4-\sqrt{11})^n q\right)}$$

$$= \frac{f\left(((2+\sqrt{3})^n + (2-\sqrt{3})^n)p - (2-\sqrt{3})^n p\right)}{g\left(((4+\sqrt{11})^n + (4-\sqrt{11})^n)q - (4-\sqrt{11})^n q\right)}$$

$$= \frac{f\left(rp - (2-\sqrt{3})^n p\right)}{g\left(sq - (4-\sqrt{11})^n q\right)}, \quad \text{where } r,s \in \mathbb{Z}$$

$$= \frac{f\left(-(2-\sqrt{3})^n p\right)}{g\left(-(4-\sqrt{11})^n q\right)}.$$

Therefore, we get

$$\lim_{n\to\infty} \frac{f\left((2+\sqrt{3})^n p\right)}{g\left((4+\sqrt{11})^n q\right)}$$

$$= \lim_{n\to\infty} \frac{f\left(-(2-\sqrt{3})^n p\right)}{-(2-\sqrt{3})^n p} \cdot \frac{(2-\sqrt{3})^n p}{(4-\sqrt{11})^n q} \cdot \frac{-(4-\sqrt{11})^n q}{g\left(-(4-\sqrt{11})^n q\right)}$$

$$= \frac{p}{q} \lim_{n\to\infty} \frac{f\left(-(2-\sqrt{3})^n p\right)}{-(2-\sqrt{3})^n p} \cdot \left(\frac{2-\sqrt{3}}{4-\sqrt{11}}\right)^n \cdot \frac{-(4-\sqrt{11})^n q}{g\left(-(4-\sqrt{11})^n q\right)} = 0.$$

Note that $0 < 4 - \sqrt{11} < 1$, $0 < 2 - \sqrt{3} < 1$ and $0 < \dfrac{2-\sqrt{3}}{4-\sqrt{11}} < 1$.

98. *Let $f : \mathbb{R} \to \mathbb{R}$ be a function such that for every $x, y \in \mathbb{R}$ we have*

$$f(x+y) = f(x) + f(y). \tag{2.15}$$

Prove that

(1) If f is continuous at x_0, then f is continuous in every point of $\mathbb{R}$;
(2) If f is continuous in one point of $\mathbb{R}$, then $f(x) = ax$, where $a = f(1)$.

Solution. (1) If f is continuous at x_0, then $\lim_{h\to 0}\left(f(x_0+h) - f(x_0)\right) = 0$. By Equation (2.15), for every $x \in \mathbb{R}$ we have

$$f(x+h) - f(x) = f(h) = f(x_0+h) - f(x_0).$$

Thus, $\lim_{h\to 0}\left(f(x+h) - f(x)\right) = 0$, this yield that f is continuous in every arbitrary point x.

(2) Similar to Problem (58), we can see that $f(x) = xf(1)$, for all $x \in \mathbb{Q}$. By using (1), we can suppose that f is continuous in every point of $\mathbb{R}$. Let $x \in \mathbb{R}$ be an arbitrary point. There exists a sequence of rational numbers $\{r_n\}_{n=0}^{\infty}$ such that $r_n \to x$. Hence, we have

$$f(x) = f\left(\lim_{n \to \infty} r_n\right) = \lim_{n \to \infty} f(r_n) = \lim_{n \to \infty} r_n f(1) = xf(1).$$

99. *Let $n \in \mathbb{N}$. Find all continuous functions $f : \mathbb{R}^+ \to \mathbb{R}^+$ such that*

$$f(x) + f(y) = f\left(\sqrt[n]{x^n + y^n}\right), \quad \text{for all } x, y \in \mathbb{R}^+. \tag{2.16}$$

Solution. Since x, y are positive, it follows that (2.16) is equivalent to

$$f\left(\sqrt[n]{x}\right) + f\left(\sqrt[n]{y}\right) = f\left(\sqrt[n]{x+y}\right).$$

If we define $g(x) = \sqrt[n]{x}$, for all $x \in \mathbb{R}^+$, then we get

$$f \circ g(x) + f \circ g(y) = f \circ g(x + y).$$

Now, similar to Problem (98), we conclude that $f \circ g(x) = ax$. Since g is one to one and onto, it follows that $f(x) = ax^n$, where $a = f(1)$.

100. *Let f and g be two continuous functions on an interval I, $f(x) \neq 0$, for all $x \in I$, and $(f(x))^2 = (g(x))^2$, for all $x \in I$. Prove that*

$$\left(\forall x \in I, f(x) = g(x)\right) \text{ or } \left(\forall x \in I, f(x) = -g(x)\right).$$

Solution. Suppose this proposition is false. So there exist $x_1, x_2 \in I$ such that $f(x_1) = g(x_1)$ and $f(x_2) = -g(x_2)$. Since f is continuous on I and $f(x) \neq 0$ for all $x \in I$, we conclude that $f(x_1)$ and $f(x_2)$ have the same sign. Hence, $g(x_1)$ and $g(x_2)$ have opposite signs. So, by the intermediate value theorem, there exists c between x_1 and x_2 such that $g(c) = 0$. Since $(f(x))^2 = (g(x))^2$, for all $x \in I$, it follows that $f(c) = 0$. This is a contradiction. So, we were wrong to assume the proposition was false.

101. *Suppose that $f : [0, 1] \to [0, 1]$ is a continuous function. Prove that f has a fixed point in $[0, 1]$, i.e., there is at least one real number $x_0 \in [0, 1]$ such that $f(x_0) = x_0$.*

Solution. Let $g(x) = f(x) - x$. Then, $g(0) = f(0) \geq 0$ and $g(1) = f(1) - 1 \leq 0$. If $f(0) = 0$, then 0 is a fixed point and if $f(1) = 1$, then 1 is a fixed point. Otherwise, we have $g(1) < 0 < g(0)$. Now, by intermediate value theorem, there is a point $x_0 \in (0, 1)$ such that $g(x_0) = 0$ which implies that $f(x_0) = x_0$.

102. *Let $f, g : [0, 1] \to [0, 1]$ be two continuous functions. Prove that there exist $x_0, y_0 \in [0, 1]$ such that $f(x_0) = y_0$ and $g(y_0) = x_0$.*

Solution. We define $h : [0, 1] \to [0, 1]$ by $h(x) = g(f(x)) - x$. Then, we have

$$h(0) = g(f(0)) \geq 0,$$
$$h(1) = g(f(1)) - 1 \leq 0.$$

Consequently, by the intermediate value theorem, there is $x_0 \in [0, 1]$ such that $h(x_0) = 0$ or $g(f(x_0)) = x_0$. Now, it is enough to put $y_0 = f(x_0)$.

103. *Suppose that f is a continuous function on $[0, 2]$ such that $f(0) = f(2)$. Show that there exist real numbers x, $y \in [0, 2]$ such that $f(x) = f(y)$ and $|y - x| = 1$.*

Solution. We define $g(x) = f(x) - f(x - 1)$. Then, we have

$$g(1) = f(1) - f(0),$$
$$g(2) = f(2) - f(1) = f(0) - f(1) = -\left(f(1) - f(0)\right).$$

Hence, $g(1)$ and $g(2)$ have different signs. Now, by the intermediate value theorem, there exists $\xi \in (1, 2)$ such that $g(\xi) = 0$. This means that $f(\xi) = f(\xi - 1)$. Now, it is enough to take $y = \xi$ and $x = \xi - 1$.

104. *Let f be a continuous function on $[a, b]$ and let α and β be two positive real numbers. Show that there exists $c \in [a, b]$ such that $\alpha f(a) + \beta f(b) = (\alpha + \beta) f(c)$.*

Solution. Since f is continuous on $[a, b]$, it follows that there are $p, q \in [a, b]$ such that $f(p)$ and $f(q)$ are the maximum and minimum of f, respectively. So, we have $f(q) \leq f(a) \leq f(p)$ and $f(q) \leq f(b) \leq f(p)$. Since α and β are positive, it follows that

$$\alpha f(q) \leq \alpha f(a) \leq \alpha f(p) \tag{2.17}$$

and

$$\beta f(q) \leq \beta f(b) \leq \beta f(p). \tag{2.18}$$

From (2.17) and (2.18) we get

$$(\alpha + \beta) f(q) \leq \alpha f(a) + \beta f(b) \leq (\alpha + \beta) f(p),$$

or equivalently

$$f(q) \leq \frac{\alpha f(a) + \beta f(b)}{\alpha + \beta} \leq f(p).$$

Now, by the intermediate value theorem, there is $c \in [a, b]$ such that

$$\frac{\alpha f(a) + \beta f(b)}{\alpha + \beta} = f(c).$$

105. *Let $f : [a, b] \to \mathbb{R}$ be a continuous function and $x_1, \ldots, x_n \in [a, b]$. Prove that there exists $\xi \in [a, b]$ such that*

$$f(\xi) = \frac{f(x_1) + \cdots + f(x_n)}{n}.$$

Solution. Since f is continuous on the closed interval $[a, b]$, it takes its maximum and minimum on this interval. Suppose that

$$f(x_i) = \min\{f(x_1), \ldots, f(x_n)\} \text{ and } f(x_j) = \max\{f(x_1), \ldots, f(x_n)\},$$

for some $1 \le i, j \le n$. Thus, $f(x_i) \le f(x_k) \le f(x_j)$, for all $1 \le k \le n$. This implies that $nf(x_i) \le f(x_1) + \cdots + f(x_n) \le nf(x_j)$ or

$$f(x_i) \le \frac{f(x_1) + \cdots + f(x_n)}{n} \le f(x_j).$$

Now, by the intermediate value theorem, there exists ξ between x_i and x_j such that
$$f(\xi) = \frac{f(x_1) + \cdots + f(x_n)}{n}.$$

106. *Let $f : [a, b] \to \mathbb{R}$ be a continuous function and $x_1, \ldots, x_n \in [a, b]$. Prove that there exists $\xi \in [a, b]$ such that*

$$\frac{1}{2}n(n + 1)f(\xi) = f(x_1) + 2f(x_2) + 3f(x_3) + \cdots + nf(x_n).$$

Solution. Similar to Problem (105), suppose that

$$f(x_i) = \min\{f(x_1), \ldots, f(x_n)\} \text{ and } f(x_j) = \max\{f(x_1), \ldots, f(x_n)\},$$

for some $1 \le i, j \le n$. Then, we have

$$f(x_i) + 2f(x_i) + \cdots + nf(x_i) \le f(x_1) + 2f(x_2) + \cdots + nf(x_n)$$
$$\le f(x_j) + 2f(x_j) + \cdots + nf(x_j),$$

and so we obtain

$$(1 + 2 + \cdots n)f(x_i) \le f(x_1) + 2f(x_2) + \cdots + nf(x_n) \le (1 + 2 + \cdots n)f(x_j).$$

Since $1 + 2 + \cdots n = \dfrac{n(n + 1)}{2}$, it follows that

$$\frac{n(n + 1)}{2}f(x_i) \le f(x_1) + 2f(x_2) + \cdots + nf(x_n) \le \frac{n(n + 1)}{2}f(x_j).$$

This implies that

$$f(x_i) \leq \frac{2}{n(n+1)}\Big(f(x_1) + 2f(x_2) + 3f(x_3) + \cdots + nf(x_n)\Big) \leq f(x_j).$$

Now, by the intermediate value theorem, there exists ξ between x_i and x_j such that

$$f(\xi) = \frac{2}{n(n+1)}\Big(f(x_1) + 2f(x_2) + 3f(x_3) + \cdots + nf(x_n)\Big).$$

This completes the proof.

107. *Show that there is a positive number a such that $3^a = a^5$.*

Solution. We define $f(x) = 3^x - x^5$. Clearly, f is continuous on $\mathbb{R}$. Moreover, we have $f(0) = 1 > 0$ and $f(2) = 9 - 32 = -23 < 0$. Now, by the intermediate value theorem, there is $a \in (0, 2)$ such that $f(a) = 0$. This implies that $3^a - a^5 = 0$ or $3^a = a^5$.

108. *Let $f : \mathbb{R} \to \mathbb{R}$ be a continuous function and $\lim\limits_{x \to \pm\infty} \dfrac{f(x)}{x} = 0$. Prove that there exists $c \in \mathbb{R}$ such that $c - f(c) = 0$.*

Solution. Since $\lim\limits_{x \to \infty} \dfrac{f(x)}{x} = 0$, then for $\epsilon = 1$, we can find a number $M_1 > 0$ such that $\left| \dfrac{f(x)}{x} \right| < 1$ whenever $x > M_1$. Since $x > 0$, it follows that $\dfrac{|f(x)|}{x} < 1$ or $-x < f(x) < x$, for all $x > M_1$. Hence

$$x - f(x) > 0 \text{ whenever } x > M_1 > 0. \tag{2.19}$$

Similarly, since $\lim\limits_{x \to -\infty} \dfrac{f(x)}{x} = 0$, then for $\epsilon = 1$, we can find a number $M_2 > 0$ such that $\left| \dfrac{f(x)}{x} \right| < 1$ whenever $x < -M_2$. Since $x < 0$, it follows that $\dfrac{|f(x)|}{-x} < 1$ or $x < f(x) < -x$, for all $x < -M_2$. Hence

$$x - f(x) < 0 \text{ whenever } x < -M_2 < 0. \tag{2.20}$$

Now, by (2.19), (2.20) and the intermediate value theorem, there exists $c \in \mathbb{R}$ such that $c - f(c) = 0$.

109. *Let f be a continuous function on an interval I and the range of f is contained in rational numbers. Prove that f is a constant function.*

Solution. Suppose that there exist $a, b \in I$ such that $f(a) \neq f(b)$. Without loss of generality, we may assume that $f(a) < f(b)$ and so there exists an irrational number k such that $f(a) < k < f(b)$. Since f is continuous, by the intermediate value theorem there exists c between a and b such that $f(c) = k$. This is a contradiction.

110. *Prove that a polynomial f of odd degree has at least one real root.*

Solution. Let $f(x) = a_0 + a_1 x + \cdots + a_n x^n$, where $a_n \neq 0$ and n is odd. We may suppose that $a_n = 1$, otherwise we would work with $(\frac{1}{a_n})f$. We have

$$\lim_{x \to \infty} f(x) = \infty \quad \text{and} \quad \lim_{x \to -\infty} f(x) = -\infty.$$

Since f is continuous, it follows that $f(x) < 0$ for some x and $f(x') > 0$ for some other x'. Now by the intermediate value theorem, there exists a real number ξ such that $f(\xi) = 0$.

111. *Let $f : [0, 1] \to \mathbb{R}$ be a continuous function such that $f(0) = f(1)$ and $n \in \mathbb{N}$. Prove that there is $\xi \in [0, 1]$ such that $f(\xi) = f(\xi + \frac{1}{n})$.*

Solution. If $n = 1$, then $\xi = 0$ is the answer. Assume that $n \neq 1$. We consider the function $g(x) = f(x) - f(x + \frac{1}{n})$. Then, we have

$$g(0) = f(0) - f\left(\frac{1}{n}\right),$$
$$g\left(\frac{1}{n}\right) = f\left(\frac{1}{n}\right) - f\left(\frac{2}{n}\right),$$
$$g\left(\frac{2}{n}\right) = f\left(\frac{2}{n}\right) - f\left(\frac{3}{n}\right),$$
$$\vdots$$
$$g\left(\frac{n-1}{n}\right) = f\left(\frac{n-1}{n}\right) - f(1).$$

If we add the above equalities together, we get

$$g(0) + g\left(\frac{1}{n}\right) + g\left(\frac{2}{n}\right) + \cdots + g\left(\frac{n-1}{n}\right) = 0.$$

If one of the numbers $g(0)$, $g(\frac{1}{n})$, $g(\frac{2}{n}), \ldots, g(\frac{n-1}{n})$ is equal to 0, the proof of our statement completes. In otherwise, at least one of them must be positive and one of them negative. Let $a, b \in \{0, \frac{1}{n}, \frac{2}{n}, \ldots, \frac{n-1}{n}\}$ such that $g(a) < 0$ and $g(b) > 0$. Then, by the intermediate value theorem there is ξ between a and b such that $g(\xi) = 0$. This implies that $f(\xi) = f(\xi + \frac{1}{n})$.

112. *A function f is said to satisfy the Lipschitz condition of order α on an interval $[a, b]$ if*

$$|f(x) - f(x_0)| < k|x - x_0|^\alpha,$$

for all x, x_0 in the interval. Show that f is continuous at x_0 if $\alpha > 0$.

Solution. Let $\alpha > 0$. We observe that

$$-k|x - x_0|^\alpha < f(x) - f(x_0) < k|x - x_0|^\alpha.$$

Since

$$\lim_{x \to x_0} k|x - x_0|^\alpha = \lim_{x \to x_0} -k|x - x_0|^\alpha = 0,$$

it follows that $\lim_{x \to x_0} (f(x) - f(x_0)) = 0$ or $\lim_{x \to x_0} f(x) = f(x_0)$, i.e., f is continuous at x_0.

113. *A real valued function f defined on (a, b) is said to be convex if*

$$f(\lambda x + (1 - \lambda)y) \le \lambda f(x) + (1 - \lambda)f(y),$$

whenever $x, y \in (a, b)$ and $0 < \lambda < 1$. Geometrically, this inequality means that the line segment between $(x, f(x))$ and $(y, f(y))$, which is the chord from x to y, lies above the graph of f. Prove that every convex function is continuous.

Solution. We show that f satisfies the Lipschitz condition on any closed interval $[x_0, y_0] \subset (a, b)$. We choose $\epsilon > 0$ such that $x_0 - \epsilon, y_0 + \epsilon \in (a, b)$. Suppose that $x, y \in [x_0, y_0]$ and $x \ne y$. We pick

$$z = y + \frac{\epsilon}{|y - x|}(y - x) \tag{2.21}$$

and

$$\lambda = \frac{|y - x|}{\epsilon + |y - x|}. \tag{2.22}$$

Clearly, $0 < \lambda < 1$ and $z \in [x_0 - \epsilon, y_0 + \epsilon]$. From (2.22), we obtain

$$\epsilon = \frac{|y - x|(1 - \lambda)}{\lambda}. \tag{2.23}$$

We put ϵ from (2.23) into (2.21). Then, we get $z = \frac{1}{\lambda}y - \frac{(1 - \lambda)}{\lambda}x$, which implies that $y = \lambda z + (1 - \lambda)x$. Since f is convex, it follows that

$$f(y) = f(\lambda z + (1 - \lambda)x) \le \lambda f(z) + (1 - \lambda)f(x).$$

So, $f(y) - f(x) \le \lambda(f(z) - f(x))$. Since f is bounded on $[x_0, y_0]$ with bound M, it follows that

$$f(y) - f(x) \le 2\lambda M$$
$$= 2\frac{|y - x|}{\epsilon + |y - x|}M$$
$$< 2\frac{|y - x|}{\epsilon}M.$$

If we set $\dfrac{2M}{\epsilon} = k$, then $f(y) - f(x) < k|y - x|$. Similarly, we obtain $f(x) - f(y) < k|y - x|$. Therefore, $|f(x) - f(y)| < k|y - x|$.

2.3 Exercises

Easier Exercises

By using the definition of limit, prove that

1. $\lim\limits_{x \to 1} (x^2 + x + 1) = 3,$

2. $\lim\limits_{x \to a} \sqrt[3]{x} = \sqrt[3]{a},$

3. $\lim\limits_{x \to 5} \dfrac{5}{x - 4} = 5.$

By using the definition of limit, prove that

4. $\lim\limits_{x \to -4} \left| \dfrac{5 - x}{4 + x} \right| = \infty,$

5. $\lim\limits_{x \to 2} \dfrac{3}{(x - 2)^2} = \infty,$

6. $\lim\limits_{x \to 3} \dfrac{-1}{(x - 3)^2} = -\infty.$

7. Give a definition for each of the following:

   (a) $\lim\limits_{x \to \infty} f(x) = -\infty,$

   (b) $\lim\limits_{x \to -\infty} f(x) = \infty,$

   (c) $\lim\limits_{x \to -\infty} f(x) = -\infty.$

8. We have
$$\lim\limits_{x \to \infty} \left(\frac{1}{x} - x^3 \right) = -\infty \text{ and } \lim\limits_{x \to -\infty} \left(\frac{1}{x} - x^3 \right) = \infty.$$

   How would you define these statements precisely and prove them?

9. Find the vertical and horizontal asymptotes of the graph of the equation $xy^2 - 2y^2 - 4x = 0$, and draw a sketch of the graph.

10. Define the *sign function* $sgn : \mathbb{R} \to \mathbb{R}$ by
$$sgn(x) = \begin{cases} 1 & \text{if } x > 0 \\ 0 & \text{if } x = 0 \\ -1 & \text{if } x < 0. \end{cases}$$

   Prove that $\lim\limits_{x \to 0} sgn(x)$ does not exist.

11. If
$$f(x) = \begin{cases} x & \text{if } x < 1 \\ 2 + (x - 1)^2 & \text{if } x \geq 1, \end{cases}$$

   prove that $\lim\limits_{x \to 1} f(x)$ does not exist. Find the value of the given limit:

12. $\lim\limits_{x\to-1} \dfrac{2x^2 - x - 3}{x^3 + 2x^2 + 6x + 5}$,

13. $\lim\limits_{x\to4} \dfrac{2x^3 - 11x^2 + 10x + 8}{3x^3 - 17x^2 + 16x + 16}$,

14. $\lim\limits_{x\to5} \dfrac{4 - \sqrt{3x+1}}{x^2 - 7x + 10}$,

15. $\lim\limits_{x\to3} \dfrac{\sqrt{x+13} - 2\sqrt{x+1}}{x^2 - 9}$,

16. $\lim\limits_{x\to0} \dfrac{(1+3x)^3 - (1+2x)^4}{x^2}$,

17. $\lim\limits_{x\to0} \dfrac{\sqrt[3]{x+1} - 1}{x}$.

Find the one-sided limit if it exists:

18. $\lim\limits_{x\to0^+} \dfrac{x + x^2}{|x|}$,

19. $\lim\limits_{x\to1^-} \dfrac{[x] - 1}{x^2 - 1}$,

20. $\lim\limits_{x\to2^+} \dfrac{[x] - 1}{[x] - x}$,

21. $\lim\limits_{x\to1^+} \dfrac{[x^2] - [x]^2}{x^2 - 1}$.

Evaluate the value of the given limit:

22. $\lim\limits_{x\to-\infty} \left(3x + \dfrac{1}{x^2}\right)$,

23. $\lim\limits_{x\to\infty} \left(\sqrt{x^2 + x} - x\right)$,

24. $\lim\limits_{x\to-\infty} \dfrac{\sqrt{x^4 + 1}}{2x^2 - 3}$,

25. $\lim\limits_{x\to-\infty} \left(\sqrt[3]{x^3 + x} - \sqrt[3]{x^3 + 1}\right)$,

26. $\lim\limits_{x\to\infty} \sqrt{x + \sqrt{x + \sqrt{x}}} - \sqrt{x})$,

27. $\lim\limits_{x\to0} \dfrac{1 + \sqrt{x} + \sqrt[3]{x}}{1 + \sqrt[3]{x} + \sqrt[4]{x}}$.

Evaluate the limit if it exists:

28. $\lim\limits_{x\to0} \sin\left(\cot x\right) \sin x$,

29. $\lim\limits_{x\to0} \dfrac{\sin\left(\sin(\sin x)\right)}{\sin(\sin x)}$,

30. $\lim\limits_{x\to\frac{\pi}{2}} \dfrac{\frac{\pi}{2} - x}{\cos x}$,

31. $\lim\limits_{x\to\frac{\pi}{3}} \dfrac{\sin(x - \frac{\pi}{3})}{1 - 2\cos x}$,

32. $\lim\limits_{x\to\infty} \dfrac{e^x}{x}$,

33. $\lim\limits_{x\to0} \sin\left(x \sin(\dfrac{1}{x})\right)$,

34. $\lim\limits_{x\to0} \dfrac{\sqrt{1 - \cos x^2}}{1 - \cos x}$,

35. $\lim\limits_{x\to0} \dfrac{\cos(\sin x) - \cos x}{x^4}$,

36. $\lim\limits_{x\to0} \dfrac{3\tan 4x - 12\tan x}{3\sin 4x - 12\sin x}$,

37. $\lim\limits_{x\to\frac{\pi}{2}} \dfrac{\tan 3x}{\tan x}$.

38. If $\lim\limits_{x\to0} \dfrac{f(x)}{x^2} = 1$, show that $\lim\limits_{x\to0} \dfrac{f(x)}{x} = 0$.

Find the asymptotes (vertical, horizontal or/and oblique) for the given function:

39. $f(x) = \dfrac{x^3 + 1}{x^2 - 9}$,

40. $f(x) = \dfrac{2x^2 - 4x + 5}{x - 3}$,

41. $f(x) = \sqrt{x^2 + 4x}$,

42. $f(x) = \dfrac{2x + 5}{\sqrt{x^2 - 2x - 3}}$.

43. Prove that $f(x) = |x|$ is a continuous function on $\mathbb{R}$.

44. Prove that if f is continuous at x_0 and g is discontinuous at x_0, then $f + g$ is discontinuous at x_0.

45. If the function g is continuous at x_0 and the function f is discontinuous at x_0, is it possible for the quotient of the two functions, $\frac{f}{g}$, to be continuous at x_0? Prove your answer.

Find all points x_0 at which the function is continuous:

46. $(1 - \sin^2 x)^{-\frac{1}{2}}$,

47. $\ln(\sin x + \cos x)$,

48. $\dfrac{1}{\sqrt{2 \cosh x - 1}}$,

49. $\cot\left(1 - e^{-x^2}\right)$,

50. $\sin\left(\dfrac{1}{\cos x}\right)$,

51. $\tan(x + \ln x)$,

52. $\dfrac{1}{x - \ln x}$,

53. $\sin(e^{-x^2})$,

54. $\sin^{-1}(\cos x)$.

55. In which point the following function is continuous?

$$f(x) = \begin{cases} e^{-\frac{1}{x^2}} & \text{if } x \neq 0 \\ 0 & \text{if } x = 0. \end{cases}$$

56. Find the values of the constants a and b that make the following function continuous on $\mathbb{R}$, and draw a sketch of the graph of f:

$$f(x) = \begin{cases} 6x + 6a & \text{if } x < -3 \\ 3ax - 7b & \text{if } -3 \leq x \leq 3 \\ x - 12b & \text{if } 3 \leq x. \end{cases}$$

57. Show that the inverse of the discontinuous function $f(x) = (1 + x^2)sgn(x)$, $(x \in \mathbb{R})$ is a continuous function.

58. If $f : [0, 1] \to \mathbb{R}$ is continuous, prove that $g = \min\{f(x), 0\}$ is also continuous.

59. Prove that if $f : \mathbb{R} \to \mathbb{R}$ and $g : \mathbb{R} \to \mathbb{R}$ are continuous and $f(a) < g(a)$, then there exists an interval $I = (a - \delta, a + \delta)$ such that $f(z) < g(x)$, for all $x \in I$.

60. Let $f : [0, \infty) \to \mathbb{R}$ be continuous and $\lim\limits_{x \to \infty} f(x)$ exists. Prove that f is bounded.

61. Let $f : \mathbb{R} \to \mathbb{R}$ be continuous and periodic. Does it imply that f is bounded?

62. If $f : \mathbb{R} \to [0, \infty)$ is a one to one correspondence, prove that f is not continuous.

63. Show that the equation $x^6 + 2x - 2 = 0$ has exactly two real roots.

64. Show that the equation $x^2 - 10 = x - \sin x$ has a real solution.

65. Show that the equation $1 - \dfrac{x^2}{4} = \cos x$ has at least three real solution.

66. Show that the equation $\cos \pi x = x$ has one and only one root between 0 and $\dfrac{1}{2}$.

67. Prove that $x2^x = 1$ for some x in $(0, 1)$.

68. Give an example of a function f on $[0, 1]$ which is not continuous but it satisfies the intermediate value theorem.

69. Let $f : \mathbb{R} \to \mathbb{R}$ be a continuous function such that $f(x + 2\pi) = f(x)$, for all $x \in \mathbb{R}$. Show that there exists $\xi \in \mathbb{R}$ such that $f(\xi + \pi) = f(\xi)$.

70. If n is a positive integer and if $a > 0$ be a real number, prove that there exists exactly one positive real number b such that $b^n = a$.
71. Let f be a convex function in $(-\infty, \infty)$ and assume that $\lim_{x \to -\infty} f(x) = \infty$. Is it possible that $\lim_{x \to \infty} f(x) = -\infty$?
72. Let f be a convex function in $(-\infty, \infty)$ and assume that $\lim_{x \to -\infty} f(x) = 0$. Is it possible that $\lim_{x \to \infty} f(x) = -\infty$?

Harder Exercises

73. Prove that $\lim_{x \to a}$ does not exist (finite) if for some $\epsilon_0 > 0$, every deleted neighborhood of a contains points x_1 and x_2 such that

$$|f(x_1) - f(x_2)| \geq \epsilon_0.$$

74. Prove that $\lim_{x \to 0} \left| \tan(\frac{1}{x}) \right|$ does not exist.
75. Let

$$f(x) = \begin{cases} 0 & \text{if } x \text{ is rational} \\ 1 & \text{if } x \text{ is irrational} \end{cases}$$

and let a be an arbitrary real number. Prove that $\lim_{x \to a} f(x)$ does not exist.
76. Let

$$f(x) = \begin{cases} 1 & \text{if } x = \frac{1}{n}, \text{ where } n \text{ is a positive integer} \\ 0 & \text{otherwise.} \end{cases}$$

Show that

(a) If $a \neq 0$, then $\lim_{x \to a} f(x) = 0$;
(b) $\lim_{x \to 0} f(x)$ does not exist.

77. Let $a > 1$ and $k > 0$. Prove that $\lim_{x \to \infty} \dfrac{a^{\sqrt{x}}}{x^k} = \infty$.

Let m and n be positive integers. Find the value of the given limit:

78. $\lim_{x \to 1} \left(\dfrac{n}{x^n - 1} - \dfrac{m}{x^m - 1} \right)$,

79. $\lim_{x \to 1} \dfrac{1 + x^2 + x^3 + \cdots + x^n - n}{x - 1}$.

80. If $f(x) \geq 0$ for $a < x < x_0$ and $\lim_{x \to x_0^-} f(x)$ exists, prove that $\lim_{x \to x_0^-} f(x) \geq 0$. Conclude that from this if $f(x) \geq g(x)$ for $a < x < x_0$, then

$$\lim_{x \to x_0^-} f(x) \geq \lim_{x \to x_0^-} g(x)$$

if both limits exist.

81. If f and g are continuous at x_0 in $\mathbb{R}$, prove that $\min\{f, g\}$ and $\max\{f, g\}$ are continuous at x_0.

82. Let $f : [0, 1] \to \mathbb{R}$ be a one to one correspondence. If f is continuous, prove that f^{-1} is also continuous.

83. A function f is defined as follows:

$$f(x) = \begin{cases} 2 \cos x & \text{if } x \leq c \\ ax^2 + b & \text{if } x > c, \end{cases}$$

where a, b and c are constant. If b and c are given, find all values of a (if any exist) for which f is continuous at the point c.

84. Let f be a continuous function defined on $[a, b]$, where a and b are rational numbers and

$$f(x) = \begin{cases} x & \text{if } x \text{ is rational} \\ a + b - x & \text{if } x \text{ is irrational.} \end{cases}$$

At what points of $[a, b]$ is f continuous?

85. Suppose that $f : [0, \pi] \to \mathbb{R}$ is defined by

$$f(x) = \begin{cases} x \sin(\frac{1}{x}) - \frac{1}{x} \cos(\frac{1}{x}) & \text{if } x \neq 0 \\ 0 & \text{if } x = 0. \end{cases}$$

Is f continuous?

86. Suppose that $f : [0, \infty) \to \mathbb{R}$ is defined by $f(x) = [x^2] \sin \pi x$. Show that f is continuous at each $x \neq \sqrt{n}$, where $n \in \mathbb{N}$.

87. Let f be a function defined on $\mathbb{R}$ as follows:

$$f(x) = \lim_{n \to \infty} \frac{x}{1 + (2 \sin x)^{2n}}.$$

Examine f for continuity at all points $x \in \mathbb{R}$.

88. Let f be a function defined on $[a, b)$ and define

$$F(x) = \max_{a \leq t \leq x} f(t), \quad (a \leq x \leq b).$$

(How do we know F is well defined?) Prove that F is continuous on $[a, b)$.

89. Let $f, g : \mathbb{R} \to \mathbb{R}$ be continuous such that given any two points $a < b$, there exists a real number c such that $a < c < b$ and $f(c) = g(c)$. Show that $f(x) = g(x)$, for all $x \in \mathbb{R}$.

90. If the domain of f is the set of all real numbers and f is continuous at 0, prove that if $f(a + b) = f(a) \cdot f(b)$ for all $a, b \in \mathbb{R}$, then f is continuous at every point.

91. Let $f : \mathbb{R} \rightarrow \mathbb{R}$ be a continuous function such that $f(x) = f(x^2)$, for all $x \in \mathbb{R}$. Prove that f is constant.

92. Determine necessary and sufficient conditions on a pair of sets A and B so that they will have the property that there exists a continuous function $f : \mathbb{R} \rightarrow \mathbb{R}$ such that $f(x) = 0$ for all $x \in A$ and $f(x) = 1$ for all $x \in B$.

93. Let $f : [0, 2] \rightarrow \mathbb{R}$ be a continuous function and $f(0) = f(2)$. Show that there exist real numbers $a_1, a_2, b_1, b_2 \in [0, 2]$ such that $a_2 - a_1 = 1$, $b_2 - b_1 = \dfrac{1}{2}$, $f(a_1) = f(a_2)$ and $f(b_1) = f(b_2)$.

94. Prove that the equation $|x^{31} + x^8 + 20| = x^{32}$ has at least one positive real root.

95. Show that the equation $\sin x = e^x$ has infinitely many solutions.

96. Given any two triangles in the plane, show that there is one line that bisects both of them.

97. A hiker begins a backpacking trip at 6 am on Saturday morning, arriving at camp at 6 pm that evening. The next day, the hiker returns on the same trial leaving at 6 am in the morning and finishing at 6 pm. Show that there is some place on the trial that the hiker visited at the same time of the day both coming and going.

98. For which real number c does there exist a continuous real function f satisfying $f(f(x)) = cx^9$, for all $x \in \mathbb{R}$.

99. Let $f : [0, 1] \rightarrow \mathbb{R}$ be a continuous function. Show that there exists $x_0 \in [0, 1]$ such that

$$f(x_0) = \frac{1}{3}\left(f(\frac{1}{4}) + f(\frac{1}{2}) + f(\frac{3}{4})\right).$$

100. If a function f (not necessary monotonic) has left hand limits and right hand limits at every point of an open interval I, prove that f must be continuous except on a countable set.

101. If f is a one to one continuous function on $[a, b]$, prove that f is strictly monotone on $[a, b]$.

102. Suppose that f is a continuous function on $[0, 1]$ with the property that for each $y \in \mathbb{R}$, does not exist $x \in [0, 1]$ such that $f(x) = y$ or exists exactly one $x \in [0, 1]$ such that $f(x) = y$. Prove that f is strictly monotone on $[0, 1]$.

103. Suppose that f is a function on $[0, 1]$ with the property that for each $y \in \mathbb{R}$, does not exist $x \in [0, 1]$ such that $f(x) = y$ or exists exactly two $x_1, x_2 \in [0, 1]$ such that $f(x_1) = f(x_2) = y$.

    (a) Prove that f is not continuous on $[0, 1]$.
    (b) Give an example of f with the above property.
    (c) Prove that each function with the above property will have infinite discontinuities on $[0, 1]$.

104. Let A be a subset of $\mathbb{R}$. Give an example of a strictly increasing function f defined on A such that f^{-1} is not continuous on $f(A)$.

105. Prove that every increasing convex function of a convex function is convex.

106. If f is convex in (a, b) and $a < s < t < u < b$, show that

$$\frac{f(t) - f(s)}{t - s} \leq \frac{f(u) - f(s)}{u - s} \leq \frac{f(u) - f(t)}{u - t}.$$

107. Let f be a continuous real function defined in (a, b) such that

$$f\left(\frac{x + y}{2}\right) \leq \frac{f(x) + f(y)}{2},$$

for all $x, y \in (a, b)$. Prove that f is convex.

Chapter 3
Derivatives

3.1 Basic Concepts and Theorems

Definition of Derivative

Let f be a real valued function defined on an open interval containing x. We say that f is *differentiable* at x, or that f has a *derivative* at x, denoted by $f'(x)$, if the limit $f'(x) = \lim\limits_{\Delta x \to 0} \dfrac{f(x + \Delta x) - f(x)}{\Delta x}$ exists and is finite. Other symbols that are used instead of $f'(x)$ are $D_x f(x)$ and $\dfrac{d}{dx} f(x)$. If a is a particular number in the domain of f, then

$$f'(a) = \lim_{\Delta x \to 0} \frac{f(a + \Delta x) - f(a)}{\Delta x} \tag{3.1}$$

In (3.1) let $a + \Delta x = x$, then $\Delta x \to 0$ is equivalent to $x \to a$. So, we get the following formula $f'(a) = \lim\limits_{x \to a} \dfrac{f(x) - f(a)}{x - a}$.

Tangent line: Let f be a continuous function and $P(a, f(a))$ be a fixed point. We wish to define the slop of the tangent line to the graph of f at P. Let I be the open interval that contains a and on which f is defined. Let $Q(a + \Delta x, f(a + \Delta x))$ be another point on the graph of f such that $a + \Delta x$ is also in I. Draw a line through P and Q (see Fig. 3.1). Any line through two points on a curve is called a *secant line*. So, the slope of the secant line PQ is given by

$$m_{PQ} = \frac{f(a + \Delta x) - f(a)}{\Delta x}.$$

Now, we move the point Q along the graph of f toward P. This is equivalent to stating that Δx approaches zero. This discussion leads to the following definition.

Suppose that f is a continuous function at a. The *tangent line* to the graph of f at $P(a, f(a))$ is the line through P having slop

© The Author(s), under exclusive license to Springer Nature Singapore Pte Ltd. 2020
B. Davvaz, *Examples and Problems in Advanced Calculus: Real-Valued Functions*,
https://doi.org/10.1007/978-981-15-9569-1_3

Fig. 3.1 Secant line

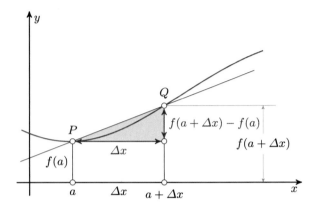

$$m = \lim_{\Delta x \to 0} m_{PQ} = \lim_{\Delta x \to 0} \frac{f(a + \Delta x) - f(a)}{\Delta x}$$

if this limits exists and is finite. Therefore, the graph $y = f(x)$ has a non-vertical tangent line at $P(a, f(a))$ if and only if f is differentiable at a, and the slop of tangent line is then equal to $f'(a)$. Hence, an equation of the tangent line to the graph of $y = f(x)$ at the point $(a, f(a))$ is

$$y = f'(a)(x - a) + f(a).$$

Note that if $f'(a) = 0$, then the tangent line is the horizontal line $y = f(a)$.

Normal line: An equation of the *normal* to the graph of $y = f(x)$ at the point $(a, f(a))$ is

$$y = -\frac{1}{f'(a)}(x - a) + f(a)$$

if $f'(a) \neq 0$ and $x = a$ if $f'(a) = 0$.

Tangent line approximation: For a differentiable function f, the equation of the tangent line to f at $x = a$ can be used to approximate $f(x)$ for x near a. Therefore, we can write

$$f(x) \approx f(a) + f'(a)(x - a),$$

for x near a (where $\approx$ denotes approximately equal to). We call the linear function

$$L(x) = f(a) + f'(a)(x - a),$$

the *linear approximation*, or *tangent line approximation* of f at $x = a$.

Differential: If $y = f(x)$, then the *differential of x*, denoted by dx, is given by $dx = \Delta x$, where Δx is an arbitrary increment of x, and x is any number in the domain of f'. The *differential of y*, denoted by dy, is given by $dy = f'(x)dx$.

One sided derivatives: Similar to one sided limits we can define *one sided deriva-*

tives. The *derivative from the right* of f at a denoted by $f'_+(a)$ and is defined by $f'_+(a) = \lim\limits_{x \to a^+} \dfrac{f(x) - f(a)}{x - a}$ and the *derivative from the left* of f at a denoted by $f'_-(a)$ and is defined by $f'_-(a) = \lim\limits_{x \to a^-} \dfrac{f(x) - f(a)}{x - a}$. Therefore, $f'(a)$ exists if and only if $f'_+(a)$ and $f'_-(a)$ both exist and are equal, in this case $f'(a) = f'_+(a) = f'_-(a)$. If $f'_+(a)$ and $f'_-(a)$ both exist but are not equal, then the curve $y = f(x)$ is said to have a *corner* at the point $(a, f(a))$.

Differentiability implies continuity: If f is differentiable at a, then f is continuous at a. Note that the converse is not true in general, for instance, $f(x) = |x|$ is continuous at 0 but $f'_+(0) \neq f'_-(0)$.

Derivative Rules

Suppose that f and g are differentiable at x. Then, $f \pm g$, fg and $\dfrac{f}{g}$ are differentiable at x and

(1) $(f \pm g)'(x) = f'(x) \pm g'(x)$,
(2) $(fg)'(x) = f'(x)g(x) + f(x)g'(x)$,
(3) $\left(\dfrac{f}{g}\right)'(x) = \dfrac{f'(x)g(x) - f(x)g'(x)}{\big(g(x)\big)^2}$, where $g(x) \neq 0$.

A list of derivative of elementary functions is presented in Table 3.1.

Chain rule: Let f be differentiable at x and g be differentiable at $f(x)$. If $h = g \circ f$, then h is differentiable at x and

$$h'(x) = g'\big(f(x)\big)f'(x).$$

If $y = f(x)$ is a differentiable function, then its derivative $f'(x)$ is also a function. If f' is also differentiable, then we can differentiate f' to get a new function of x denoted by f''. So, $f'' = (f')'$. The function f'' is called the *second derivative* of f because it is the derivative of the first derivative. It is written in several ways: $f''(x)$, y'', $\dfrac{d^2y}{dx^2}$, $\dfrac{d}{dx}\left(\dfrac{dy}{dx}\right)$, $D_x^2 f(x)$. If f'' is differentiable, its derivative, $f'''(x) = y''' = \dfrac{d^3y}{dx^3}$, is the *third derivative* of f. The names continue as you imagine. We denote the nth derivative of f by $f^{(n)}$. The *nth derivative* of the function f, where $n \in \mathbb{N}$, is the derivative of $f^{(n-1)}$. So far, we were dealing with given function $y = f(x)$. These functions are called *explicit functions*. But there are equations involving y's and x's can not be necessarily solved for y. In this case, y is called an *implicit function* of x. In order to find y', when y is an implicit function of x, we need a method called *implicit differentiation*.

Method of implicit differentiation: Given an equation involving the variables x

Table 3.1 A list of derivative of elementary functions

$f(x)$	$f'(x)$	$f(x)$	$f'(x)$
c	0	a^x	$a^x \ln a$
x^n	nx^{n-1}	$\ln|x|$	$\dfrac{1}{x}$
$\sin x$	$\cos x$	$\log_a |x|$	$\dfrac{1}{x \ln a}$
$\cos x$	$-\sin x$	$\sinh x$	$\cosh x$
$\tan x$	$\sec^2 x$	$\cosh x$	$\sinh x$
$\cot x$	$-\csc^2 x$	$\tanh x$	$\operatorname{sech}^2 x$
$\sec x$	$\sec x \tan x$	$\coth x$	$-\operatorname{csch}^2 x$
$\csc x$	$-\csc x \cot x$	$\operatorname{sech} x$	$-\operatorname{sech} x \tanh x$
$\sin^{-1} x$	$\dfrac{1}{\sqrt{1-x^2}}$	$\operatorname{csch} x$	$-\operatorname{csch} x \coth x$
$\cos^{-1} x$	$\dfrac{-1}{\sqrt{1-x^2}}$	$\sinh^{-1} x$	$\dfrac{1}{\sqrt{x^2+1}}$
$\tan^{-1} x$	$\dfrac{1}{1+x^2}$	$\cosh^{-1} x$	$\dfrac{1}{\sqrt{x^2-1}}$
$\cot^{-1} x$	$\dfrac{-1}{1+x^2}$	$\tanh^{-1} x$	$\dfrac{1}{1-x^2}$
$\sec^{-1} x$	$\dfrac{1}{|x|\sqrt{x^2-1}}$	$\coth^{-1} x$	$\dfrac{1}{1-x^2}$
$\csc^{-1} x$	$\dfrac{-1}{|x|\sqrt{x^2-1}}$	$\operatorname{sech}^{-1} x$	$\dfrac{-1}{x\sqrt{1+x^2}}$
e^x	e^x	$\operatorname{csch}^{-1} x$	$\dfrac{-1}{x\sqrt{1+x^2}}$

and y, the derivative of y is found using implicit differentiation as follows: (1) Apply $\dfrac{d}{dx}$ to both sides of the equation. Remember that in the process of applying the derivative rules, y' will appear, possibly more than once; (2) Solve for y'.

There are an important class of problems in which we are given the rate of change on one quantity, usually with respect to time, and are asked to find the rate of change of another related quantity. The method of solving such problems involving *related rates* closely resembles the technique of implicit differentiation, but there are now two dependent variables.

Some Important Theorems

The following theorems are of the most important theorems in calculus.

Derivative of an inverse function: Let f be continuous and one to one in a neighborhood of the point x, and suppose that f has a finite non-zero derivative at x, equal to $f'(x)$. Then, f^{-1} has a derivative at the point $y = f(x)$, equal to

$$\left(f^{-1}\right)'(y) = \frac{1}{f'(x)}, \tag{3.2}$$

where $x = f^{-1}(y)$.

Rolle's theorem: If f is a continuous function on $[a, b]$ which is differentiable in (a, b) and $f(a) = f(b)$, then there exists $c \in (a, b)$ such that $f'(c) = 0$.

Mean value theorem: If f is a continuous function on $[a, b]$ which is differentiable in (a, b), then there exists $c \in (a, b)$ such that

$$f(b) - f(a) = f'(c)(b - a).$$

Cauchy's mean value theorem: If f and g are two continuous functions on $[a, b]$ which are differentiable in (a, b), then there exists $c \in (a, b)$ such that

$$\left(f(b) - f(a)\right)g'(c) = \left(g(b) - g(a)\right)f'(c).$$

The following theorem is frequently useful in the evaluation of limits.

L'Hospital's rule for indeterminate form $\frac{0}{0}$: Let f and g be functions that are differentiable in a deleted neighborhood of a and $g'(x) \neq 0$ for all $x \neq a$. If $\lim_{x \to a} f(x) = \lim_{x \to a} g(x) = 0$ and

$$\lim_{x \to a} \frac{f'(x)}{g'(x)} = L,$$

then

$$\lim_{x \to a} \frac{f(x)}{g(x)} = L.$$

L'Hospital's rule also holds if either x increase without bound or x decreases without bound.

L'Hospital's rule for indeterminate form $\frac{\infty}{\infty}$: Let f and g be functions that are differentiable in a deleted neighborhood of a and $g'(x) \neq 0$ for all $x \neq a$. If $\lim_{x \to a} f(x) = \infty$ or $-\infty$ and $\lim_{x \to a} g(x) = \infty$ or $-\infty$, and if

$$\lim_{x \to a} \frac{f'(x)}{g'(x)} = L,$$

then

$$\lim_{x \to a} \frac{f(x)}{g(x)} = L.$$

L'Hospital's rule is valid if all the limits are right hand limits or the limits are left hand limits.

Taylor's theorem: Let f be a real function on $[a, b]$, n be a natural number, $f^{(n-1)}$

is continuous on $[a, b]$ and $f^{(n)}(t)$ exists for all $t \in (a, b)$. Suppose that α and β are two distinct points of $[a, b]$, and define

$$P(t) = \sum_{k=0}^{n-1} \frac{f^{(k)}(\alpha)}{k!} (t - \alpha)^k.$$

Then, there exists x between α and β such that

$$f(\beta) = P(\beta) + \frac{f^{(n)}(x)}{n!} (\beta - \alpha)^n.$$

For $n = 1$, this is just the mean value theorem. In general, Taylor's theorem shows that f can be approximated by a polynomial of degree $n - 1$.

Maximum and Minimum Values of a Function

A function f is said to have a *local maximum value* at c if there exists a neighborhood of c, on which f is defined, such that $f(x) \leq f(c)$ for all x in this neighborhood. A function f is said to have a *local minimum value* at c if there exists a neighborhood of c, on which f is defined, such that $f(x) \geq f(c)$ for all x in this neighborhood. If the function f has either a local maximum or a local minimum value at c, then f is said to have a *local extremum* at c.

Test for absolute extremum: Let f be a continuous function on $[a, b]$ and suppose f has local extremum at $c_1, c_2, \ldots, c_n$ of (a, b) and only at these points. Then, $M = \max\{f(a), f(c_1), f(c_2), \ldots, f(c_n), f(b)\}$ is the absolute maximum of f on $[a, b]$ and $m = \min\{f(a), f(c_1), f(c_2), \ldots, f(c_n), f(b)\}$ is the absolute minimum of f on $[a, b]$.

Fermat's theorem: If f has a local maximum or minimum at c, then $f'(c)$ either fails to exist or $f'(c) = 0$.

A *critical point* of a function f is a point c in the domain of f such that $f'(c)$ does not exist or $f'(c) = 0$.

Monotonicity theorem: Suppose that f is differentiable in (a, b).

(1) If $f'(x) \geq 0$ for all $x \in (a, b)$, then f is increasing.
(2) If $f'(x) \leq 0$ for all $x \in (a, b)$, then f is decreasing.
(3) If $f'(x) = 0$ for all $x \in (a, b)$, then f is constant.

First derivative test for a local extremum: Suppose that c is a critical point of a function f.

(1) If f' changes sign from positive to negative at c, then f has a local maximum at c.
(2) If f' changes sign from negative to positive at c, then f has a local minimum at c.

Second derivative test for a local extremum: Suppose that c is a critical point of a function f and $f''(c)$ exists and is non-zero.

(1) If $f''(c) < 0$, then f has a local maximum at c.
(2) If $f''(c) > 0$, then f has a local minimum at c.

Concavity and Points of Inflection

Let f be a differentiable function on an interval I and suppose f' is monotonic on I. Then, we say that f is *concave upward* on I if f' is increasing on I, and *concave downward* on I if f' is decreasing on I.

Concavity test: Let f has a continuous derivative f' on an interval I, and suppose that f'' exists and has the same sign at every interior point of I.

(1) If f'' is positive at every interior point of I, then f is concave upward on I.
(2) If f'' is negative at every interior point of I, then f is concave downward on I.

A function f is said to have *inflection point* at c or $(c, f(c))$ is a reflection point of the graph of f, if f changes its concavity at c. Let f be a differentiable function on an open interval containing c. If $(c, f(c))$ is a reflection point of the graph of f and $f''(c)$ exists, then $f''(c) = 0$.

Third derivative test for an inflection point: If $f''(c) = 0$ and if $f'''(c)$ exists and non-zero, then f has an inflection point at c.

If the reader is interested to see the proof of theorems that are presented in this chapter, we refer him/her to [1–16].

3.2 Problems

114. *Let $f(x) = \sin x$. Using the definition of derivative, calculate $f'(x)$.*

Solution. By the definition, we have

$$
\begin{aligned}
f'(x) &= \lim_{\Delta x \to 0} \frac{f(x + \Delta x) - f(x)}{\Delta x} = \lim_{\Delta x \to 0} \frac{\sin(x + \Delta x) - \sin(x)}{\Delta x} \\
&= \lim_{\Delta x \to 0} \frac{\sin x \cos \Delta x + \cos x \sin \Delta x - \sin x}{\Delta x} \\
&= \lim_{\Delta x \to 0} \frac{\sin x (\cos \Delta x - 1) + \cos x \sin \Delta x}{\Delta x} \\
&= \sin x \lim_{\Delta x \to 0} \frac{\cos \Delta x - 1}{\Delta x} + \cos x \lim_{\Delta x \to 0} \frac{\sin \Delta x}{\Delta x} \\
&= \sin x \cdot (0) + \cos x \cdot (1) = \cos x.
\end{aligned}
$$

115. *Interprets*

$$\lim_{x \to 0} \frac{\cosh^n x - 1}{x}, \quad for \ n \in \mathbb{N},$$

as a derivative and evaluate the limit.

Solution. Let $f(x) = \cosh^n x$. Then, $f'(0) = \lim_{x \to 0} \dfrac{f(x) - f(0)}{x - 0}$. Since $\cosh x = \dfrac{e^x + e^x}{2}$, it follows that $f(0) = 1$. So, $f'(0) = \lim_{x \to 0} \dfrac{\cosh^n x - 1}{x}$. On the other hand, $f'(x) = n \cosh^{n-1} x \sinh x$, and so $f'(0) = 0$.

116. *Estimate $\sqrt[3]{25}$ by using the tangent line approximation.*

Solution. If Δx is small, $\Delta y = f(x + \Delta x) - f(x) \approx f'(x)\Delta x$. Suppose that $f(x) = \sqrt[3]{x}$. Then, we obtain

$$\sqrt[3]{x + \Delta x} - \sqrt[3]{x} \approx \frac{1}{3} x^{-\frac{2}{3}} \Delta x.$$

If $x = 27$ and $\Delta x = -2$, then

$$\sqrt[3]{27 - 2} - \sqrt[3]{27} \approx \frac{1}{3}(27)^{-\frac{2}{3}}(-2).$$

This implies that $\sqrt[3]{25} \approx 3 - \dfrac{2}{27}$ or 2.926.

117. *A function f is defined as follows:*

$$f(x) = \begin{cases} x^2 & if \ x \le c \\ ax + b & if \ x > c, \end{cases}$$

where a, b, c are constants. Find values of a and b (in terms of c) such that $f'(c)$ exists.

Solution. The derivative $f'(c)$ exists if and only if $f'_+(c)$ and $f'_-(c)$ both exist and are equal. We have

$$f'_+(c) = \lim_{x \to c^+} \frac{f(x) - f(c)}{x - c} = \lim_{x \to c^+} \frac{ax + b - c^2}{x - c}$$

and

$$f'_-(c) = \lim_{x \to c^-} \frac{f(x) - f(c)}{x - c} = \lim_{x \to c^-} \frac{x^2 - c^2}{x - c} = 2c.$$

So, $\lim_{x \to c^+} \dfrac{ax + b - c^2}{x - c} = 2c$ or $\lim_{x \to c^+} \dfrac{ax + b - c^2}{2cx - 2c^2} = 1$. This implies that $b = -c^2$ and $a = 2c$.

118. *For what values of the positive integer n does the curve*

$$f(x) = \begin{cases} b & if \ x < a \\ b + (x - a)^n & if \ x \geq a \end{cases}$$

have a corner at the point (a, b)?

Solution. We compute the right and left-hand derivatives at $x = a$. We obtain

$$f'_+(a) = \lim_{x \to a^+} \frac{f(x) - f(a)}{x - a} = \lim_{x \to a^+} \frac{b + (x - a)^n - b}{x - a} = \lim_{x \to a^+} (x - a)^{n-1}$$
$$= \begin{cases} 1 & if \ n = 1 \\ 0 & if \ n \neq 1 \end{cases}$$

and

$$f'_-(a) = \lim_{x \to a^-} \frac{f(x) - f(a)}{x - a} = \lim_{x \to a^-} \frac{b - b}{x - a} = 0.$$

So, the curve f has a corner at (a, b) only for $n = 1$.

119. *Find the values of a and b such that*

$$f(x) = \begin{cases} a \sin x + b \cos x & if \ x < 0 \\ x^2 + 1 & if \ x \geq 0 \end{cases}$$

is differentiable at $x = 0$.

Solution. First, f must be continuous at $x = 0$. We have

$$f(0) = 1 = \lim_{x \to 0^+} f(x) \text{ and } \lim_{x \to 0^-} f(x) = \lim_{x \to 0^-} (a \sin x + b \cos x) = b,$$

and so we get $b = 1$.

On the other hand, the left and right derivations at $x = 0$ must be equal. We have

$$f'_+(0) = \lim_{x \to 0^+} \frac{f(x) - f(0)}{x - 0} = \lim_{x \to 0^+} \frac{x^2 + 1 - 1}{x} = 0$$

and

$$f'_-(0) = \lim_{x \to 0^-} \frac{f(x) - f(0)}{x - 0} = \lim_{x \to 0^-} \frac{a \sin x + \cos x - 1}{x}$$
$$= \lim_{x \to 0^-} (a \cos x - \sin x) = a.$$

Thus, we conclude that $a = 0$.

120. *(1) Let $f : \mathbb{R} \to \mathbb{R}$ be a function and f' exists in a neighborhood of $x = a$ and $\lim\limits_{x \to a^+} f'(x) = L$. Prove that $f'(a) = L$.*

(2) Determine whether

$$f(x) = \begin{cases} 1 & if \ x > 0 \\ 0 & if \ x \le 0 \end{cases}$$

is derivative of any function.

Solution. By the mean value theorem, there exists $x_h \in (a, a + h)$ such that $f'(x_h) = \dfrac{f(a + h) - f(a)}{h}$. So, we find that

$$f'(a) = \lim_{h \to 0} \frac{f(a + h) - f(a)}{h} = \lim_{h \to 0^+} \frac{f(a + h) - f(a)}{h} = \lim_{h \to 0^+} f'(x_h) = L.$$

(2) No, because $\lim\limits_{x \to 0^+} f(x) = 1 \ne f(0)$.

121. *Let g be a continuous function and $g(a) \ne 0$. If $f(x) = |x - a|g(x)$, show that $f'(a)$ does not exist.*

Solution. We evaluate the left and right derivative of f at a. We have

$$f'_+(a) = \lim_{x \to a^+} \frac{f(x) - f(a)}{x - a} = \lim_{x \to a^+} \frac{|x - a|g(x)}{x - a} = g(a)$$

$$f'_-(a) = \lim_{x \to a^-} \frac{f(x) - f(a)}{x - a} = \lim_{x \to a^-} \frac{|x - a|g(x)}{x - a} = -g(a).$$

So, $f'_+(a) \ne f'_-(a)$.

122. *Let*

$$f(x) = cx - \frac{x^3}{1 + x^2},$$

where c is constant. If $c \ge \dfrac{9}{8}$, prove that $f'(x) \ge 0$ for all $x \in \mathbb{R}$.

Solution. By derivative rules we obtain

$$f'(x) = c - \frac{3x^2(1 + x^2) - 2x \cdot x^3}{(1 + x^2)^2} = \frac{c + (2c - 3)x^2 + (c - 1)x^4}{(1 + x^2)^2}.$$

The numerator of the last fraction is equal to

$$c + (2c - 3)x^2 + (c - 1)x^4 = (c - 1)\left(x^2 + \frac{2c - 3}{2(c - 1)}\right)^2 + \left(c - \frac{(2c - 3)^2}{4(c - 1)}\right)$$

$$= (c - 1)\left(x^2 + \frac{2c - 3}{2(c - 1)}\right)^2 + \frac{8c - 9}{4(c - 1)}.$$

Now, if $c \geq \dfrac{9}{8}$, then $c - 1 \geq 0$ and $\dfrac{8c - 9}{4(c - 1)} \geq 0$. This completes the our proof.

123. *Let f and g be two differentiable functions.*

(1) If $h(x) = f(x)g(x)$, $h(1) = 20$, $f(1) = 5$, $g'(1) = 3$ and $h'(1) = 12$, find $f'(1)$.
(2) If $k(x) = f^2(g(x))$, $g(1) = 1$, $g'(1) = 2$, $f(1) = 3$ and $f'(1) = 6$, find $k'(1)$.

Solution. (1) Since $h(1) = f(1)g(1)$, it follows that $20 = 5g(1)$. Hence $g(1) = 4$. On the other hand, since $h'(1) = f'(1)g(1) + f(1)g'(1)$, it follows that $12 = 4f'(1) + 5 \cdot 3$. This implies that $f'(1) = -\dfrac{3}{4}$.

(2) We have $k'(x) = 2f(g(x)) \cdot f'(g(x)) \cdot g'(x)$. So, we obtain

$$k'(1) = 2f(g(1)) \cdot f'(g(1)) \cdot g'(1)$$
$$= 2f(1) \cdot f'(1) \cdot g'(1) = 2 \cdot 3 \cdot 6 \cdot 2 = 72.$$

124. *If $f(x) = \ln x$, prove that*

$$f^{(n)}(x) = (-1)^{n-1} \frac{(n-1)!}{x^n},$$

for all $x > 0$.

Solution. For the proof, we use mathematical induction. If $n = 1$, then $f'(x) = \dfrac{1}{x}$. Now, suppose that the statement is true for $n = k$, i.e.,

$$f^{(k)}(x) = (-1)^{k-1} \frac{(k-1)!}{x^k}. \tag{3.3}$$

In order to prove the statement for $n = k + 1$, we take derivative from both sides of (3.3). Then, we get

$$f^{(k+1)}(x) = (-1)^{k-1}(k-1)!(-k)x^{-k-1} = (-1)^k \frac{k!}{x^{k+1}}.$$

This complete the proof.

125. *For every $n \in \mathbb{N}$, prove that*

$$\frac{d^n}{dx^n}\left(x^{n-1} \ln x\right) = \frac{(n-1)!}{x}. \tag{3.4}$$

Solution. We apply mathematical induction. If $n = 1$, we have $\dfrac{d}{dx}(\ln x) = \dfrac{1}{x}$ and so (3.4) is true. Now, suppose that (3.4) holds for n, we prove (3.4) for $n + 1$. Indeed, we have

$$\frac{d^{n+1}}{dx^{n+1}}(x^n \ln x) = \frac{d^n}{dx^n}\left(\frac{d}{dx}(x^n \ln x)\right) = \frac{d^n}{dx^n}\left(\frac{d}{dx}(x \cdot x^{n-1} \ln x)\right)$$

$$= \frac{d^n}{dx^n}\left(x^{n-1} \ln x + x\frac{d}{dx}(x^{n-1} \ln x)\right)$$

$$= \frac{d^n}{dx^n}\left(x^{n-1} \ln x + x\left((n-1)x^{n-2} \ln x + x^{n-2}\right)\right)$$

$$= \frac{d^n}{dx^n}\left(x^{n-1} \ln x\right) + \frac{d^n}{dx^n}\left((n-1)x^{n-1} \ln x + x^{n-1}\right)$$

$$= \frac{(n-1)!}{x} + \frac{d^n}{dx^n}\left((n-1)x^{n-1} \ln x + x^{n-1}\right)$$

$$= \frac{(n-1)!}{x} + \frac{d^n}{dx^n}\left((n-1)x^{n-1} \ln x\right) + \frac{d^n}{dx^n}\left(x^{n-1}\right)$$

$$= \frac{(n-1)!}{x} + (n-1)\frac{d^n}{dx^n}\left(x^{n-1} \ln x\right)$$

$$= \frac{(n-1)!}{x} + (n-1)\frac{(n-1)!}{x}$$

$$= \frac{n!}{x}.$$

This completes the proof.

126. *Prove that*

(1) $\dfrac{d^n}{dx^n} \cos ax = a^n \cos\left(ax + \dfrac{n\pi}{2}\right)$.

(2) $\dfrac{d^n}{dx^n}\left(\cos^4 x + \sin^4 x\right) = 4^{n-1} \cos\left(4x + \dfrac{n\pi}{2}\right)$.

Solution. (1) We use mathematical induction. If $n = 1$, then

$$\frac{d}{dx}\cos ax = -a\sin ax = a\cos\left(ax + \frac{\pi}{2}\right),$$

so (1) is true for $n = 1$. Now, suppose that (1) is true for n. We prove it for $n + 1$. We have

$$\frac{d^{n+1}}{dx^{n+1}}\cos ax = \frac{d}{dx}\left(a^n \cos\left(ax + \frac{n\pi}{2}\right)\right)$$
$$= a^n\left(-a\sin\left(ax + \frac{n\pi}{2}\right)\right)$$
$$= a^{n+1} \cos\left(ax + \frac{(n+1)\pi}{2}\right).$$

Thus, (1) holds for all $n \in \mathbb{N}$.

(2) We have

$$\frac{d}{dx}\left(\cos^4 x + \sin^4 x\right) = -4\cos^3 x \cdot \sin x + 4\sin^3 x \cdot \cos x$$
$$= -4\sin x \cdot \cos x \left(\cos^2 x - \sin^2 x\right)$$
$$= -2\sin(2x) \cdot \cos(2x)$$
$$= -\sin(4x)$$
$$= \cos\left(4x + \frac{\pi}{2}\right).$$

Now, by using part (1), our proof completes.

127. *Use Problem (8) to establish Leibniz's rule*

$$\frac{d^n}{dx^n}\left(f(x)g(x)\right) = \sum_{k=0}^{n} \binom{n}{k} f^{(n-k)}(x) g^{(k)}(x)$$

for the nth derivative of the product of two n-times differentiable functions f and g.

Solution. We prove the formula by mathematical induction. It evidently holds for $n = 1$. If we suppose that it holds for n, its validity for $n + 1$ is proved as follows:

$$\frac{d^{n+1}}{dx^{n+1}}\left(f(x)g(x)\right) = \frac{d}{dx}\left(\sum_{k=0}^{n} \binom{n}{k} f^{(n-k)}(x) g^{(k)}(x)\right)$$

$$= \sum_{k=0}^{n} \binom{n}{k} f^{(n-k)}(x) g^{(k+1)}(x) + \sum_{k=0}^{n} \binom{n}{k} f^{(n+1-k)}(x) g^{(k)}(x)$$

$$= \sum_{k=1}^{n+1} \binom{n}{k-1} f^{(n+1-k)}(x) g^{(k)}(x) + \sum_{k=0}^{n} \binom{n}{k} f^{(n+1-k)}(x) g^{(k)}(x)$$

$$= \binom{n}{n} f(x) g^{(n+1)}(x) + \sum_{k=1}^{n} \binom{n}{k-1} f^{(n+1-k)}(x) g^{(k)}(x)$$

$$+ \sum_{k=1}^{n} \binom{n}{k} f^{(n+1-k)}(x) g^{(k)}(x) + \binom{n}{0} f^{(n+1)}(x) g(x)$$

$$= f(x) g^{(n+1)}(x) + \sum_{k=1}^{n} \left(\binom{n}{k-1} + \binom{n}{k}\right) f^{(n+1-k)}(x) g^{(k)}(x)$$

$$+ f^{(n+1)}(x) g(x)$$

$$= \binom{n+1}{n+1} f(x) g^{(n+1)}(x) + \sum_{k=1}^{n} \binom{n+1}{k} f^{(n+1-k)}(x) g^{(k)}(x)$$

$$+ \binom{n+1}{0} f^{(m+1)}(x) g(x)$$

$$= \sum_{k=0}^{n+1} \binom{n+1}{k} f^{(n+1-k)}(x) g^{(k)}(x).$$

128. *Find the equation of the tangent lines to the ellipse* $25x^2 + 9y^2 = 72$ *that are parallel to the line* $5x - 3y = 1$.

Solution. By implicit differentiation, we have $50x + 18yy' = 0$, or equivalently $y' = -\dfrac{25x}{9y}$. The slope of the line $5x - 3y = 1$ is $\dfrac{5}{3}$. So, for the tangent line to be parallel to $5x - 3y = 1$, we must have $-\dfrac{25x}{9y} = \dfrac{5}{3}$ or

$$x = -\frac{3}{5}y. \qquad (3.5)$$

By substituting (3.5) into the equation of the ellipse, we get

$$72 = 25\left(\frac{9}{25}y^2\right) + 9y^2$$
$$= 18y^2.$$

So, $y = \pm 2$. Since $x = -\dfrac{3}{5}y$, it follows that the points of tangency are $(-\dfrac{6}{5}, 2)$ and $(\dfrac{6}{5}, -2)$. Therefore, the required equations are $y - 2 = \dfrac{5}{3}(x + \dfrac{6}{5})$ and $y + 2 = \dfrac{5}{3}(x - \dfrac{6}{5})$, or equivalently

$$y = \frac{5}{3}x + 4 \text{ and } y = \frac{5}{3}x - 4.$$

129. *Let* $f(x) = x^5 + 2x^3 + 2$ *and* $g(x) = f\left(3f^{-1}(x)\right)$. *Find the equation of the tangent line to* $y = g(x)$ *at* $x = 5$.

Solution. The equation of tangent line is $y - g(5) = g'(5)(x - 5)$. So, we must find $g(5)$ and $g'(5)$. We have

$$g(5) = f\left(3f^{-1}(5)\right) = f(3 \cdot 1) = f(3) = 3^5 + 2 \cdot 3^3 + 2 = 299.$$

The derivative of g is equal to

$$g'(x) = 3(f^{-1})'(x) \cdot f'\left(3f^{-1}(x)\right).$$

By the inverse derivative formula, we have

$$(f^{-1})'(5) = \frac{1}{f'\left(f^{-1}(5)\right)} = \frac{1}{f'(1)} = \frac{1}{11}.$$

Consequently, we get

$$g'(5) = \frac{3}{11} f'(3) = \frac{3}{11} \cdot 459 = \frac{1377}{11}.$$

Therefore, the desired tangent line has the equation $y - 299 = \dfrac{1377}{11}(x - 5)$, or equivalently,

$$y = \frac{1377}{11}x - \frac{3596}{11}.$$

130. *The graph of the equation*

$$x = a \sinh^{-1} \sqrt{\frac{a^2}{y^2} - 1} - \sqrt{a^2 - y^2}$$

is called a tractrix. Prove that the slope of the curve at any point (x, y) is $-\dfrac{y}{\sqrt{a^2 - y^2}}$.

Solution. The slope of the curve at point (x, y) is equal to the derivative at (x, y). So, by implicit differentiation, we get

$$1 = \frac{a}{\sqrt{\dfrac{a^2}{y^2} - 1 + 1}} \cdot \frac{\dfrac{-2a^2 y'}{y^3}}{2\sqrt{\dfrac{a^2}{y^2} - 1}} + \frac{yy'}{\sqrt{a^2 - y^2}}$$

$$= \frac{a}{\sqrt{\dfrac{a^2}{y^2}}} \cdot \frac{-a^2 y'}{y^3 \sqrt{\dfrac{a^2}{y^2} - 1}} + \frac{yy'}{\sqrt{a^2 - y^2}}$$

$$= \frac{-a^2 y'}{y\sqrt{a^2 - y^2}} + \frac{yy'}{\sqrt{a^2 - y^2}}.$$

Thus, we have

$$y\sqrt{a^2 - y^2} = -a^2 y' + y^2 y'$$
$$= -y'(a^2 - y^2).$$

This implies that

$$y' = \frac{-y}{\sqrt{a^2 - y^2}}.$$

131. *Prove that $\lim\limits_{n \to \infty} n\left(\sqrt[n]{x} - 1\right) = \ln x$, for all $x > 0$.*

Solution. By substituting $x = e^{nh}$ we get $n = \dfrac{\ln x}{h}$ and so

$$n\left(\sqrt[n]{x} - 1\right) = \frac{\ln x}{h}\left(\sqrt[n]{e^{nh}} - 1\right) = \frac{\ln x}{h}\left(e^h - 1\right).$$

Consequently, we have

$$\lim_{n \to \infty} n\left(\sqrt[n]{x} - 1\right) = \lim_{h \to 0} \frac{\ln x}{h}\left(e^h - 1\right) = \ln x \lim_{h \to 0} \frac{e^h - 1}{h}.$$

We already know that $\lim_{h \to 0} \dfrac{e^h - 1}{h} = 1$, this completes the proof.

132. *Let f and g be two differentiable functions such that $f'(x) = g(x)$ and $g'(x) = -f(x)$, for all $x \in \mathbb{R}$. Furthermore, suppose that $f(0) = 0$ and $g(0) = 1$. Prove that*

$$(f(x))^2 + (g(x))^2 = 1.$$

Solution. We consider the functions F and G where $F(x) = (f(x))^2$ and $G(x) = -(g(x))^2$. Then, we have

$$F'(x) = 2f(x)f'(x) = 2f(x)g(x),$$
$$G'(x) = -2g(x)g'(x) = 2f(x)g(x),$$

and so $F'(x) = G'(x)$. Therefore, we have $F(x) = G(x) + C$, or equivalently $(f(x))^2 = -(g(x))^2 + C$. Now, by assumption $f(0) = 0$ and $g(0) = 1$, we get $C = 1$.

133. *Find all differentiable functions f and g on $(0, \infty)$ such that for every $x > 0$,*

$$f'(x) = -\frac{g(x)}{x} \text{ and } g'(x) = -\frac{f(x)}{x}.$$

Solution. We have

$$\left(x\big(f(x) + g(x)\big)\right)' = xf'(x) + xg'(x) + f(x) + g(x) = 0,$$
$$\left(\frac{f(x) - g(x)}{x}\right)' = \frac{xf'(x) - xg'(x) - f(x) + g(x)}{x^2} = 0.$$

Thus, there are two real constant c_1 and c_2 such that $f(x) + g(x) = \dfrac{2c_1}{x}$ and $f(x) - g(x) = 2c_2 x$. Therefore, we conclude that

$$f(x) = \frac{c_1}{x} + c_2 x \text{ and } g(x) = \frac{c_1}{x} - c_2 x.$$

134. *Let $f : \mathbb{R} \to \mathbb{R}$ be a function.*

(1) If there exists a fixed number M such that $|f(x)| \leq M|x|^2$, for all $x \in \mathbb{R}$, prove that f is differentiable at $x = 0$ and $f'(0) = 0$.

(2) If $|f(x)| \leq |x|$, determine whether $f'(0) = 0$.

Solution. (1) Since $|f(0)| \leq M \cdot 0^2 = 0$, it follows that $f(0) = 0$. Now, we have

$$\frac{|f(x) - f(0)|}{|x - 0|} = \frac{|f(x)|}{|x|} \leq M|x|.$$

Therefore, $\lim\limits_{x \to 0} \dfrac{|f(x) - f(0)|}{|x - 0|} = 0$. Consequently, we have $f'(0) = 0$.

(2) If we consider $f(x) = x$, then $f'(x) = 1$, for all $x \in \mathbb{R}$.

135. *Let f be a differentiable function at a. Find*

$$\lim_{x \to a} \frac{a^n f(x) - x^n f(a)}{x - a},$$

where n is a natural number.

Solution. We have

$$\frac{a^n f(x) - x^n f(a)}{x - a} = \frac{a^n f(x) - a^n f(a) + a^n f(a) - x^n f(x)}{x - a}$$

$$= a^n \frac{f(x) - f(a)}{x - a} - f(a)\frac{x^n - a^n}{x - a}.$$

Since $f(x)$ and x^n are differentiable at a, we get

$$\lim_{x \to a} \frac{a^n f(x) - x^n f(a)}{x - a} = a^n f'(a) - f(a)na^{n-1}.$$

136. *If $x \in [0, \frac{1}{2}]$, show that*

$$(1 - x)e^{(x+x^2)} \geq 1.$$

Solution. We define $f(x) = (1 - x)e^{(x+x^2)}$. Then, we have

$$f'(x) = -e^{(x+x^2)} + (1 - x)(1 + 2x)e^{(x+x^2)} = x(1 - 2x)e^{(x+x^2)}.$$

Since $x \in [0, \frac{1}{2}]$, it follows that $f'(x) \geq 0$, and so f is increasing on $[0, \frac{1}{2}]$. Since $0 \leq x$, it follows that $f(0) \leq f(x)$. This completes the proof.

137. *Let f be a real differentiable function at a. Let $\ell(x)$ be the tangent line to f at a. Prove that ℓ is the unique linear function with the property*

$$\lim_{x \to a} \frac{f(x) - \ell(x)}{x - a} = 0.$$

Solution. We put $h = x - a$. Then, we have

$$\lim_{x \to a} \frac{f(x) - \ell(x)}{x - a} = \lim_{h \to 0} \frac{f(a + h) - \left(f(a) + f'(a)h \right)}{h}$$
$$= \lim_{h \to 0} \frac{f(x + h) - f(a)}{h} - f'(a) = 0.$$

Now, assume that there is another function $L(x)$ that satisfies the property. Then, we have

$$f(a) - L(a) = \lim_{x \to a} \left(f(x) - L(x) \right) = \lim_{x \to a} \frac{f(x) - L(a)}{x - a} \cdot (x - a) = 0.$$

So, $L(x) = f(a) + m(x - a)$, where m is the slope and

$$m = L'(a) = \lim_{h \to 0} \frac{L(a + h) - L(a)}{h}$$
$$= \lim_{h \to 0} \frac{L(a + h) - f(a + h) + f(a + h) - L(a)}{h}$$
$$= \lim_{h \to 0} \frac{L(a + h) - f(a + h)}{h} + \lim_{h \to 0} \frac{f(a + h) - L(a)}{h}$$
$$= 0 + f'(a) = f'(a).$$

Therefore, $L = \ell$.

138. *If $f(x) = 3x + 2 \cos x$, find $(f^{-1})'(2)$.*

Solution. Since $f'(x) = 3 - 2 \sin x > 0$, it follows that f is increasing, and so it is one to one. We have $f(0) = 2$ and so $f^{-1}(2) = 0$. Therefore,

$$(f^{-1})'(2) = \frac{1}{f'\left(f^{-1}(2) \right)} = \frac{1}{f'(0)} = \frac{1}{3}.$$

139. *Let f be a continuous function and one to one in a neighborhood of the point x and suppose that f has a finite second derivative at the point x, equal to $f''(x)$. Show that f^{-1} has a second derivative at the point $y = f(x)$, equal to*

$$\left(f^{-1} \right)''(y) = -\frac{f''(x)}{\left(f'(x) \right)^3}.$$

Solution. By (3.2) and chain rule we have

$$\left(f^{-1}\right)''(y) = \frac{d}{dy}\left(f^{-1}\right)'(y) = \frac{d}{dy}\left(\frac{1}{f'(x)}\right)$$
$$= \frac{d}{dx}\left(\frac{1}{f'(x)}\right) \cdot \frac{dx}{dy}$$
$$= \frac{-f''(x)}{(f'(x))^2} \cdot \frac{1}{f'(x)}$$
$$= -\frac{f''(x)}{(f'(x))^3}.$$

140. *Let f be a real function such that*

$$|f(x) - f(y)| \leq (x - y)^2,$$

for all $x, y \in \mathbb{R}$. Show that f is a constant function.

Solution. Let $x > y$. Then, we have $-(x-y)^2 \leq f(x) - f(y) \leq (x-y)^2$, and so

$$-(x - y) \leq \frac{f(x) - f(y)}{x - y} \leq (x - y).$$

By Sandwich theorem, we conclude that $\lim\limits_{x \to y^+} \dfrac{f(x) - f(y)}{x - y} = 0$ or $f'_+(y) = 0$. Similarly, we obtain $f'_-(y) = 0$. Thus, $f'(y) = 0$, for all $y \in \mathbb{R}$. This implies that f is constant.

141. *Find all polynomials f of degree not exceeding 2 which commute with their first derivatives, i.e., for which*

$$f(f'(x)) = f'(f(x)),$$

for all $x \in \mathbb{R}$.

Solution. If $f(x) = a_0 + a_1 x + a_2 x^2$, then $f'(x) = a_1 + 2a_2 x$. So, we have

$$f(f'(x)) = a_0 + a_1(a_1 + 2a_2 x) + a_2(a_1 + 2a_2 x)^2$$
$$= a_0 + a_1^2 + 2a_1 a_2 x + a_1^2 a_2 + 4a_2^3 x^2 + 4a_1 a_2^2 x$$
$$= (a_0 + a_1^2 + a_1^2 a_2) + (2a_1 a_2 + 4a_1 a_2^2)x + 4a_2^3 x^2$$

and

$$f'(f(x)) = a_1 + 2a_2(a_0 + a_1 x + a_2 x^2)$$
$$= (a_1 + 2a_0 a_2) + 2a_1 a_2 x + 2a_2^2 x^2.$$

Consequently, we must have

$$a_0 + a_1^2 + a_1^2 a_2 = a_1 + 2a_0 a_2,$$
$$2a_1 a_2 + 4a_1 a_2^2 = 2a_1 a_2,$$
$$4a_2^3 = 2a_2^2.$$

From the last equation, we obtain $a_2 = 0$ or $a_2 = \dfrac{1}{2}$. If $a_2 = 0$, then $a_0 + a_1^2 = a_1$ and a_1 arbitrary. If we take $a_1 = a$, then $a_0 = a - a^2$, and so

$$f(x) = (a - a^2) + ax.$$

If $a_2 = \dfrac{1}{2}$, then we get $a_1 = 0$ and a_0 arbitrary. If we take $a_0 = a$, then

$$f(x) = a + \frac{1}{2}x^2.$$

142. *Find the polynomial function f such that*

$$f(f'(x)) = 2 + 6x^2 - 27x^4 + 27x^6. \tag{3.6}$$

Solution. Let the degree of f be n. Then, the degree of f' is equal to $n - 1$. So, we must have $n(n - 1) = 6$, which implies that $n = 3$. Suppose that $f(x) = a_0 + a_1 x + a_2 x^2 + a_3 x^3$. By (3.6), we conclude that f' contains only even power of x. So, $a_2 = 0$ and $f(x) = a_0 + a_1 x + a_3 x^3$. Hence, $f'(x) = a_1 + 3a_3 x^2$. Therefore, $f(f'(x)) = a_0 + a_1(a_1 + 3a_3 x^2) + a_3(a_1 + 3a_3 x^2)^3$. This implies that

$$f(f'(x)) = (a_0 + a_1^2 + a_1^3 a_3) + (3a_1 a_3 + 9a_1^2 a_3^2)x^2 + 27a_1 a_3^3 x^4 + 27a_3^4 x^6. \tag{3.7}$$

By (3.6) and (3.7), we conclude that

$$a_0 + a_1^2 + a_1^3 a_3 = 2,$$
$$3a_1 a_3 + 9a_1^2 a_3^2 = 6,$$
$$27a_1 a_3^3 = -27,$$
$$27a_3^4 = 27.$$

Then, we obtain $(a_3 = 1, a_1 = -1, a_0 = 2)$ or $(a_3 = -1, a_1 = 1, a_0 = 2)$. Therefore, $f(x) = 2 - x + x^3$ or $f(x) = 2 + x - x^3$.

143. *Prove that for $x \geq 1$, $\cosh^{-1} x = \ln\left(x + \sqrt{x^2 - 1}\right)$.*

Solution. We put $f(x) = \cosh^{-1} x$ and $g(x) = \ln\left(x + \sqrt{x^2 - 1}\right)$. Then, $f'(x) = g'(x) = \dfrac{1}{\sqrt{x^2 - 1}}$. So, $f(x) - g(x) = c$, where c is a constant. Take $x = 1$, then $\cosh^{-1}(1) - \ln\left(1 + \sqrt{1 - 1}\right) = c$. So, $\cosh c = 1$, which implies that $c = 0$ and this completes the proof of our statement.

144. *Prove that for $x > 0$, $\tan^{-1} x + \tan^{-1} \dfrac{1}{x} = \dfrac{\pi}{2}$.*

Solution. We define $f(x) = \tan^{-1} x + \tan^{-1} \dfrac{1}{x}$. Then, $f'(x) = 0$. So, f is a constant function., i.e., $\tan^{-1} x + \tan^{-1} \dfrac{1}{x} = c$, for all $x > 0$. Now, take $x = 1$, then $\tan^{-1} 1 + \tan^{-1} 1 = c$. This implies that $\dfrac{\pi}{4} + \dfrac{\pi}{4} = c$ and so $c = \dfrac{\pi}{2}$.

145. *Suppose that a and c are real numbers, $c > 0$, and f is defined on $[-1, 1]$ by*

$$f(x) = \begin{cases} x^a \sin(x^{-c}) & if \ x \neq 0 \\ 0 & if \ x = 0. \end{cases}$$

Prove that the following statements:

(1) f is continuous if and only if $a > 0$.
(2) $f'(0)$ exists if and only if $a > 1$.

Solution. (1) Since x^a and $\sin(x^{-c})$ are continuous on $I = [-1, 1] - \{0\}$, it follows that their products is continuous on I. So, it is enough to investigate the continuity of f at $x = 0$.

Let $a > 0$. If $\epsilon > 0$, then

$$|x^a \sin(x^{-c})| = |x^a| \cdot |\sin(x^{-c})| \leq |x^a| = |x|^a.$$

If we consider $\delta = \epsilon^{\frac{1}{a}}$, then

$$|x^a \sin(x^{-c})| < \epsilon, \ \text{whenever} \ |x| < \delta.$$

Therefore, $\lim\limits_{x \to 0} f(x) = 0$, i.e., f is continuous at $x = 0$.

Now, let f be continuous at $x = 0$. Then,

$$\lim_{x \to 0} x^a \sin(x^{-c}) = f(0) = 0. \tag{3.8}$$

Moreover,

$$|\sin(x^{-c})| \leq 1. \tag{3.9}$$

From (3.8) and (3.9), we conclude that $a > 0$.

If not, then $a = 0$ or $a < 0$. If $a = 0$, then similar to problem (62) we conclude that $\lim\limits_{x \to 0} \sin(x^{-c})$ does not exist, and this is a contradiction. If $a < 0$, choosing $\epsilon = 1$, we can find a number $\delta > 0$ such that

$$|x^a \sin(x^{-c})| < 1, \tag{3.10}$$

whenever $0 < |x| < \delta$. Let n be a positive integer such that

$$x_0 = \left(\frac{1}{2n\pi + \frac{\pi}{2}}\right)^{c^{-1}} < \delta.$$

Then, $\sin(x_0^{-c}) = \sin(2n\pi + \frac{\pi}{2}) = 1$. Now, from (3.10) we get $|x_0^a| < 1$. Since $a < 0$, it follows that $\lim_{n\to\infty} x_0^a = \infty$ and this is a contradiction.

(2) We have

$$f'(0) = \lim_{x\to 0} \frac{x^a \sin(x^{-c})}{x} = \lim_{x\to 0} x^{a-1} \sin(x^{-c}). \tag{3.11}$$

If $a > 1$, then $a - 1 > 0$. Hence, $f'(0)$ exists.

If $f'(0)$ exists, then the limits in (3.11) must exists, but this limit exists if $a - 1 > 0$.

146. *Give an example of a function f such that f' exists on* [0, 1] *but is not continuous on* [0, 1].

Solution. We define $f : [0, 1] \to \mathbb{R}$ by

$$f(x) = \begin{cases} x^2 \sin(\frac{1}{x}) & \text{if } x \in (0, 1] \\ 0 & \text{if } x = 0. \end{cases}$$

Then, $f'(x) = 2x \sin(\frac{1}{x}) - \cos(\frac{1}{x})$ for $x > 0$ and

$$f'(0) = \lim_{x\to 0} \frac{x^2 \sin(\frac{1}{x}) - 0}{x - 0} = \lim_{x\to 0} x \sin(\frac{1}{x}) = 0.$$

Therefore, f' exists on [0, 1] but is not continuous at 0, since f' does not have a limit at 0.

147. *Suppose that*

$$f(x) = \begin{cases} x^2 \tanh(\frac{1}{x}) & \text{if } x \neq 0 \\ 0 & \text{if } x = 0. \end{cases}$$

(1) Find f'(x).
(2) Investigate the continuity of f' at x = 0.

Solution. (1) If $x \neq 0$, then

$$f'(x) = 2x \tanh(\frac{1}{x}) + x^2(-\frac{1}{x^2})\left(1 - \tanh^2(\frac{1}{x})\right)$$

$$= 2x \tanh(\frac{1}{x}) + \tanh^2(\frac{1}{x}) - 1.$$

If $x = 0$, then

$$f'(0) = \lim_{x \to 0} \frac{f(x) - f(0)}{x - 0} = \lim_{x \to 0} x \tanh(\frac{1}{x}) = 0,$$

since tanh is bounded on $\mathbb{R}$. Therefore,

$$f'(x) = \begin{cases} 2x \tanh(\frac{1}{x}) + \tanh^2(\frac{1}{x}) - 1 & \text{if } x \neq 0 \\ 0 & \text{if } x = 0. \end{cases}$$

(2) We have

$$\lim_{x \to 0^+} f'(x) = \lim_{x \to 0^+} \left(2x \tanh(\frac{1}{x}) + \tanh^2(\frac{1}{x}) - 1\right) = 0 + 1 - 1 = 0,$$

$$\lim_{x \to 0^-} f'(x) = \lim_{x \to 0^-} \left(2x \tanh(\frac{1}{x}) + \tanh^2(\frac{1}{x}) - 1\right) = 0 + (-1)^2 - 1 = 0.$$

Hence, we have $\lim_{x \to 0} f'(x) = f'(0)$. Consequently, f' is continuous at $x = 0$.

148. *Suppose that*

$$f(x) = \begin{cases} e^{-\frac{1}{x^2}} \sin(\frac{1}{x}) & \text{if } x \neq 0 \\ c & \text{if } x = 0. \end{cases}$$

For which values of real number c is f differentiable?

Solution. If $x \neq 0$, then f is differentiable at x, since f is a combination and product of differentiable functions. So, we consider the case $x = 0$. Since

$$-e^{-\frac{1}{x^2}} \leq e^{-\frac{1}{x^2}} \sin(\frac{1}{x}) \leq e^{-\frac{1}{x^2}}$$

and

$$\lim_{x \to 0} -e^{-\frac{1}{x^2}} = \lim_{x \to 0} e^{-\frac{1}{x^2}} = 0,$$

we obtain

$$\lim_{x \to 0} f(x) = \lim_{x \to 0} e^{-\frac{1}{x^2}} \sin(\frac{1}{x}) = 0.$$

Thus, f is continuous at $x = 0$ if and only if $c = 0$. Now, we have

$$\lim_{x \to 0^+} \frac{f(x) - f(0)}{x - 0} = \lim_{x \to 0^+} \frac{e^{-\frac{1}{x^2}} \sin(\frac{1}{x})}{x} = \lim_{u \to \infty} \frac{e^{-u^2} \sin u}{\frac{1}{u}} = \lim_{u \to \infty} \frac{u \sin u}{e^{u^2}}.$$

Since

$$-\frac{|u|}{e^{u^2}} \le \frac{u \sin u}{e^{u^2}} \le \frac{|u|}{e^{u^2}}$$

and

$$\lim_{u \to \infty} -\frac{|u|}{e^{u^2}} = \lim_{u \to \infty} \frac{|u|}{e^{u^2}} = 0,$$

we conclude that

$$\lim_{x \to 0^+} \frac{f(x) - f(0)}{x - 0} = \lim_{u \to \infty} \frac{u \sin u}{e^{u^2}} = 0.$$

Analogously, we get

$$\lim_{x \to 0^-} \frac{f(x) - f(0)}{x - 0} = 0.$$

Hence, if $c = 0$, then f is differentiable at $x = 0$. Therefore, we obtain

$$f'(x) = \begin{cases} \frac{1}{x^2}\left(\frac{2}{x}\sin(\frac{1}{x}) - \cos(\frac{1}{x})\right)e^{-\frac{1}{x^2}} & \text{if } x \ne 0 \\ 0 & \text{if } x = 0. \end{cases}$$

149. *Let $f : (0, \infty) \to \mathbb{R}$ be a differentiable function such that $f'(x) = \dfrac{1}{x}$ and $f(1) = 0$. Prove that $f(xy) = f(x) + f(y)$.*

Solution. Suppose that $x \in (0, \infty)$ is arbitrary and $y \in (0, \infty)$ is a fixed number. We define $g(x) = f(xy) - f(x)$. Then, we have

$$g'(x) = yf'(xy) - f'(x) = y \cdot \frac{1}{xy} - \frac{1}{x} = 0.$$

So, g is a constant function. If $g(x) = c$, then $c = f(xy) - f(x)$, for all $x \in (0, \infty)$. Take $x = 1$, then $c = f(y) - f(1) = f(y)$. Therefore, $f(y) = f(xy) - f(x)$ or $f(xy) = f(x) + f(y)$.

150. *Let f be a real function such that for every $x, y \in \mathbb{R}$ the following conditions hold:*

(1) $f(x + y) = f(x)f(y)$;

(2) $f(x) = 1 + xg(x)$, *where* $\lim_{x \to 0} g(x) = 1$.

Prove that $f'(x)$ exists and $f'(x) = f(x)$, for all $x \in \mathbb{R}$.

Solution. We obtain $f(0) = 1 + 0g(0) = 1$ and

$$f'(0) = \lim_{x \to 0} \frac{f(x) - f(0)}{x - 0} = \lim_{x \to 0} \frac{1 + xg(x) - 1}{x} = \lim_{x \to 0} g(x) = 1.$$

Now, for any arbitrary x, we have

$$f'(x) = \lim_{h \to 0} \frac{f(x + h) - f(x)}{h} = \lim_{h \to 0} \frac{f(x)f(h) - f(x)}{h} = \lim_{h \to 0} \frac{f(x)(f(h) - 1)}{h}$$
$$= \lim_{h \to 0} \frac{f(x)(f(h) - f(0))}{h} = f(x) \lim_{h \to 0} \frac{f(h) - f(0)}{h} = f(x)f'(0) = f(x).$$

151. *For every $x \geq 0$ prove that*

$$1 - \frac{x^2}{2} \leq \cos x.$$

Solution. Writing $f(x) = \cos x - (1 - \frac{x^2}{2})$, we have $f'(x) = -\sin x + x$ and $f''(x) = -\cos x + 1$. If $0 \leq x < \frac{\pi}{2}$, then $f''(x) \geq 0$ and so f' is increasing. Hence, $f'(x) \geq f'(0) = 0$. Clearly, if $x \geq \frac{\pi}{2}$, then $f'(x) \geq 0$. Thus, for every $x \geq 0$, we have $f'(x) \geq 0$. This implies that f is increasing. So, $f(x) \geq f(0) = 0$. Therefore, we obtain $1 - \frac{x^2}{2} \leq \cos x$.

152. *Prove that*

$$e^\pi > \pi^e.$$

Solution. Since the function $\ln$ is increasing, it is enough to prove that $\ln e^\pi > \ln \pi^e$ or $\pi \ln e > e \ln \pi$, or equivalently $\frac{\ln \pi}{\pi} < \frac{\ln e}{e}$. Now, we define $f(x) = \frac{\ln x}{x}$, for all $x > e$. Then, we have $f'(x) = \frac{1 - \ln x}{x^2}$. Since $e < x$, it follows that $1 < \ln x$. Therefore, $f'(x) < 0$ which implies that f is monotone decreasing. Since $e < \pi$, it follows that $f(\pi) < f(e)$ or $\frac{\ln \pi}{\pi} < \frac{1}{e}$. So, $e \ln \pi < \pi$. This implies that $\ln \pi^e < \ln e^\pi$. Therefore, we conclude that $\pi^e < e^\pi$.

153. *Suppose that g is a real function on $\mathbb{R}$, with bounded derivative. Fix $\epsilon > 0$, and define $f(x) = x + \epsilon g(x)$. Prove that f is one to one if ϵ is small enough.*

Solution. Suppose that f is not one to one. Hence, there exist $x_1 \neq x_2$ with $f(x_1) = f(x_2)$. If $x_1 < x_2$, then by Rolle's theorem there exists $t \in (x_1, x_2)$ such that

$$f'(t) = 0. \tag{3.12}$$

We have $f'(t) = 1 + \epsilon g'(t)$ and $|g'| \leq M$, for some $M > 0$. If ϵ is small enough, we conclude that

$$f'(t) > 0. \tag{3.13}$$

Now, (3.12) and (3.13) are contradictory.

154. *Let $a > 0$ and f be a continuous function on $[a, b]$, differentiable in (a, b) and $f(a) = f(b) = 0$. Prove that there exists $c \in (a, b)$ such that*

$$f'(c) = \frac{f(c)}{c}.$$

Solution. We consider the function g on $[a, b]$ as follows: $g(x) = \dfrac{f(x)}{x}$. Clearly, g is continuous on $[a, b]$, differentiable in (a, b) and $g(a) = g(b) = 0$. By Rolle's theorem there exists $c \in (a, b)$ such that $g'(c) = 0$. So, we have $\dfrac{cf'(c) - f(c)}{c^2} = 0$ which implies that $cf'(c) - f(c) = 0$. Thus, $f'(c) = \dfrac{f(c)}{c}$.

155. *Let f be a continuous function on $[a, b]$ and $f'(x) = 1$, for all $x \in (a, b)$. Prove that $f(x) = x - a + f(a)$, for all $x \in [a, b]$.*

Solution. Let $x \in [a, b]$ be arbitrary. Then, f is continuous on $[a, x]$ and differentiable in (a, x). Hence, by the mean value theorem there is $c \in (a, x)$ such that $f(x) - f(a) = f'(c)(x - a)$. Thus, we obtain $f(x) = x - a + f(a)$.

156. *Show that the difference of square roots of two consecutive natural numbers greater than m^2 is less than $\dfrac{1}{2m}$.*

Solution. Let $n \in \mathbb{N}$. We consider $f(x) = \sqrt{x}$ on $[n, n + 1]$. By the mean value theorem, there is $\xi \in (n, n + 1)$ such that $f(n + 1) - f(n) = f'(\xi)$ or $\sqrt{n + 1} - \sqrt{n} = \dfrac{1}{2\sqrt{\xi}}$. Now, if $n > m^2$, then $\xi > m^2$, which implies that $\dfrac{1}{2\sqrt{\xi}} < \dfrac{1}{2m}$. Therefore, we obtain

$$\sqrt{n + 1} - \sqrt{n} < \frac{1}{2m}.$$

157. *Give an example of a function that is continuous on $[-a, a]$ and for which the conclusion of the mean value theorem does not hold.*

Solution. Suppose that $f(x) = |x|$. Then, $\dfrac{f(a) - f(-a)}{a - (-a)} = \dfrac{a - a}{2a} = 0$. Since

$$f'(x) = \begin{cases} 1 & \text{if } x > 0 \\ -1 & \text{if } x < 0 \end{cases}$$

and $f'(0)$ does not exist, there is no number $\xi \in (-a, a)$ such that $f'(\xi) = 0$.

158. *Let f be a continuous function on $[a, b]$ and differentiable in (a, b) such that $f(a) = f(b) = 0$. Show that for every real number λ, there exists $\xi \in (a, b)$ such that $f'(\xi) = \lambda f(\xi)$.*

Solution. We consider $g(x) = e^{-\lambda x} f(x)$. Then, $g'(x) = e^{-\lambda x}(-\lambda f(x) + f'(x))$. On the other hand, by Rolle's theorem there exists $\xi \in (a, b)$ such that $g'(\xi) = 0$. Therefore, $-\lambda f(\xi) + f'(\xi) = 0$.

159. *Let f be a continuous function on $[a, b]$ and differentiable in (a, b). If for every $x \in [a, b]$, $f(x)$ is non-zero, show that there exists $c \in (a, b)$ such that*

$$\frac{f(a)}{f(b)} = e^{\dfrac{(a - b)f'(c)}{f(c)}}.$$

Solution. We define $g(x) = \ln |f(x)|$, for all $x \in [a, b]$. Then, g is continuous on $[a, b]$ and differentiable in (a, b). So, by the mean value theorem, there exists $c \in (a, b)$ such that $g(b) - g(a) = (b - a)g'(c)$. This implies that

$$\ln \left| \frac{f(a)}{f(b)} \right| = (a - b)\frac{f'(c)}{f(c)}.$$

Since f is continuous and $f(x) \neq 0$ for each x, it follows that

$$\left| \frac{f(a)}{f(b)} \right| = \frac{f(a)}{f(b)}.$$

This completes the our proof.

160. *Let f be a continuous function on $[a, b]$ and differentiable in (a, b) such that $f'(x) \neq 1$, for all $x \in (a, b)$. Show that there is at most one $x_0 \in [a, b]$ such that $f(x_0) = x_0$.*

Solution. Assume that f has two fixed points, i.e., $f(x_1) = x_1$ and $f(x_2) = x_2$ for some $x_1, x_2 \in [a, b]$ and $x_1 \neq x_2$. By the mean value theorem, there is $c \in (a, b)$ such that $f(x_2) - f(x_1) = f'(c)(x_2 - x_1)$, which implies that $x_2 - x_1 = f'(c)(x_2 - x_1)$. So, we get $f'(c) = 1$, a contradiction.

161. *Let f be a differentiable function that satisfies $f'(x) = f(\frac{x}{2})$ for all $x \in \mathbb{R}$ and $f(0) > 0$. If $f(a) = 0$ for some $a > 0$, prove that there exists $c \in (0, a)$ such that $f(c) = 0$.*

Solution. Applying the mean value theorem on $[0, a]$, there exists $b \in (0, a)$ such that $f(a) - f(0) = af'(b)$. This implies that $f'(b) = -\dfrac{f(0)}{a} < 0$. Since $f(\frac{b}{2}) = f'(b)$, we deduce that $f(\frac{b}{2}) < 0$. Now, we apply the intermediate value theorem on $[0, \frac{b}{2}]$. Then, we find $c \in (0, \frac{b}{2}) \subseteq (0, a)$ such that $f(c) = 0$.

162. *Let f and g be two continuous on $[a, b]$ and differentiable in (a, b) such that $f(a) = f(b) = 0$. Show that there exists $\xi \in (a, b)$ such that $f'(\xi) + f(\xi)g'(\xi) = 0$.*

Solution. We consider $h(x) = f(x)e^{g(x)}$. Then, h is continuous on $[a, b]$ and differentiable in (a, b). So, by the mean value theorem there exists $\xi \in (a, b)$ such that $h(b) - h(a) = h'(\xi)(b - a)$. Hence, $h'(\xi) = 0$. But $0 = h'(\xi) = (f'(\xi) + f(\xi)g'(\xi))e^{g(\xi)}$. This implies that $f'(\xi) + f(\xi)g'(\xi) = 0$.

163. *Let f be a continuous function on $[a, b]$, f'' exists on (a, b), $f(a) = f(b) = 0$ and there exists $c \in (a, b)$ such that $f(c) > 0$. Prove that there exists $\xi \in (a, b)$ such that $f''(\xi) < 0$.*

Solution. Suppose that $f''(x) \geq 0$, for all $x \in (a, b)$. So, f' is increasing. Since f is continuous on $[a, c]$ and $[c, d]$, and is differentiable in (a, c) and (c, b), then there exist $t_1 \in (a, c)$ and $t_2 \in (c, b)$ such that

$$f'(t_1) = \frac{f(c)}{c - a} \quad \text{and} \quad f'(t_2) = \frac{f(c)}{c - b}.$$

This implies that
$$f'(t_2) < f'(t_1). \tag{3.14}$$

On the other hand, since f' is increasing and $t_1 < t_2$, we have

$$f'(t_2) \geq f'(t_1). \tag{3.15}$$

Now, (3.14) and (3.15) are contradiction.

164. *Suppose that f is defined and differentiable for all $x > 0$ and $f'(x) \to 0$ as $x \to \infty$. Put $g(x) = f(x + 1) - f(x)$. Prove that $g(x) \to 0$ as $x \to \infty$.*

Solution. Since f is differentiable, it follows that f is continuous too. By applying the mean value theorem on $[x, x + 1]$, we observe that there is $t \in (x, x + 1)$ such that $f(x + 1) - f(x) = f'(t)(x + 1 - x)$. Hence, $g(x) = f'(t)$. Therefore, we obtain

$$\lim_{x \to \infty} g(x) = \lim_{x \to \infty} f'(t), \text{ where } x < t < x + 1.$$

This implies that $g(x) \to 0$ as $x \to \infty$.

165. *Let $f : [0, 2] \to \mathbb{R}$ be a continuous function and f', f'' exist on $(0, 2)$. Let $f(0) = 0$, $f(1) = 1$ and $f(2) = 2$. Show that there exists $\xi \in (0, 2)$ such that $f''(\xi) = 0$.*

Solution. Let $g(x) = f(x) - x$. Then, $g(0) = g(1) = g(2) = 0$. Now, by Rolle's theorem there exist $c_1 \in (0, 1)$ and $c_2 \in (1, 2)$ such that $g'(c_1) = g'(c_2) = 0$. Again, by Rolle's theorem there exists $\xi \in (c_1, c_2)$ such that $g''(\xi) = 0$. This implies that $f''(\xi) = 0$.

166. *Let f be a differentiable function on $[0, 2]$, $f(0) = 0$ and $f(1) = f(2) = 1$. Prove that there exists $\xi \in (0, 2)$ such that $f'(\xi) = \dfrac{1}{1000}$.*

Solution. Since f is differentiable in $[1, 2]$, by Rolle's theorem there exists $c \in (1, 2)$ such that $f'(c) = 0$. Also, since f is differentiable in $[0, 1]$, by the mean value theorem there exists $d \in (0, 1)$ such that $f(1) - f(0) = f'(d)(1 - 0)$. So, $f'(d) = 1$. Now, we observe that

$$f'(c) < \frac{1}{1000} < f'(d).$$

Now, let $g(t) = f(t) - \dfrac{1}{1000}t$. Then, $g'(c) < 0$, so that $g(t_1) < g(c)$ for some $t_1 \in (a, b)$, and $g'(d) > 0$, so that $g(t_2) < g(d)$ for some $t_2 \in (a, b)$. Hence, g attains its minimum on $[d, c]$ at some point ξ such that $d < \xi < c$. Hence, $g'(\xi) = 0$. Therefore, we conclude that $f'(\xi) = \dfrac{1}{1000}$.

167. *Let $f(x) = x^5 - x^3 - x$. Show that*

$$\left| \frac{f(x_1) - f(x_2)}{x_1 - x_2} \right| \leq \frac{3}{2},$$

for all $x_1, x_2 \in [-1, 1]$ and $x_1 \neq x_2$.

Solution. Suppose that $x_1, x_2 \in [-1, 1]$ and $x_1 \neq x_2$ are arbitrary. Then, f is continuous on $[x_1, x_2]$ and differentiable in (x_1, x_2). So, by the mean value theorem there exists $\xi \in (x_1, x_2)$ such that $\dfrac{f(x_1) - f(x_2)}{x_1 - x_2} = f'(\xi)$. If M is the maximum of $|f'(x)|$ on $[-1, 1]$, then we get

$$\left| \frac{f(x_1) - f(x_2)}{x_1 - x_2} \right| \leq M. \tag{3.16}$$

On the other hand, $f'(x) = 5x^4 - 3x^2 - 1$ and $f''(x) = 20x^3 - 6x$. If $f'(x) = 0$, then we conclude that $0, \sqrt{\dfrac{3}{10}}$ and $-\sqrt{\dfrac{3}{10}}$ are critical points of $f''(x)$. Now, we find that

$$M = \max \left\{ |f'(-1)|, \ |f'(1)|, \ |f'(0)|, \ |f'(\sqrt{\tfrac{3}{10}})|, \ |f'(-\sqrt{\tfrac{3}{10}})| \right\} \tag{3.17}$$
$$= \max \left\{ 1, \frac{29}{20} \right\} = \frac{29}{20} \leq \frac{3}{2}.$$

By (3.16) and (3.17) the proof of desired inequality is completed.

168. *Let f be a twice differentiable function such that $f(0) = 1$, $f'(0) = -1$, $f(1) = 2$, $f'(1) = 5$ and for every $0 \leq x \leq 1$, $f''(x) \geq 0$. Prove that $f(x) \geq \dfrac{1}{3}$, for all $0 \leq x \leq 1$.*

Solution. Suppose that a, b are two arbitrary elements in $[0, 1]$ and $a < b$. By applying the mean value theorem for f' on $[a, b]$, we get $c \in (a, b)$ such that $f'(b) - f'(a) = (b - a) f''(c)$. Since $f''(c) \geq 0$, it follows that $f'(a) \leq f'(b)$.
   So, for every $0 < x < 1$, we have

$$-1 = f'(0) \leq f'(x) \leq f'(1) = 5.$$

Now, we apply the mean value theorem for f on $[0, x]$ and $[x, 1]$, respectively. Then, there are $\alpha \in (0, x)$ and $\beta \in (x, 1)$ such that

$$f(x) - f(0) = x f'(\alpha) \text{ and } f(1) - f(x) = (1 - x) f'(\beta).$$

Since $f'(\alpha) \geq -1$ and $f'(\beta) \leq 5$, we conclude that $f(x) \geq -x + 1$ and $f(x) \geq 5x - 3$. Consequently, we obtain $f(x) \geq \dfrac{1}{3}$.

169. *Let f, g be two differentiable functions at a such that $f(a) = g(a)$ and $f'(a) < g'(a)$. Show that there exists $\delta > 0$ such that $f(a + b) < g(a + b)$, whenever $0 < b < \delta$.*

Solution. Suppose that $h(x) = g(x) - f(x)$. Then, $h(a) = 0$ and $h'(a) = g'(a) - f'(a) > 0$. Let $h'(a) > \epsilon > 0$. Then, there is $\delta > 0$ such that

$$\left| \frac{h(a+b) - h(a)}{b} - h'(a) \right| < \epsilon, \text{ whenever } 0 < |b| < \delta.$$

This implies that

$$-\epsilon < \frac{h(a+b) - h(a)}{b} - h'(a) < \epsilon.$$

So, if $0 < b < \delta$, then

$$0 \le h'(a) - \epsilon < \frac{h(a+b) - h(a)}{b}.$$

Consequently, we get $0 < h(a+b) - h(a)$, which implies that $f(a+b) < g(a+b)$.

170. *Show that the formula in the mean value theorem can be written as follows:*

$$\frac{f(x+h) - f(x)}{h} = f'(x + ch),$$

where $0 < c < 1$ and x, h are real numbers.

Solution. Since f is differentiable in $[x, x+h]$, by the mean value theorem there exists $\xi \in (x, x+h)$ such that $f(x+h) - f(x) = f'(\xi)(x + h - x)$. Hence, $f'(\xi) = \dfrac{f(x+h) - f(x)}{h}$. Since $x < \xi < x + h$, it follows that $\xi = x + a$, where $0 < a < h$. We pick $c = \dfrac{a}{h}$. Then, $a = ch$ and $0 < c < 1$. So, we have $\xi = x + ch$ and $\dfrac{f(x+h) - f(x)}{h} = f'(x + ch)$.

171. *Let f be a continuous function on $[a - h, a + h]$ and differentiable in $(a - h, a + h)$. Prove that there exists a real number c between 0 and 1 such that*

$$f(a+h) - 2f(a) + f(a-h) = h\Big(f'(a + ch) - f'(a - ch) \Big).$$

Solution. We consider the function F on $[0, 1]$ as follows:

$$F(x) = f(a + hx) + f(a - hx),$$

for all $x \in [0, 1]$. Hence, F is continuous on $[0, 1]$ and differentiable in $(0, 1)$. Then, by the mean value theorem, there exists c between 0 and 1 such that

$F(1) - F(0) = (1 - 0)F'(c)$. This implies that $f(a + h) + f(a - h) - 2f(a) = h\Big(f'(a + hc) - f'(a - hc)\Big)$.

172. *Suppose that f is defined in a neighborhood of x and $f''(x)$ exists. Prove that*

$$\lim_{h \to 0} \frac{f(x + h) + f(x - h) - 2f(x)}{h^2} = f''(x).$$

Solution. By assumptions, we can suppose that f' is defined in $(x - \delta, x + \delta)$. We consider

$$F(h) = f(x + h) + f(x - h) - 2f(x) \text{ and } G(h) = h^2.$$

Now, by L'Hospital's rule, we have

$$\begin{aligned}
\lim_{h \to 0} \frac{F(h)}{G(h)} &= \lim_{h \to 0} \frac{F'(h)}{G'(h)} \\
&= \lim_{h \to 0} \frac{f'(x + h) - f'(x - h)}{2h} \\
&= \lim_{h \to 0} \left(\frac{f'(x + h) - f'(x)}{2h} + \frac{f'(x) - f'(x - h)}{2h} \right) \\
&= \frac{f''(x)}{2} + \frac{f''(x)}{2} = f''(x).
\end{aligned}$$

173. *Give an example in which the limit in Problem (172) exists but f'' does not exists.*

Solution. We define

$$f(x) = \begin{cases} \dfrac{x^2}{2} & \text{if } x > 0 \\ \dfrac{-x^2}{2} & \text{if } x \leq 0. \end{cases}$$

Then, $f'(x) = |x|$, which is not differentiable at $x = 0$, but we have

$$\lim_{h \to 0} \frac{f(h) + f(-h) - 2f(0)}{h^2} = \lim_{h \to 0} \frac{h^2 - (-h)^2}{2h^2} = 0.$$

174. *Let f be a real function such that $f^{(4)}(x)$ exists, for all $x \in [a, b]$ and $f'(x_0) = f''(x_0) = f'''(x_0) = 0$, for some $x_0 \in (a, b)$. Prove that*

(1) f has a local maximum at x_0 if $f^{(4)}(x_0) < 0$.
(2) f has a local minimum at x_0 if $f^{(4)}(x_0) > 0$.

Solution. (1) Suppose that $f^{(4)}(x_0) < 0$. Then, by the second derivative test for a local extremum, we conclude that x_0 is a local maximum for f''. On the other hand, since $f''(x_0) = 0$, it follows that $f''(x) \leq 0$, for all x in a neighborhood I

of x_0. This implies that f' is a decreasing function on I. Now, for $x > x_0$ we have $f'(x) < f'(x_0) = 0$ and for $x < x_0$ we have $f'(x) > f'(x_0) = 0$. Therefore, x_0 is a local maximum for f.

(2) The proof is similar to the proof of (1).

175. *Prove that*

$$|\sin y - \sin x| \le |y - x|, \quad for\ all\ x, y \in \mathbb{R}.$$

Solution. If $x = y$, then the inequality holds. Suppose that $x < y$. We consider $f(x) = \sin x$. Then, f is continuous on $[x, y]$ and differentiable in (x, y). Hence, by the mean value theorem, there exists $c \in (x, y)$ such that $f'(c) = \dfrac{f(y) - f(x)}{y - x}$. On the other hand, $f'(c) = \cos c$. Thus, we have

$$\left| \frac{f(y) - f(x)}{y - x} \right| = |\cos c| \le 1.$$

Therefore, $|\sin y - \sin x| \le |y - x|$.

176. *Prove that*

$$|\tan y - \tan x| \le 4|y - x|, \quad for\ all\ x, y \in [-\frac{\pi}{3}, \frac{\pi}{3}].$$

Solution. If $x = y$, then the inequality holds. Suppose that $x < y$. We consider $f(x) = \tan x$. Then, f is continuous on $[x, y]$ and differentiable in (x, y). Hence, by the mean value theorem, there exists $c \in (x, y)$ such that $f'(c) = \dfrac{f(y) - f(x)}{y - x}$. Since $f'(c) = 1 + \tan^2 c$, it follows that

$$1 + \tan^2 c = \frac{f(y) - f(x)}{y - x}.$$

Since $|c| \le \dfrac{\pi}{3}$, it follows that $|\tan c| \le \tan \dfrac{\pi}{3} = \sqrt{3}$. So, we have

$$\left| \frac{f(y) - f(x)}{y - x} \right| = |1 + \tan^2 c| \le 4.$$

Therefore, we obtain $|\tan y - \tan x| \le 4|y - x|$.

177. *Prove that*

$$|\sin^{-1} y - \sin^{-1} x| \ge |y - x|, \quad for \; all \; x, y \in [-1, 1].$$

Solution. It is clear if $x = y$. Let $x < y$. We define $f(x) = \sin^{-1} x$. Then, f is continuous on $[x, y]$ and differentiable in (x, y). Hence, by the mean value theorem, there exists $c \in (x, y)$ such that $f'(c) = \dfrac{f(y) - f(x)}{y - x}$. Moreover, $f'(c) = \dfrac{1}{\sqrt{1 - c^2}}$.

Therefore, for all $x, y \in (-1, 1)$ and $x \ne y$, we obtain

$$\frac{|\sin^{-1} y - \sin^{-1} x|}{|y - x|} = f'(c) = \frac{1}{\sqrt{1 - c^2}} \ge 1.$$

This completes the proof of our statement.

178. *By applying the mean value theorem, prove the following inequality:*

$$\frac{x}{\sin x} < \frac{\tan x}{x},$$

for all $x \in (0, \dfrac{\pi}{2})$.

Solution. It is enough to prove that $\tan x \sin x - x^2 > 0$. We define $f(x) = \tan x \sin x - x^2$. Then, we have

$$f'(x) = \sec^2 x \sin x + \tan x \cos x - 2x = \sin x (\sec^2 x + 1) - 2x,$$
$$f''(x) = (\sqrt{\sec x} - \sqrt{\cos x})^2 + 2 \sin^2 x \sec^3 x > 0,$$

for all $x \in (0, \dfrac{\pi}{2})$. Since $f''(x) > 0$, it follows that $f'(x)$ is strictly increasing on $(0, \dfrac{\pi}{2})$. So, $f'(x) > f'(0)$, for all $x \in (0, \dfrac{\pi}{2})$. This implies that $f'(x) > 0$, and so f is an increasing function. For all $x \in (0, \dfrac{\pi}{2})$, we obtain $f(x) > f(0) = 0$. Therefore, $\tan x \sin x - x^2 > 0$, for all $x \in (0, \dfrac{\pi}{2})$.

179. *If $x, y \in [\dfrac{\pi}{2}, \dfrac{3\pi}{2}]$, prove that*

$$\cos y - \cos x \ge (x - y) \sin x.$$

Solution. Let $f(x) = \cos x$. Then, by Taylor's theorem, there exists ξ between x and y such that

$$f(y) = f(x) + f'(x)(y - x) + \frac{(y - x)^2}{2} f''(\xi).$$

Since $f''(\xi) = -\cos\xi > 0$, it follows that $f(y) - f(x) \geq (y - x)f'(x)$. This implies that $\cos y - \cos x \geq (x - y)\sin x$.

180. *If $0 \leq p \leq 1$ and $a, b > 0$, prove that*

$$(a + b)^p \leq a^p + b^p.$$

Solution. Let $a < b$ and define $f(x) = x^p$. Then, f is continuous on $[b, a + b]$ and differentiable in $(b, a + b)$. By the mean value theorem, there is $\xi \in (b, a + b)$ such that $f(a + b) - f(b) = (a + b - b)f'(\xi)$. So, we get

$$p\xi^{p-1}a = (a + b)^p - b^p. \tag{3.18}$$

Since $0 \leq 1 - p \leq 1$ and $a < \xi$, we have $a^{1-p} \leq \xi^{1-p}$ or $\xi^{p-1} \leq a^{p-1}$. Hence,

$$p\xi^{p-1} \leq a^{p-1}. \tag{3.19}$$

By (3.18) and (3.19), we conclude that $(a + b)^p - b^p \leq aa^{p-1}$, which implies that $(a + b)^p \leq a^p + b^p$.

181. *For $\alpha \geq 1$ and $x > -1$, prove that*

$$(1 + x)^\alpha \geq 1 + \alpha x. \tag{3.20}$$

Inequality (3.20) is called Bernoulli's inequality.

Solution. For $x > -1$, we define $f(x) = (1 + x)^\alpha$. Note that if $x = 0$, we have equality. So, we consider the following two cases:

(1) $0 < x$;
(2) $-1 < x < 0$.

If $0 < x$, then f is continuous on $[0, x]$ and is differentiable in $(0, x)$. So, by the mean value theorem there exists $c_1 \in (0, x)$ such that $f(x) - f(0) = f'(c_1)(x - 0)$. This implies that

$$(1 + x)^\alpha - 1 = \alpha x(1 + c_1)^{\alpha-1}. \tag{3.21}$$

Since $c_1 > 0$ and $\alpha - 1 \geq 0$, it follows that $(1 + c_1)^{\alpha-1} > 1$, and so $\alpha x(1 + x)^{\alpha-1} > \alpha x$. Now, by (3.21), we conclude that $(1 + x)^\alpha - 1 = \alpha x(1 + c_1)^{\alpha-1} > \alpha x$.

If $-1 < x < 0$, then f is continuous on $[x, 0]$ and is differentiable in $(x, 0)$. So, by the mean value theorem there exists $c_2 \in (x, 0)$ such that $f(0) - f(x) = f'(c_2)(0 - x)$. This implies that

$$1 - (1 + x)^\alpha = -\alpha x(1 + c_2)^{\alpha-1}. \tag{3.22}$$

Since $x < c_2 < 0$, it follows that $(1+x)^{\alpha-1} < (1+c_2)^{\alpha-1} < 1$, and so $-\alpha x(1+c_2)^{\alpha-1} < -\alpha x$. Now, by (3.22), we conclude that $1 + \alpha x < (1+x)^\alpha$.

182. *Let $0 < a < b$. Prove that*

$$\frac{b-a}{b} < \ln\frac{b}{a} < \frac{b-a}{a}.$$

Solution. We define $f(x) = \ln x$. Since f is continuous on $[a, b]$ and differentiable in (a, b), by the mean value theorem there exists $c \in (a, b)$ such that $f'(c) = \dfrac{\ln b - \ln a}{b-a}$. On the other hand, $f'(c) = \dfrac{1}{c}$. Since $a < c < b$, it follows that $\dfrac{1}{b} < \dfrac{1}{c} < \dfrac{1}{a}$. This means that $\dfrac{1}{b} < \dfrac{\ln b - \ln a}{b-a} < \dfrac{1}{a}$, which implies that

$$\frac{b-a}{b} < \ln b - \ln a < \frac{b-a}{a}$$

This completes the proof of our statement.

183. *For every $x > 0$, prove that*

$$\frac{x}{x+1} < \ln(1+x) < x.$$

Solution. Suppose that $x > 0$. We define $f(t) = \ln(1+t)$, for all $t \in [0, x]$. Clearly, f is continuous on $[0, x]$ and differentiable in $(0, x)$. So, by the mean value theorem there exists $c \in (0, x)$ such that $f(x) - f(0) = xf'(c)$. This implies that

$$\ln(1+x) = \frac{x}{1+c}.$$

Since $0 < c < x$, it follows that $\dfrac{1}{1+x} < \dfrac{1}{1+c} < 1$. Hence, we have

$$\frac{x}{1+x} < \frac{x}{1+c} < x.$$

This completes the proof of our statement.

184. *Let $x \geq 0$. Prove that*

$$\frac{x}{1+x^2} \leq \tan^{-1} x \leq x.$$

Solution. It is obvious if $x = 0$. Let $x \neq 0$. We define $f(t) = \tan^{-1} t$. Since f is continuous on $[0, x]$ and differentiable in $(0, x)$, by the mean value theorem there exists $c \in (0, x)$ such that $f'(c) = \dfrac{\tan^{-1} x - \tan^{-1} 0}{x - 0} = \dfrac{\tan^{-1} x}{x}$. Moreover, $f'(c) = \dfrac{1}{1 + c^2} \leq 1$. Thus, we obtain $\tan^{-1} x \leq x$. Since $0 < c < x$, it follows that $\dfrac{1}{1 + x^2} < \dfrac{1}{1 + c^2}$, which implies that $\dfrac{1}{1 + x^2} < \dfrac{\tan^{-1} x}{x}$. So, $\dfrac{x}{1 + x^2} < \tan^{-1} x$.

185. *Show that the equation $x^5 - 5x + 1 = 0$ has three real roots, but no more than three.*

Solution. Writing $f(x) = x^5 - 5x + 1$, we have $f(-2) = -21$, $f(-1) = 5$, $f(1) = -3$ and $f(2) = 23$. Therefore, by the intermediate value theorem, there exist $r_1 \in (-2, -1)$, $r_2 \in (-1, 1)$ and $r_3 \in (1, 2)$ such that $f(r_1) = f(r_2) = f(r_3) = 0$. These are three roots of the equation. Now, suppose that there is another root. Let c_1, c_2, c_3 and c_4 be four different roots of the equation such that $c_1 < c_2 < c_3 < c_4$. Hence, $f(c_1) = f(c_2) = f(c_3) = f(c_4) = 0$. So, by Rolle's theorem, there exist $t_1 \in (c_1, c_2)$, $t_2 \in (c_2, c_3)$ and $t_3 \in (c_3, c_4)$ such that $f'(t_1) = f'(t_2) = f'(t_3) = 0$. But this is impossible, since the equation $f'(x) = 5x^4 - 5 = 0$ has only two real roots, 1 and -1.

186. *Let n be a positive even integer and a, b be fixed real numbers. Show that the equation $x^n + ax + b = 0$ has at most two real roots.*

Solution. We consider $f(x) = x^n + ax + b$. Then, f is a continuous and differentiable function on $\mathbb{R}$. Suppose that f has three real roots x_1, x_2 and x_3 such that $x_1 < x_2 < x_3$. Hence, $f(x_1) = f(x_2) = f(x_3) = 0$. By Rolle's theorem, there exist $c_1 \in (x_1, x_2)$ and $c_2 \in (x_2, x_3)$ such that $f'(c_1) = f'(c_2) = 0$. But $f'(x) = nx^{n-1} + a$ and $n - 1$ is odd. So, f' has exactly one real root, i.e., $\sqrt[n-1]{\dfrac{-a}{n}}$, which contradicts with f has more than two roots.

187. *Let f be a non-negative function such that f', f'', f''' exist on $(0, 1)$ and let $f(x_1) = f(x_2) = 0$ for $0 < x_1 < x_2 < 1$. Show that there exists $\xi \in (0, 1)$ such that $f'''(\xi) = 0$.*

Solution. We have

$$f'_+(x_1) = \lim_{\Delta x \to 0^+} \frac{f(x_1 + \Delta x) - f(x_1)}{\Delta x} = \lim_{\Delta x \to 0^+} \frac{f(x_1 + \Delta x)}{\Delta x} \geq 0,$$

$$f'_-(x_1) = \lim_{\Delta x \to 0^-} \frac{f(x_1 + \Delta x) - f(x_1)}{\Delta x} = \lim_{\Delta x \to 0^-} \frac{f(x_1 + \Delta x)}{\Delta x} \leq 0.$$

Since $f'(x_1)$ exists, it follows that $f'(x_1) = 0$. Similarly, we observe that $f'(x_2) = 0$. On the other hand, since $f(x_1) = f(x_2) = 0$, it follows that there exists $c \in (x_1, x_2)$ such that $f'(c) = 0$. Now, by Rolle's theorem, there exist $\xi_1 \in (x_1, c)$ and $\xi_2 \in (c, x_2)$ such that $f''(\xi_1) = f''(\xi_2) = 0$. Again, by Rolle's theorem, there exists $\xi \in (\xi_1, \xi_2)$ such that $f'''(\xi) = 0$.

188. *Let f, g and h be continuous functions on $[a, b]$ and differentiable in (a, b). Prove that there exists $c \in (a, b)$ such that*

$$\begin{vmatrix} f'(c) & g'(c) & h'(c) \\ f(a) & g(a) & h(a) \\ f(b) & g(b) & h(b) \end{vmatrix} = 0$$

Solution. We define the function F on $[a, b]$ as follows:

$$F(x) = \begin{vmatrix} f(x) & g(x) & h(x) \\ f(a) & g(a) & h(a) \\ f(b) & g(b) & h(b) \end{vmatrix}.$$

Then, F is continuous on $[a, b]$ and differentiable in (a, b). Moreover, $F(a) = F(b) = 0$. Note that if we have any matrix with two identical rows then its determinant is equal to zero. It is easy to see that

$$F'(x) = \begin{vmatrix} f'(x) & g'(x) & h'(x) \\ f(a) & g(a) & h(a) \\ f(b) & g(b) & h(b) \end{vmatrix}.$$

Now, by Rolle's theorem there exists $c \in (a, b)$ such that $F'(c) = 0$.

189. *Let $f_1, \ldots, f_n$ and $g_1, \ldots, g_n$ be continuous on $[a, b]$ and differentiable on (a, b). Suppose that $g_i(a) \neq g_i(b)$, for all $i = 1, \ldots, n$. Prove that there exists $\xi \in (a, b)$ such that*

$$\sum_{i=1}^{n} f_i'(\xi) = \sum_{i=1}^{n} g_i'(\xi) \frac{f_i(b) - f_i(a)}{g_i(b) - g_i(a)}.$$

Solution. We define the following function:

$$h(x) = \sum_{i=1}^{n} \left(\left(g_i(x) - g_i(a) \right) \frac{f_i(b) - f_i(a)}{g_i(b) - g_i(a)} - \left(f_i(x) - f_i(a) \right) \right).$$

Clearly, h is continuous on $[a, b]$ and differentiable in (a, b). Moreover, we have

$$h'(x) = \sum_{i=1}^{n} \left(g_i'(x) \frac{f_i(b) - f_i(a)}{g_i(b) - g_i(a)} - f_i'(x) \right).$$

Since $h(a) = h(b) = 0$, by Roll's theorem, there exists $\xi \in (a, b)$ such that $h'(\xi) = 0$. This completes the proof of our statement.

190. *Let $f''(x)$ exist for all $x \in [a, b]$. Prove that there exist $c, \xi \in (a, b)$ such that*

$$\begin{vmatrix} 1 & 1 & 1 \\ a & b & c \\ f(a) & f(b) & f(c) \end{vmatrix} - \frac{1}{2} f''(\xi) \begin{vmatrix} 1 & 1 & 1 \\ a & b & c \\ a^2 & b^2 & c^2 \end{vmatrix} = 0.$$

Solution. We define the function F on $[a, b]$ as follows:

$$F(x) = \begin{vmatrix} 1 & 1 & 1 \\ a & b & x \\ f(a) & f(b) & f(x) \end{vmatrix} - A \begin{vmatrix} 1 & 1 & 1 \\ a & b & x \\ a^2 & b^2 & x^2 \end{vmatrix},$$

where A is a constant such that $F(c) = 0$. We try to find A. Clearly, $F(a) = F(b) = 0$. Since F is differentiable, it follows that F on both intervals $[a, c]$ and $[c, b]$ satisfies the conditions of Rolle's theorem. So, there exist c_1 and c_2 such that $a < c_1 < c < c_2 < b$ and $F'(c_1) = F'(c_2) = 0$. Now, by Rolle's theorem, there exists $\xi \in (c_1, c_2)$ such that $F''(\xi) = 0$. Since

$$F'(x) = \begin{vmatrix} 1 & 1 & 0 \\ a & b & 1 \\ f(a) & f(b) & f'(x) \end{vmatrix} - A \begin{vmatrix} 1 & 1 & 1 \\ a & b & 1 \\ a^2 & b^2 & 2x \end{vmatrix},$$

$$F''(x) = \begin{vmatrix} 1 & 1 & 0 \\ a & b & 0 \\ f(a) & f(b) & f''(x) \end{vmatrix} - A \begin{vmatrix} 1 & 1 & 0 \\ a & b & 0 \\ a^2 & b^2 & 2 \end{vmatrix},$$

it follows that $A = \dfrac{1}{2} f''(\xi)$.

191. *Let f be a continuous and periodic function with period 2π.*

(1) Show that there is $c \in \mathbb{R}$ such that $f(c + \pi) = f(c)$.
(2) If f is differentiable, show that f' has at least two roots in every interval with length 2π.

Solution. (1) We define $g(x) = f(x + \pi) - f(x)$. Let $a \in \mathbb{R}$ be arbitrary. Then, we have

$$g(a + \pi) = f((a + \pi) + \pi) - f(a + \pi) = f(a + 2\pi) - f(a + \pi)$$
$$= f(a) - f(a + \pi) = -g(a).$$

If $g(a) = 0$, then $f(a + \pi) = f(a)$ and our statement is proved. But if $g(a) \neq 0$, then by the intermediate value theorem there is c between a and $a + \pi$ such that $g(c) = 0$, which implies that $f(c + \pi) = f(c)$.

(2) Let $[a, a + 2\pi]$ be an arbitrary interval.

If $f(a) = f(a + \pi)$, then $f(a) = f(a + \pi) = f(a + 2\pi)$. Hence, by Rolle's theorem, there are numbers c_1 and c_2, where $a < c_1 < a + \pi + c_2 < a + 2\pi$, such that $f'(c_1) = f'(c_2) = 0$.

If $f(a) \neq f(a + \pi)$, then by the first part, there is $c \in [a, a + \pi]$ such that $f(c) = f(c + \pi)$. So, by Rolle's theorem there is $\xi_1 \in (c, c + \pi)$ such that $f'(\xi_1) = 0$. Since $a < c < \xi_1 < c + \pi < a + 2\pi$, it follows that $\xi \in (a, a + 2\pi)$. Since f is periodic, we have $f(c + \pi) = f(c + 2\pi)$. Again, by Rolle's theorem, there is $\xi_2 \in (c + \pi, c + 2\pi)$ such that $f'(\xi_2) = 0$. If $\xi_2 \in (a, a + 2\pi)$, then our statement is proved. In otherwise, $a + 2\pi < \xi_2 < c + 2\pi$. Hence, $a < \xi_2 - 2\pi < c < \xi_1$ and $f'(\xi_2 - 2\pi) = 0$.

192. *Prove that*

$$\frac{d^n}{dx^n}(x^2 - 1)^n = 0$$

has n roots, all different and lying between -1 *and* 1.

Solution. Let $f(x) = (x^2 - 1)^n$. Since the degree of $f^{(n)}(x)$ is n, it follows that $f^{(n)}$ has at most n roots. By using Leibniz's rule (see Problem (127)), if $k < n$, then we have

$$\frac{d^k}{dx^k} f(x) = \frac{d^k}{dx^k}\left((x - 1)^n \cdot (x + 1)^n\right)$$
$$= \sum_{i=0}^{k} \binom{k}{i} \frac{d^i}{dx^i}(x - 1)^n \cdot \frac{d^{k-i}}{dx^{k-i}}(x + 1)^n.$$

Therefore, 1 and -1 are roots of $f^{(k)}(x)$, for all $k < n$.

Since $f(1) = f(-1)$, by Rolle's theorem there is $a_0 \in (-1, 1)$ such that $f'(a_0) = 0$.

Now, we have $f'(-1) = f'(a_0) = f'(1) = 0$. So, by Rolle's theorem there are $b_1 \in (-1, a_0)$ and $b_2 \in (a_0, 1)$ such that $f''(b_1) = f''(b_2) = 0$.

Again, we have $f''(-1) = f''(b_1) = f''(b_2) = f''(1) = 0$. Then, by Rolle's theorem there are $c_1 \in (-1, b_1)$, $c_2 \in (b_1, b_2)$ and $c_3 \in (b_2, 1)$ such that $f'''(c_1) = f'''(c_2) = f'''(c_3) = 0$.

We continue the above process. Finally, in the last step, we get distinct numbers $\xi_1, \xi_2, \ldots, \xi_{n-1}$ in $(-1, 1)$ such that

$$f^{(n-1)}(-1) = f^{(n-1)}(\xi_1) = f^{(n-1)}(\xi_2) = \ldots = f^{(n-1)}(\xi_{n-1}) = f^{(n-1)}(1) = 0.$$

Another application of Rolle's theorem, yields distinct numbers $d_1, d_2, \ldots, d_n$ in $(-1, 1)$ such that

$$f^{(n)}(d_1) = f^{(n)}(d_2) = \ldots = f^{(n)}(d_n) = 0.$$

193. *Let f and g be two functions such that $f(x)g'(x) - g(x)f'(x) \neq 0$. Prove that between any two consecutive roots of f, the function g has one root, but no more than one.*

Solution. Let x_1 and x_2 be two consecutive roots of f such that $x_1 < x_2$. Then, $f(x_1) = f(x_2) = 0$.

First, we suppose that g does not have any root between x_1 and x_2. This means that $g(x) \neq 0$, for all $x \in (x_1, x_2)$. We define the function h as follows:

$$h(x) = \frac{f(x)}{g(x)}.$$

Then, we have

$$h'(x) = \frac{f'(x)g(x) - f(x)g'(x)}{g^2(x)} \neq 0.$$

On the other hand, since $h(x_1) = h(x_2) = 0$, it follows that there exists $c_1 \in (x_1, x_2)$ such that $h'(c_1) = 0$, this is a contradiction.

Now, suppose that g has two roots z_1 and z_2 between x_1 and x_2. Then, $g(z_1) = g(z_2) = 0$. We define the function k as follows:

$$k(x) = \frac{g(x)}{f(x)}.$$

Then, we have

$$k'(x) = \frac{g'(x)f(x) - g(x)f'(x)}{f^2(x)} \neq 0.$$

On the other hand, since $k(z_1) = k(z_2) = 0$, it follows that there exists $c_2 \in (z_1, z_2)$ such that $k'(c_2) = 0$, this is a contradiction again.

194. *Suppose that*

(1) f is a continuous function on $[0, \infty)$,
(2) $f'(x)$ exists for each $x > 0$,
(3) $f(0) = 0$,
(4) f' is monotonically increasing.

Put

$$g(x) = \frac{f(x)}{x}, \quad (x > 0)$$

and prove that g is also monotonically increasing.

Solution. Let $x > 0$ be arbitrary. Since f is continuous on $[0, x]$ and differentiable in $(0, x)$, by the mean value theorem there exists $t \in (0, x)$ such that
$$f'(t) = \frac{f(x) - f(0)}{x - 0} = \frac{f(x)}{x}.$$ Since $0 < t < x$, it follows that $f'(t) < f'(x)$.
Hence, $\frac{f(x)}{x} < f'(x)$, which implies that $f(x) < xf'(x)$. Therefore,
$$g'(x) = \frac{xf'(x) - f(x)}{x^2} > 0,$$ for al $x > 0$. This implies that g is monotonically increasing.

195. *Suppose that*

(1) f and g are continuous functions on $[0, \infty)$ and $g(x) \neq 0$, for all $x > 0$,
(2) $f'(x)$ and $g'(x)$ exist for each $x > 0$ and $g'(x) \neq 0$,
(3) $f(0) = g(0) = 0$,
(4) $\frac{f'}{g'}$ is monotonically increasing on $(0, \infty)$.

Put
$$h(x) = \frac{f(x)}{g(x)} \quad (x > 0)$$

and prove that h is also monotonically increasing.

Solution. Let $x > 0$ be arbitrary. Since f and g are continuous on $[0, x]$ and differentiable in $(0, x)$, by Cauchy's mean value theorem there exists $t \in (0, x)$ such that
$$(f(x) - f(0))g'(t) = (g(x) - g(0))f'(t).$$

Hence, $f(x)g'(t) = g(x)f'(t)$. Since $0 < t < x$, it follows that $\frac{f'(t)}{g'(t)} < \frac{f'(x)}{g'(x)}$,
which implies that $\frac{f(x)}{g(x)} < \frac{f'(x)}{g'(x)}$ or $f'(x)g(x) - g'(x)f(x) > 0$. Therefore,
$$h'(x) = \frac{f'(x)g(x) - g'(x)f(x)}{(g(x))^2} > 0,$$ for all $x > 0$. This implies that h is monotonically increasing.

196. *(1) Let f be a real differentiable function such that $f(a) = 0$ for some a and $f'(x) > f(x)$, for all $x \in \mathbb{R}$. Show that $f(x) > 0$, for all $x > a$.*

(2) If c is a real positive number, show that the equation $ce^x = 1 + x + \frac{x^2}{2}$ has a unique real root.

Solution. (1) We consider $h(x) = e^{-x} f(x)$. Then, we have $h'(x) = e^{-x}(f'(x) - f(x)) > 0$, which implies that h is strictly increasing. So, for every $x > a$ we get $h(x) > h(a) = 0$. This implies that $f(x) > 0$.

(2) If c is a real positive number, show that the equation $ce^x = 1 + x + \dfrac{x^2}{2}$ has a unique real root.

Solution. (2) We define $f(x) = ce^x - 1 - x - \dfrac{x^2}{2}$. It is not difficult to see that

$$\lim_{x \to -\infty} f(x) = -\infty \quad \text{and} \quad \lim_{x \to \infty} f(x) = \infty.$$

Hence, by the intermediate value theorem, f has at least one real root. Obviously, $f'(x) > f(x)$, for all $x \in \mathbb{R}$. Now, by using part (1), we conclude that f does not have another real root.

197. *Assume that the following limit*

$$\lim_{x \to 0^+} \left(\frac{1}{\sqrt{\sin x}} - \frac{1}{\sqrt{x}} \right)$$

exists, evaluate it by using L'Hospital's rule.

Solution. Suppose that $L = \lim\limits_{x \to 0^+} \left(\dfrac{1}{\sqrt{\sin x}} - \dfrac{1}{\sqrt{x}} \right)$. Then, we have

$$\lim_{x \to 0^+} \frac{\sqrt{x} - \sqrt{\sin x}}{\sqrt{x \sin x}} = \frac{0}{0}.$$

So, we can apply L'Hospital's rule. Hence, we get

$$L = \lim_{x \to 0^+} \frac{\dfrac{1}{2\sqrt{x}} - \dfrac{\cos x}{2\sqrt{\sin x}}}{\dfrac{\sin x + x \cos x}{2\sqrt{x \sin x}}} = \lim_{x \to 0^+} \frac{\sqrt{\sin x} - \sqrt{x} \cos x}{\sin x + x \cos x} = \frac{0}{0}.$$

Again, we use L'Hospital's rule. So, we obtain

$$
\begin{aligned}
L &= \lim_{x \to 0^+} \frac{\dfrac{\cos x}{2\sqrt{\sin x}} - \dfrac{1}{2\sqrt{x}} \cos x + \sqrt{x} \sin x}{\cos x + \cos x - x \sin x} \\
&= \lim_{x \to 0^+} \frac{\dfrac{\cos x}{2} \left(\dfrac{1}{\sqrt{\sin x}} - \dfrac{1}{\sqrt{x}} \right) + \sqrt{x} \sin x}{2 \cos x - x \sin x} \\
&= \frac{\frac{1}{2}L + 0}{2 - 0} \\
&= \frac{L}{4}.
\end{aligned}
$$

Therefore, we conclude that $L = 0$.

198. *Show that*

$$\lim_{x \to \infty} \left(1 + \frac{a}{x}\right)^x = e^a.$$

Solution. To resolve this indeterminacy of the form 1^∞, we make the substitution $t = \frac{a}{x}$, obtaining

$$\lim_{x \to \infty} \left(1 + \frac{a}{x}\right)^x = \lim_{t \to 0^+} \left(1 + t\right)^{\frac{a}{t}} = \lim_{t \to 0^+} e^{\frac{a}{t} \ln(1+t)}$$
$$= e^{\lim_{t \to 0^+} \frac{a}{t} \ln(1+t)} = e^L.$$

Now, by using L'Hospital's rule, we obtain

$$L = \lim_{t \to 0^+} a \frac{\ln(1+t)}{t} = a \cdot 1 = a.$$

This completes the proof.

199. *Find* $\lim_{x \to 0^+} (\ln \frac{1}{x})^x$ *if it exists.*

Solution. We have

$$\lim_{x \to 0^+} (\ln \frac{1}{x})^x = \lim_{x \to 0^+} e^{x \ln(\ln \frac{1}{x})} = e^{\lim_{x \to 0^+} x \ln(\ln \frac{1}{x})}.$$

Suppose that $L = \lim_{x \to 0^+} x \ln(\ln \frac{1}{x})$. Then, we obtain

$$L = \lim_{x \to 0^+} \frac{\ln(\ln \frac{1}{x})}{\frac{1}{x}} = \lim_{y \to \infty} \frac{\ln(\ln y)}{y}$$
$$= \lim_{y \to \infty} \frac{1}{y} \cdot \frac{1}{\ln y} = 0.$$

Therefore, $\lim_{x \to 0^+} (\ln \frac{1}{x})^x = e^0 = 1$.

200. *Find c so that*

$$\lim_{x \to \infty} \left(\frac{x + c}{x - c}\right)^x = 4.$$

Solution. We can write

$$\lim_{x\to\infty}\left(\frac{x+c}{x-c}\right)^x = \lim_{x\to\infty} e^{x\ln(\frac{x+c}{x-c})} = e^{\lim_{x\to\infty} x\ln(\frac{x+c}{x-c})} = e^L.$$

Then, by using L'Hospital's rule, we have

$$L = \lim_{x\to\infty}\frac{\ln(\frac{x+c}{x-c})}{\frac{1}{x}} = \lim_{x\to\infty}\frac{2cx^2}{x^2-c^2} = 2c.$$

Hence, $e^{2c} = 4$, which implies that $2c = \ln 2^2$. Thus, $c = \ln 2$.

201. *Show that the following function is continuous on $\mathbb{R}$:*

$$f(x) = \begin{cases} (e^x - 1)\ln x & if\ x > 0 \\ 0 & if\ x = 0 \\ |x|^{\cos x} & if\ x < 0. \end{cases}$$

Solution. Clearly for all $x > 0$ and $x < 0$, f is continuous. We show that f is continuous at $x = 0$. We have

$$\lim_{x\to 0^+} f(x) = \lim_{x\to 0^+}(e^x - 1)\ln x = \lim_{x\to 0^+}\left(\frac{e^x-1}{x}\right)(x\ln x).$$

By using L'Hospital's rule, we obtain

$$\lim_{x\to 0^+}\frac{e^x-1}{x} = 1 \text{ and } \lim_{x\to 0^+} x\ln x = \lim_{x\to 0^+}\frac{\ln x}{\frac{1}{x}} = \lim_{x\to 0^+}\frac{\frac{1}{x}}{-\frac{1}{x^2}} = 0.$$

So, we conclude that $\lim_{x\to 0^+} f(x) = 0$. Also, we get

$$\lim_{x\to 0^-} f(x) = \lim_{x\to 0^-} |x|^{\cos x} = \lim_{x\to 0^-} e^{\cos x \ln |x|} = e^{\lim_{x\to 0^-} \cos x \ln |x|} = 0.$$

Consequently, we have

$$\lim_{x\to 0^+} f(x) = \lim_{x\to 0^-} f(x) = f(0) = 0.$$

Therefore, f is continuous on $\mathbb{R}$.

202. *Let f and g be two continuous functions on* $[a, b]$ *and differentiable in* (a, b) *such that*

$$f'(x) \le g'(x),$$

for all $x \in (a, b)$. *Prove that*

$$f(b) - f(a) \le g(b) - g(a).$$

Solution. We define $h = g - f$. We apply the mean value theorem for h. Hence, there exists $\xi \in (a, b)$ such that $h(b) - h(a) = h'(\xi)(b - a)$, or equivalently

$$g(b) - f(b) - g(a) + f(a) = \big(g'(\xi) - f'(\xi)\big)(b - a). \tag{3.23}$$

By assumption $g'(\xi) - f'(\xi)$ is positive, so the left side of (3.23) is positive, too. This implies that $f(b) - f(a) \le g(b) - g(a)$.

203. *Let f be a differentiable function on* $[0, \infty)$ *and let* $\lim\limits_{x \to \infty} \big(f(x) + f'(x)\big) = 0$. *Prove that* $\lim\limits_{x \to \infty} f(x) = 0$.

Solution. If $g(x) = e^x f(x)$, then $g'(x) = e^x \big(f(x) + f'(x)\big)$. Let $\epsilon > 0$. Then, there is $M > 0$ such that $|f(x) + f'(x)| < \dfrac{\epsilon}{2}$, whenever $x > M$. By Cauchy's mean value theorem, there is $\xi \in (M, x)$ such that

$$\big(g(x) - g(M)\big)e^\xi = \big(e^x - e^M\big)g'(\xi),$$

or equivalently,

$$g(x) - g(M) = \big(f(\xi) + f'(\xi)\big)(e^x - e^M).$$

Thus, we have

$$|g(x) - g(M)| < \frac{\epsilon}{2}|e^x - e^M|.$$

This implies that $|g(x)| < \dfrac{\epsilon}{2}(e^x - e^M) + |g(M)|$. Thus, we get

$$|f(x)| < \frac{\epsilon}{2}(1 - e^{M-x}) + |f(M)|e^{M-x} < \frac{\epsilon}{2} + |f(M)|e^{M-x}.$$

Since $|f(M)|e^{M-x} \to 0$ as $x \to \infty$, it follows that there is $N > 0$ such that $|f(M)|e^{M-x} < \dfrac{\epsilon}{2}$, whenever $x > N$. Now, we pick $K = \max\{M, N\}$, then $|f(x)| < \epsilon$, whenever $x > K$. This completes the proof.

204. *Let f be a continuous function on $[a, b]$ and differentiable in (a, b), where $a \geq 0$. Show that there exist $c_1, c_2 \in (a, b)$ such that*

$$f'(c_1) = \frac{a+b}{2c_2} f'(c_2).$$

Solution. By the mean value theorem, there is $c_1 \in (a, b)$ such that

$$f'(c_1) = \frac{f(b) - f(a)}{b - a}. \tag{3.24}$$

Now, we apply Cauchy's mean value theorem for functions $f(x)$ and $g(x) = x^2$. Then, there is $c_2 \in (a, b)$ such that

$$\Big(f(b) - f(a)\Big)(2c_2) = (b^2 - a^2) f'(c_2),$$

or equivalently,

$$\frac{f(b) - f(a)}{b - a} = \frac{a+b}{2c_2} f'(c_2). \tag{3.25}$$

From (3.24) and (3.25), we conclude that $f'(c_1) = \dfrac{a+b}{2c_2} f'(c_2)$.

205. *Suppose that f is a real function, two times differentiable on $[a, b]$ such that $f'(a) = f'(b) = 0$. Show that*

$$2\Big(f(b) - f(a)\Big) = \Big(\frac{b-a}{2}\Big)^2 \Big(f''(\xi_1) - f''(\xi_2)\Big).$$

for some $\xi_1, \xi_2 \in (a, b)$.

Solution. We apply Taylor's theorem on $[a, \frac{a+b}{2}]$ and $[\frac{a+b}{2}, b]$. Then, we get

$$f\Big(\frac{a+b}{2}\Big) = f(a) + \Big(\frac{b-a}{2}\Big)^2 \cdot \frac{1}{2} f''(\xi_1) \tag{3.26}$$

and

$$f\Big(\frac{a+b}{2}\Big) = f(b) + \Big(\frac{b-a}{2}\Big)^2 \cdot \frac{1}{2} f''(\xi_2), \tag{3.27}$$

for some $\xi_1, \xi_2 \in (a, b)$. From (3.26) and (3.27), we get

$$f(a) + \Big(\frac{b-a}{2}\Big)^2 \cdot \frac{1}{2} f''(\xi_1) = f(b) + \Big(\frac{b-a}{2}\Big)^2 \cdot \frac{1}{2} f''(\xi_2),$$

and we done.

206. *Suppose that* f *is a real function, three times differentiable on* $[-1, 1]$ *such that* $f(-1) = 0$, $f(0) = 0$, $f(1)=1$ *and* $f'(0) = 0$. *Prove that* $f^{(3)}(x) \geq 3$ *for some* $x \in (-1, 1)$.

Solution. We apply Taylor's theorem on $[0, 1]$ and $[-1, 0]$. Then, there exist $s \in (0, 1)$ and $t \in (-1, 0)$ such that

$$f(1) = f(0) + f'(0)(1 - 0) + \frac{f''(0)}{2!}(1 - 0)^2 + \frac{f^{(3)}(s)}{3!}(1 - 0)^3$$
$$f(-1) = f(0) + f'(0)(-1 - 0) + \frac{f''(0)}{2!}(-1 - 0)^2 + \frac{f^{(3)}(t)}{3!}(-1 - 0)^3.$$

Thus, we observe that

$$1 = \frac{f''(0)}{2!} + \frac{f^{(3)}(s)}{3!}, \tag{3.28}$$

$$0 = \frac{f''(0)}{2!} - \frac{f^{(3)}(t)}{3!}. \tag{3.29}$$

From (3.28) and (3.29), we conclude that

$$6 = f^{(3)}(s) + f^{(3)}(t).$$

Hence, $f^{(3)}(s) \geq 3$ or $f^{(3)}(t) \geq 3$.

207. *Let* f *be a continuous function on* $[a, b]$ *and differentiable in* (a, b). *Let* $f'(x) \leq M$, *for all* $x \in (a, b)$ *and* $f(b) - f(a) = M(b - a)$. *Prove that* $f(x) = f(a) + M(x - a)$, *for all* $x \in [a, b]$.

Solution. We define $g(x) = f(x) - f(a) - M(x - a)$. Then, $g(a) = g(b) = 0$ and $g'(x) = f'(x) - M \leq 0$. So,

$$g'(x) \leq 0, \text{ for all } x \in [a, b]. \tag{3.30}$$

Now, let $x \in [a, b]$ be an arbitrary element. Then, g is continuous on $[a, x]$ and $[x, b]$, and is differentiable in (a, x) and (x, b). By the mean value theorem, there exist $\xi_1 \in (a, x)$ and $\xi_2 \in (x, b)$ such that

$$g(x) - g(a) = g'(\xi_1)(x - a), \tag{3.31}$$

$$g(b) - g(x) = g'(\xi_2)(b - x). \tag{3.32}$$

From (3.31) and (3.30), we obtain $g(x) = g'(\xi_1)(x - a) \leq 0$. So,

$$g(x) \leq 0. \tag{3.33}$$

From (3.32) and (3.30), we obtain $-g(x) = g'(\xi_2)(b - x) \leq 0$. Thus,

$$g(x) \geq 0. \tag{3.34}$$

By (3.33) and (3.34), we conclude that $g(x) = 0$. This completes the proof.

208. *Let f be a continuous function on $[0, 1]$, differentiable in $(0, 1)$ and $f(0) = 0$. If $|f'(x)| \leq |f(x)|$, for all $x \in (0, 1)$, show that f is the zero function.*

Solution. Since $|f|$ is continuous on $[a, b]$, it attains its maximum, named M. So, we have $|f(x)| \leq M$, for all $x \in (0, 1)$. Now, assume that $x \in [0, 1]$ is arbitrary. By the mean value theorem, there is $c_1 \in (0, x)$ such that $f(x) - f(0) = f'(c_1)(x - 0)$ or $f(x) = xf'(c_1)$. This implies that

$$|f(x)| = x|f'(c_1)| \leq x|f(c_1)|. \tag{3.35}$$

Again, we apply mean value theorem. Then, there exists $c_2 \in (0, c_1)$ such that $f(c_1) - f(0) = f'(c_2)(c_1 - 0)$, and so

$$|f(c_1)| \leq x|f'(c_2)| \leq x|f(c_2)|. \tag{3.36}$$

Now, by (3.35) and (3.36), we get

$$|f(x)| \leq x^2|f(c_2)|.$$

If we repeat the above process, then we obtain a sequence $\{c_n\}$ such that

$$|f(x)| \leq x^n|f(c_n)| \leq x^n M, \quad \text{for all } n \in \mathbb{N}.$$

Since $x^n M \to 0$ as $n \to \infty$, it follows that $f(x) = 0$, for all $x \in (0, 1)$. The continuity of f on $[0, 1]$ implies that $f(x) = 0$, for all $x \in [0, 1]$.

209. *Suppose that f is differentiable on $[a, b]$, $f(a) = 0$, and there is a real number A such that $|f'(x)| \leq A|f(x)|$ on $[a, b]$. Prove that $f(x) = 0$, for all $x \in [a, b]$.*

Solution. Fix $x_0 \in [a, b]$. Consider the closed interval $[a, x_0]$. Suppose that

$$M_0 = \max\{|f(x)| \ : \ a \leq x \leq x_0\},$$
$$M_1 = \max\{|f'(x)| \ : \ a \leq x \leq x_0\}.$$

For any such x, f is continuous on $[a, x]$ and is differentiable in (a, x), by the mean value theorem there exists $\xi \in (a, x)$ such that $f(x) - f(a) = (x - a)f'(\xi)$. So,

$$|f(x)| \leq (x_0 - a)|f'(\xi)| \leq (x_0 - a)M_1 \leq A(x_0 - a)M_0.$$

Hence, $M_0 \le A(x_0 - a)M_0$. If we pick $a < x_0 < b$ such that $\dfrac{1 + Aa}{A} > x_0$, then $A(x_0 - a) < 1$. Thus, $M_0 = 0$ if $A(x_0 - a) < 1$. That is $f = 0$ on $[a, x_0]$. Proceed.

210. *Let* f *be a real function with finite derivative on* (a, b) *and* $\lim\limits_{x \to b^-} f(x) = \infty$. *Prove that* $\lim\limits_{x \to b^-} f'(x)$ *does not exist or it is infinite.*

Solution. Suppose that $\lim\limits_{x \to b^-} f'(x) = L$, where L is a real number. Let $\epsilon = 1$. Then, there exists $\delta > 0$ such that $|f'(x) - L| < 1$, whenever $0 < b - x < \delta$.

Let y be a real number such that $0 < b - y < \delta$. Then,

$$L - 1 < f'(y) < L + 1. \tag{3.37}$$

Pick a positive number M such that

$$M > f(y) + (|L| + 1)(b - a). \tag{3.38}$$

Since $\lim\limits_{x \to b^-} f(x) = \infty$, it follows that there exists $\delta_1 > 0$ such that $f(x) > M$, whenever $0 < b - x < \delta_1$.

Suppose that $x^\star$ is a number such that $0 < b - x^\star < \delta$ and $0 < b - x^\star < \delta_1$. Then, we have

$$f(x^\star) > M. \tag{3.39}$$

Now, we apply the mean value theorem for the interval between $x^\star$ and y. Then, there exists ξ between $x^\star$ and y such that $f(x^\star) - f(y) = f'(\xi)(x^\star - y)$. So, by (3.37) and (3.38) we have

$$\begin{aligned}
f(x^\star) &= f(y) + f'(\xi)(x^\star - y) \\
&< f(y) + (|L| + 1)(x^\star - y) \\
&< f(y) + (|L| + 1)(b - a) \\
&< M.
\end{aligned}$$

Thus,

$$f(x^\star) < M. \tag{3.40}$$

Inequalities (3.39) and (3.40) are contradiction. Therefore, the left limit of f at b does not exist.

211. *Let* f *be a differentiable real function defined on* (a, b). *Prove that* f *is convex if and only if* f' *is increasing.*

Solution. Suppose that f' is increasing in (a, b). Let $x_1, x_2 \in (a, b), x_1 < x_2, 0 < \lambda < 1$ and $x_0 = \lambda x_1 + (1 - \lambda)x_2$. By the mean value theorem, there exist $c_1 \in (x_1, x_0)$

and $c_2 \in (x_0, x_2)$ such that

$$f(x_0) - f(x_1) = f'(c_1)(x_0 - x_1),$$
$$f(x_2) - f(x_0) = f'(c_2)(x_2 - x_0).$$

So, we get

$$\lambda f(x_1) = \lambda f(x_0) + \lambda f'(c_1)(x_1 - x_0), \tag{3.41}$$

$$(1 - \lambda) f(x_2) = (1 - \lambda) f(x_0) + (1 - \lambda) f'(c_2)(x_2 - x_0). \tag{3.42}$$

If we add the left sides of Eqs. (3.41) and (3.42) together, and add the right sides together, then we obtain

$$\lambda f(x_1) + (1 - \lambda) f(x_2) = f(x_0) + \lambda f'(c_1)(x_1 - x_0) + (1 - \lambda) f'(c_2)(x_2 - x_0).$$

Since $c_1 < c_2$, it follows that $f'(c_1) \leq f'(c_2)$. Hence,

$$\begin{aligned}
\lambda f(x_1) + (1 - \lambda) f(x_2) &\geq f(x_0) + \lambda f'(c_1)(x_1 - x_0) + (1 - \lambda) f'(c_1)(x_2 - x_0) \\
&= f(x_0) + f'(c_1)(\lambda x_1 + (1 - \lambda) x_2 - x_0) \\
&= f(x_0) = f(\lambda x_1 + (1 - \lambda) x_2).
\end{aligned}$$

Therefore, f is convex.

Now, let f be a convex function on (a, b). Let $s, y \in (a, b)$ and $s < y$. We show that $f'(s) \leq f'(y)$.

Suppose that $a < s < t < u < y < b$. We have $f(t) \leq \dfrac{u - t}{u - s} f(s) + \dfrac{t - s}{u - s} f(u)$, and so

$$\frac{f(t) - f(s)}{t - s} \leq \frac{f(u) - f(s)}{u - s}.$$

Analogously, we obtain

$$\frac{f(t) - f(s)}{t - s} \leq \frac{f(u) - f(s)}{u - s} \leq \frac{f(u) - f(t)}{u - t} \leq \frac{f(y) - f(t)}{y - t} \leq \frac{f(y) - f(u)}{y - u}.$$

So,

$$f'(s) = \lim_{t \to s^+} \frac{f(t) - f(s)}{t - s} \leq \frac{f(y) - f(u)}{y - u}.$$

Therefore,

$$f'(s) \leq \lim_{u \to y^-} \frac{f(y) - f(u)}{y - u} = f'(y).$$

212. *The following are two examples of convex functions.*

(1) $f(x) = -\ln x$,
(2) $g(x) = e^{ax}$.

Can you formally verify that these functions are convex?

Solution. By Problem (211), if the second derivative of a function exists and is non-negative, then it is convex. Here, we have

$$f''(x) = \frac{1}{x^2} \text{ and } g''(x) = a^2 e^{ax}.$$

So, f and g are convex.

213. *If a and b are non-negative and $0 < \alpha < 1$, prove that*

$$a^\alpha b^{1-\alpha} \leq \alpha a + (1 - \alpha)b. \tag{3.43}$$

Solution. We consider $f(x) = -\ln x$, for $x > 0$. According to Problem (212), f is convex. So, for $a, b > 0$, we have

$$-\ln(\alpha a + (1 - \alpha)b) \leq -\alpha \ln a - (1 - \alpha) \ln b.$$

This implies that
$$\ln(a^\alpha b^{1-\alpha}) \leq \ln(\alpha a + (1 - \alpha)b).$$

Since ln is an increasing function, it follows that $a^\alpha b^{1-\alpha} \leq \alpha a + (1 - \alpha)b$.

214. *Let p and q be positive real numbers such that*

$$\frac{1}{p} + \frac{1}{q} = 1.$$

If $u \geq 0$ and $v \geq 0$, prove that

$$uv \leq \frac{u^p}{p} + \frac{v^q}{q}. \tag{3.44}$$

Solution. In Problem (213), it is enough to put $a = u^p$, $b = v^q$, $\alpha = \frac{1}{p}$. Then, we get $1 - \alpha = \frac{1}{q}$ and

$$(u^p)^{\frac{1}{p}} \cdot (v^q)^{\frac{1}{q}} \leq \frac{u^p}{p} + \frac{v^q}{q}.$$

This completes the proof.

215. *Let* $\{a_1, \ldots, a_n\}$ *and* $\{b_1, \ldots, b_n\}$ *be two sets of non-negative numbers. If* $p, q > 1$ *and* $\dfrac{1}{p} + \dfrac{1}{q} = 1$, *establish Hölder's inequality:*

$$\sum_{i=1}^{n} a_i b_i \leq \left(\sum_{i=1}^{n} a_i^p \right)^{\frac{1}{p}} \cdot \left(\sum_{i=1}^{n} b_i^q \right)^{\frac{1}{q}}. \tag{3.45}$$

When $p = q = 2$, *then (3.45) is called Cauchy-Schwarz inequality.*

Solution. Suppose that

$$A = \left(\sum_{i=1}^{n} a_i^p \right)^{\frac{1}{p}} \text{ and } B = \left(\sum_{i=1}^{n} b_i^q \right)^{\frac{1}{q}}.$$

If $A = 0$ or $B = 0$, then $a_1 = \ldots = a_n = 0$ or $b_1 = \ldots = b_b = 0$ and (3.45) obviously holds. Now, suppose that $A > 0$ and $B > 0$, and apply (3.44) to $u = \dfrac{a_i}{A}$ and $v = \dfrac{b_i}{B}$. Then, we get

$$\frac{a_i}{A} \cdot \frac{b_i}{B} \leq \frac{(\frac{a_i}{A})^p}{p} + \frac{(\frac{b_i}{B})^q}{q}, \text{ for all } 1 \leq i \leq n.$$

If we add the above inequalities to each other, then we obtain

$$\frac{1}{AB} \sum_{i=1}^{n} a_i b_i \leq \frac{1}{p A^p} \sum_{i=1}^{n} a_i^p + \frac{1}{q B^q} \sum_{i=1}^{n} b_i^q$$
$$= \frac{1}{p} + \frac{1}{q} = 1.$$

Thus, we observe that $\sum_{i=1}^{n} a_i b_i \leq AB$.

216. *Use Problem (215) and establish the triangle inequality:*

$$\left(\sum_{i=1}^{n} (a_i + b_i)^2 \right)^{\frac{1}{2}} \leq \left(\sum_{i=1}^{n} a_i^2 \right)^{\frac{1}{2}} + \left(\sum_{i=1}^{n} b_i^2 \right)^{\frac{1}{2}},$$

for each two finite set $\{a_1, \ldots, a_n\}$ *and* $\{b_1, \ldots, b_n\}$ *of non-negative numbers.*

Solution. We have

$$\sum_{i=1}^{n}(a_i + b_i)^2 = \sum_{i=1}^{n}(a_i + b_i)(a_i + b_i)$$

$$= \sum_{i=1}^{n} a_i(a_i + b_i) + \sum_{i=1}^{n} b_i(a_i + b_i)$$

$$\le \left(\sum_{i=1}^{n} a_i^2\right)^{\frac{1}{2}} \cdot \left(\sum_{i=1}^{n}(a_i + b_i)^2\right)^{\frac{1}{2}} + \left(\sum_{i=1}^{n} b_i^2\right)^{\frac{1}{2}} \cdot \left(\sum_{i=1}^{n}(a_i + b_i)^2\right)^{\frac{1}{2}}$$

$$= \left(\sum_{i=1}^{n}(a_i + b_i)^2\right)^{\frac{1}{2}} \cdot \left(\left(\sum_{i=1}^{n} a_i^2\right)^{\frac{1}{2}} + \left(\sum_{i=1}^{n} b_i^2\right)^{\frac{1}{2}}\right).$$

This completes the proof of our inequality.

217. *Let* $x_1, \ldots, x_n$ *be positive numbers and* $y_k = \dfrac{1}{x_k}$. *Prove that*

$$n^2 \le \left(\sum_{k=1}^{n} x_k\right) \cdot \left(\sum_{k=1}^{n} y_k\right).$$

Solution. Let $p = q = 2$ in problem (215). Then, by Cauchy-Schwarz inequality we have

$$\left(\sum_{k=1}^{n} \sqrt{x_k} \cdot \sqrt{y_k}\right)^2 \le \left(\sum_{k=1}^{n}(\sqrt{x_k})^2\right) \cdot \left(\sum_{k=1}^{n}(\sqrt{y_k})^2\right).$$

This implies that

$$\underbrace{(1 + \cdots + 1)}_{n \text{ times}}^2 = n^2 \le \left(\sum_{k=1}^{n} x_k\right) \cdot \left(\sum_{k=1}^{n} y_k\right).$$

218. *Let* $\{x_1, \ldots, x_n\}$, $\{y_1, \ldots, y_n\}$ *and* $\{z_1, \ldots, z_n\}$ *be three sets of non-negative numbers. Prove that*

$$\left(\sum_{i=1}^{n} x_i y_y z_i\right)^4 \le \left(\sum_{i=1}^{n} x_i^4\right) \cdot \left(\sum_{i=1}^{n} y_i^2\right)^2 \cdot \left(\sum_{i=1}^{n} z_i^4\right).$$

Solution. We apply Cauchy-Schwarz inequality in Problem (215) for $a_i = x_i z_i$ and $b_i = y_i$. We get

$$\left(\sum_{i=1}^{n} x_i y_i z_i\right)^2 \le \left(\sum_{i=1}^{n}(x_i z_i)^2\right) \cdot \left(\sum_{i=1}^{n} y_i^2\right). \qquad (3.46)$$

Again, we apply Cauchy-Schwarz inequality for the numbers $a_i = x_i^2$ and $b_i = z_i^2$. Then, we have

$$\left(\sum_{i=1}^{n} x_i^2 z_i^2\right) \leq \left(\sum_{i=1}^{n} x_i^4\right)^{\frac{1}{2}} \cdot \left(\sum_{i=1}^{n} z_i^4\right)^{\frac{1}{2}}. \tag{3.47}$$

Therefore, by (3.46) and (3.47), we obtain

$$\left(\sum_{i=1}^{n} x_i y_i z_i\right)^2 \leq \left(\sum_{i=1}^{n} x_i^4\right)^{\frac{1}{2}} \cdot \left(\sum_{i=1}^{n} z_i^4\right)^{\frac{1}{2}} \cdot \left(\sum_{i=1}^{n} y_i^2\right),$$

and this completes the proof of our inequality.

219. *Let f and g be two differentiable functions on $[a, b]$. Then, the Wronskian of f and g is defined as the following determinant.*

$$W(f, g)(x) = \begin{vmatrix} f(x) & g(x) \\ f'(x) & g'(x) \end{vmatrix} = f(x)g'(x) - f'(x)g(x).$$

If f and g are linear dependent, prove that $W(f, g)(x) = 0$, for all $x \in [a, b]$.

Solution. We consider a linear combination of f and g as follows: $c_1 f(x) + c_2 g(x) = 0$. If this holds, then by taking derivative, we obtain

$$\begin{aligned} c_1 f(x) + c_2 g(x) &= 0, \\ c_1 f'(x) + c_2 g'(x) &= 0, \end{aligned} \tag{3.48}$$

for all $x \in [a, b]$. We can rewrite Eq. (3.48) in matrix form as follows:

$$\begin{bmatrix} f(x) & g(x) \\ f'(x) & g'(x) \end{bmatrix} \begin{bmatrix} c_1 \\ c_2 \end{bmatrix} = \begin{bmatrix} 0 \\ 0 \end{bmatrix}. \tag{3.49}$$

Note that our matrix of coefficient for Eq. (3.49) is just the Wronskian of the functions f and g. If the Wronskian is not zero (even in some point $x \in [a, b]$), then there is a unique solution for (3.49), namely, $c_1 = c_2 = 0$.

220. *Consider the two functions $f(x) = x^3$ and $g(x) = x^2|x|$ on the interval $[-1, 1]$.*

(1) Show that Wronskian f and g vanishes identically.
(2) Show that f and g are not linearly dependent.

Solution. (1) Suppose that $x \neq 0$. Then, we have

$$W(f, g)(x) = \begin{vmatrix} f(x) & g(x) \\ f'(x) & g'(x) \end{vmatrix} = \begin{vmatrix} x^3 & x^2|x| \\ 3x^2 & 3x|x| \end{vmatrix} = 0.$$

(2) Assume that

$$c_1 x^3 + c_2 x^2 |x| = 0. \tag{3.50}$$

Since (3.50) must be true for all $x \in [-1, 1]$, we set $x = 1$ and $x = -1$. Then, we have

$$c_1 + c_2 = 0 \text{ and } -c_1 + c_2 = 0.$$

This implies that $c_1 = c_2 = 0$, i.e., f and g are linearly independent.

221. *Let P and Q be two continuous functions. If $f(x)$ and $g(x)$ are two linearly independent solutions of*

$$y'' + P(x)y' + Q(x)y = 0,$$

then the zeros of these functions are distinct and occur alternately. In the sense that $f(x)$ vanishes exactly once between any two successive zeros of $g(x)$, and conversely.

Solution. Since $f(x)$ and $g(x)$ are linearly independent, their Wronskian $W(f, g)$ $(x) = f(x)g'(x) - f'(x)g(x) \neq 0$. Since $W(f, g)(x)$ is continuous, it must have constant sign. First, it is easy to see that f and g cannot have a common zero; for if they do, then the Wronskian will vanish at that point, which is impossible. We now assume that x_1 and x_2 are successive zeros of g and show that f vanishes between these points. The Wronskian clearly reduces to

$$\begin{vmatrix} f(x_1) & g(x_1) \\ f'(x_1) & g'(x_1) \end{vmatrix} = \begin{vmatrix} f(x_1) & 0 \\ f'(x_1) & g'(x_1) \end{vmatrix} = f(x_1)g'(x_1) \neq 0$$

and

$$\begin{vmatrix} f(x_2) & g(x_2) \\ f'(x_2) & g'(x_2) \end{vmatrix} = \begin{vmatrix} f(x_1) & 0 \\ f'(x_2) & g'(x_2) \end{vmatrix} = f(x_2)g'(x_2) \neq 0.$$

So, both factors $f(x)$ and $g'(x)$ are non-zero at each of these points. Furthermore, $g'(x_1)$ and $g'(x_2)$ must have opposite signs, because if g is increasing at x_1 it must be decreasing at x_2, and vice versa. Since the Wronskian has constant sign, it follows that $f(x_1)$ and $f(x_2)$ must also have opposite signs, and therefore, by continuity, $f(x)$ must vanish at some point between x_1 and x_2. Note that f cannot vanish more than once between x_1 and x_2; for if it does, then the same argument shows that g must vanish between these zeros of f, which contradicts the original assumption that x_1 and x_2 are successive zeros of g.

222. Let $f(x) = e^x(\cos x + \sin x)$, where $0 \leq x \leq 2\pi$. Determine the intervals in which the curve concave downward or concave upward.

Solution. We have

$$f'(x) = 2e^x \cos x,$$
$$f''(x) = 2e^x(\cos x - \sin x).$$

If $0 < x < \dfrac{\pi}{4}$ or $\dfrac{5\pi}{4} < x < 2\pi$, then $\cos x > \sin x$, and in this case $f''(x) > 0$. If $\dfrac{\pi}{4} < x < \dfrac{5\pi}{4}$, then $\cos x < \sin x$, and in this case $f''(x) < 0$. Therefore, the curve is concave downwards in $(\dfrac{\pi}{4}, \dfrac{5\pi}{4})$ and is concave upward in $(0, \dfrac{\pi}{4}) \cup (\dfrac{5\pi}{4}, 2\pi)$.

223. *The graph of the equation*

$$x^{2/3} + y^{2/3} = k^{2/3}$$

is called an astroid, where k is a constant. Show that the length of the portion of any tangent line to an astroid cut off by the coordinate axes is constant.

Solution. Since

$$\frac{2}{3\sqrt[3]{x}} + \frac{2y'}{3\sqrt[3]{y}} = 0,$$

it follows that $y' = -\sqrt[3]{\dfrac{y}{x}}$. So, the tangent line through the point (a, b) on the curve is

$y = -\sqrt[3]{\dfrac{b}{a}}(x - a) + b$. Hence, its x and y intercepts are $(a + \sqrt[3]{ab^2}, 0)$ and $(0, b + \sqrt[3]{a^2b})$. Consequently, the square of the portion of the tangent line cut off by axes is

$$\left(a + \sqrt[3]{ab^2}\right)^2 + \left(b + \sqrt[3]{a^2b}\right)^2 = a^2 + 2a\sqrt[3]{ab^2} + b\sqrt[3]{a^2b} + b^2 + 2b\sqrt[3]{a^2b} + a\sqrt[3]{ab^2}$$
$$= a^2 + 3a\sqrt[3]{ab^2} + 3b\sqrt[3]{a^2b} + b^2$$
$$= \left(\sqrt[3]{a^2} + \sqrt[3]{b^2}\right)^3 = k^2.$$

Thus, the length of the portion is k.

224. *A ladder 25 feet long is leaning against the wall of a house. The base of the ladder is pulled away from the wall at a rate of 3 feet per second. How fast is the top of the ladder moving down the wall when its base is 15 feet from the wall?*

Solution. Let t be the number of seconds in the time that has elapsed since the ladder started to slide down the wall, h be the number of feet in the distance from the ground

Fig. 3.2 A ladder 25 feet long is leaning against the wall of a house

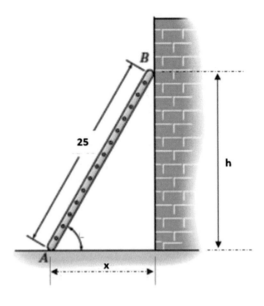

to the top of the ladder at t sec, and x be the number of feet in the distance from the bottom of the ladder to the wall at t sec, see Fig. 3.2.

Because the bottom of the ladder is pulled horizontally away from the wall at 3 ft/s, $D_t x = 3$. We wish to find $D_t h$ when $x = 15$. We have

$$h^2 = 625 - x^2. \tag{3.51}$$

Because of x and h are functions of t, we differentiate on both sides of (3.51) with respect to t and obtain $2h \, D_t h = -2x \, D_t x$, or equivalently

$$D_t h = -\frac{x}{y} D_t x. \tag{3.52}$$

When $x = 15$, it follows from (3.51) that $h = 20$. Because $D_t x = 3$, from (3.52) we get

$$D_t h \Big|_{h=20} = \frac{15}{20} \cdot 3 = -\frac{9}{4}.$$

Consequently, the top of the ladder is sliding down the wall at the rate of $2\frac{1}{4}$ ft/s when the bottom is 15 ft from the wall. The significance of the minus sign is that h is decreasing as t is increasing.

3.3 Exercises

Easier Exercises

1. Use the definition of derivative to calculate the derivative of the following functions at the indicated points.

    (a) $f(x) = x^{100}$ at $x = 1$,

    (b) $f(x) = x^2 \cos x$ at $x = 0$,

    (c) $f(x) = x^{\frac{1}{3}}$ at $x = 2$,

    (d) $f(x) = \dfrac{2x + 1}{3x - 2}$ at $x = 0$.

2. For each of the following functions defined on $\mathbb{R}$, give the set of points at which it is not differentiable.

    (a) $f(x) = |\cos x|$,
    (b) $f(x) = |x^3 - 1|$,

    (c) $f(x) = e^{|x|}$,
    (d) $f(x) = |x| + |x - 1|$.

3. Find an equation of the tangent line to the graph of $f(x) = 3x^2 - 4$ that is parallel to the line $3x + y = 4$.

4. Find an equation of each line through the point $(3, -2)$ that is tangent to the graph of $f(x) = x^2 - 7$.

5. Find an equation of each of the normal lines to the graph of $f(x) = x^3 - 4x$ that is parallel to the line $x + 8y - 8 = 0$.

6. Find an equation of the normal line to the graph of $f(x) = \dfrac{1}{x}$ that is perpendicular to the line $x + 4y = 0$.

7. Find an equation of the normal line to the graph of $x - y = \sqrt{x + y}$ at the point $(3, 1)$.

8. Find equations of the tangent and normal lines to the graph of $f(x) = 6 \sin^3 2x$ at the point $\left(\dfrac{\pi}{2}, 1\right)$.

9. Use the tangent line approximation to estimate the following quantities:

    (a) $\sqrt{35}$,
    (b) $\sin 62°$,

    (c) $\cos 121°$,
    (d) $\tan 44°$.

10. Let $e(\Delta x)$ be the error of the tangent line approximation, that is,

$$e(\Delta x) = f(a + \Delta x) - f(a) - f'(a)\Delta x.$$

    Show that

$$\lim_{\Delta x \to 0} e(\Delta x) = 0 \text{ and } \lim_{\Delta x \to 0} \frac{e(\Delta x)}{\Delta x} = 0.$$

    Thus, as $\Delta x \to 0$, the error $e(\Delta x)$ not only approaches zero but in fact approaches zero faster than Δx.

11. At what point of the graph of $x + \sqrt{xy} + y = 1$ is the tangent line parallel to the x axis?

12. Let

$$f(x) = \begin{cases} \dfrac{x+4}{3} & \text{if } x \le b \\ 2x - 1 & \text{if } b < x. \end{cases}$$

(a) Determine a value of b for which f is continuous;
(b) Is f differentiable at the value of b found in part (a)?

13. If

$$f(x) = \begin{cases} ax + b & \text{if } x \le 2 \\ 2x^2 - 1 & \text{if } 2 < x, \end{cases}$$

find the values of a and b such that $f'(2)$ exists.

14. Let

$$f(x) = \begin{cases} ax^2 + b & \text{if } x \le 1 \\ \dfrac{1}{|x|} & \text{if } 1 < x. \end{cases}$$

Find the values of a and b such that $f'(1)$ exists.

15. Let f and g be two differentiable functions on $\mathbb{R}$. If $f(0) = g(0)$ and $f'(x) \le g'(x)$ for all $x \in \mathbb{R}$, prove that $f(x) \le g(x)$ for $x \ge 0$.

16. If f is a differentiable function on $\mathbb{R}$ such that $1 \le f'(x) \le 3$ for all $x \in \mathbb{R}$ and $f(0) = 0$, prove that $x \le f(x) \le 3x$ for all $x \in \mathbb{R}$.

17. Let f be a differentiable function on $\mathbb{R}$ and $f(0) = 0$, $f(1) = 1$ and $f(2) = 1$. Show that

(a) $f'(x) = \dfrac{1}{2}$ for some $x \in (0, 2)$,

(b) $f'(x) = \dfrac{1}{7}$ for some $x \in (0, 2)$.

18. Let $h(x) = f(x)g(x)$ and suppose that $f'(x)g'(x)$ is a constant and f''' and g''' exist. Show that

$$\frac{h'''(x)}{h(x)} = \frac{f'''(x)}{f(x)} + \frac{g'''(x)}{g(x)}.$$

19. Find a general formula for y'' on the curve $x^n + y^n = a^n$, for $n \in \mathbb{N}$.

20. If $f(x) = \dfrac{1}{1 - 2x}$, prove by mathematical induction that

$$f^{(n)}(x) = \frac{2^n n!}{(1 - 2x)^{n+1}}.$$

21. If $f(x) = \sin x$, prove that $f^{(n)}(x) = \sin\left(x + \dfrac{1}{2}n\pi\right)$.

22. Let $g(x) = |f(x)|$. If $f^{(n)}(x)$ exists and $f(x) \ne 0$, prove that

$$g^{(n)}(x) = \frac{f(x)}{|f(x)|} f^{(n)}(x).$$

23. If $f(x) = x^{n-1} e^{\frac{1}{x}}$, show that

$$f^{(n)}(x) = (-1)^n \frac{f(x)}{x^{2n}}, \quad (n \in \mathbb{N}).$$

24. If $f(x) = x^2 e^{-\frac{x}{a}}$, show that

$$f^{(n)}(0) = \frac{(-1)^n n(n-1)}{a^{n-2}}, \quad (n \geq 2).$$

25. Prove that the derivative of an even function is an odd function and the derivative of an odd function is an even function, provided that these derivatives exist.
26. Prove that the derivative of a periodic function with period T is a periodic function with period T, provided that this derivative exists.

Find the absolute extreme of the given function on the specified interval:

27. $f(x) = |x^2 - 3x + 2|$ on $[-10, 10]$,

28. $f(x) = \dfrac{x}{x^2 + 1}$ on $[0, 2]$,

29. $f(x) = \sin(\sin x)$ on $[0, \pi]$,

30. $f(x) = \cos^2 x + 2 \sin^2 x$ on $[\frac{\pi}{4}, \frac{3\pi}{4}]$.

31. For what choice of constants a and b does the function

$$f(x) = \frac{ax + b}{(x - 1)(x - 4)}$$

have a strict local maximum, equal to -1 at the point $x = 2$.
32. Suppose that f is a continuous function on $[a, b]$ and f does not have a local extremum in (a, b). Prove that f is monotone on $[a, b]$.
33. Show that the function $f(x) = x^3 + ax^2 + bx + c$ always has an inflection point, regardless the values of a, b and c. For what value of a does f have an inflection point at $x = 1$.
34. The function $f(x) = \sin^3 x + \cos^3 x$ has six inflection points in the interval $[0, 2\pi]$. Find them.

Use L'Hospital rule to evaluate the limit:

35. $\displaystyle\lim_{x \to \frac{\pi}{4}} \frac{\sqrt{2}\cos x - 1}{1 - \tan^2 x}$,

36. $\displaystyle\lim_{x \to \frac{\pi}{2}^+} \left(\cos x \ln \left(x - \frac{\pi}{2} \right) \right)$,

37. $\displaystyle\lim_{x \to 0} \frac{\sin^{-1} 2x - 2 \sin^{-1} x}{x^3}$,

38. $\displaystyle\lim_{x \to 1^+} \frac{x^x - x}{1 - x + \ln x}$,

39. $\displaystyle\lim_{x \to \infty} \frac{\sqrt{x + \sqrt{x + \sqrt{x}}}}{\sqrt{x + 100}}$,

40. $\displaystyle\lim_{x \to \infty} x \left(\ln(x + \pi) - \ln x \right)$,

41. $\lim\limits_{x\to-\infty} \left(1 - \dfrac{4}{x}\right)^{x-1}$,

42. $\lim\limits_{x\to\frac{1}{2}} \left(\dfrac{x+2}{2x-1}\right)^{4x^2-1}$,

43. $\lim\limits_{x\to a} \left(\dfrac{\sin x}{\sin a}\right)^{\frac{1}{x-a}}$,

44. $\lim\limits_{x\to\infty} \left(x^2 - x^3 \sin\left(\dfrac{1}{x}\right)\right)$,

45. $\lim\limits_{x\to 0} \left(\dfrac{x}{1-\cos x} - \dfrac{2}{x}\right)$,

46. $\lim\limits_{x\to\frac{\pi}{2}} |\tan x|^{\cos x}$,

47. $\lim\limits_{x\to\infty} \dfrac{x^x}{x \ln x}$,

48. $\lim\limits_{x\to\infty} x(x+1)\left(\ln\left(1 + \dfrac{1}{x}\right)\right)^2$.

49. Let a, b and c be real numbers. Show that the equation

$$4ax^3 + 3bx^2 + 2cx = a + b + c$$

always has a root between 0 and 1.

50. Show that the equation $x^{13} + 7x^3 - 5 = 0$ has exactly one real root.

51. Let $f : [1, 3] \to \mathbb{R}$ be a continuous function that is differentiable in $(1, 3)$ with derivative $f'(x) = f(x)^2 + 4$, for all $x \in (1, 3)$. Determine whether it is true or false that $f(3) - f(1) = 5$. Justify your answer.

52. Are there any value of c for which the equation $x^4 - 4x + c = 0$ has two distinct roots in the interval $[0, 1]$? Give reasons.

53. Let $f : [a, b] \to \mathbb{R}$ be such that $f'''(x)$ exists, for all $x \in [a, b]$. Suppose that $f(a) = f(b) = f'(a) = f'(b) = 0$. Show that the equation $f'''(x) = 0$ has a solution.

Harder Exercises

54. Let $f : \mathbb{R} \to \mathbb{R}$ be a differentiable function at $x = 1$, $f(1) = 1$ and $k \in \mathbb{N}$. Show that

$$\lim\limits_{n\to\infty} n\left(f\left(1 + \dfrac{1}{n}\right) + f\left(1 + \dfrac{2}{n}\right) + \cdots + f\left(1 + \dfrac{k}{n}\right) - k\right) = \dfrac{k(k+1)}{2} f'(1).$$

55. Find the derivative of the following functions:

(a) $f(x) = \sqrt[3]{\tan^{-1}\left(\sqrt[5]{\cos^3 (\ln x)}\right)}$,

(b) $f(x) = \sin^{-1}(\tanh x)$.

(c) $f(x) = \dfrac{\sqrt{x+1}}{\sqrt{(x+2)^3} \cdot \sqrt[3]{(x+5)^2}}$,

(d) $f(x) = (\cos x)^{\sin x}$,

56. Let f' be a continuous function on $[a, b]$ and $\epsilon > 0$. Prove that there exists $\delta > 0$ such that

$$\left|\dfrac{f(t) - f(x)}{t - x} - f'(x)\right| < \epsilon,$$

whenever $0 < |t - x| < \delta$, $a \le x \le b$ and $a \le t \le b$.

57. Let f and g be differentiable functions on an open interval I and consider $a \in I$. Define h on I by

$$h(x) = \begin{cases} f(x) & \text{if } x < a \\ g(x) & \text{if } x \geq a. \end{cases}$$

Prove that h is differentiable at a if and only if both $f(a) = g(a)$ and $f'(a) = g'(a)$.

58. Suppose that the function f is defined by

$$f(x) = \begin{cases} \dfrac{g(x) - g(a)}{x - a} & \text{if } x \neq a \\ g'(x) & \text{if } x = a. \end{cases}$$

Prove that if $g'(a)$ exists, then f is continuous at a.

59. Give an example of a function $f : \mathbb{R} \to \mathbb{R}$ which is differentiable only at $x = 1$.

60. Suppose that $f : \mathbb{R} \to \mathbb{R}$ is a differentiable function at $c \in \mathbb{R}$.

    (a) If $f(c) \neq 0$, show that $|f|$ is also differentiable at c.
    (b) If $f(c) = 0$, give examples to show that $|f|$ may or may not be differentiable at c.

61. Using the mean value theorem determine

$$\lim_{x \to 0} \frac{(1 + x)^n - 1}{x}.$$

62. Suppose that

$$f(x) = \frac{\sinh x - \sin x}{x^3} \quad (x \neq 0).$$

Show that we can define $f(0)$ such that the function $f : \mathbb{R} \to \mathbb{R}$ is continuous at every point of $\mathbb{R}$.

63. Find the smallest positive integer p such that

$$\lim_{x \to 0} \frac{\sin\left(\sinh x\right) - \sinh\left(\sin x\right)}{x^p} \neq 0.$$

64. Let

$$f(x) = \sin \pi x \cdot \sin^{-1}\left(\frac{1}{x}\right).$$

    (a) Where is f differentiable?
    (b) Find an interval in which f is invertible.
    (c) If a ranges in $(0, \infty)$, compute $\lim_{x \to \infty} x^a f(x)$, if existing.

65. (a) If f is a continuous function at c and $\lim_{x \to c} f'(x)$ exists, prove that $f'(c)$ exists
    and f' is continuous at c.

    (b) Give an example to show that it is necessary to assume in (a) that f is
    continuous at c.

66. Suppose that f is a differentiable function in $(-\infty, \infty)$ and a is a critical point of
    f. Let $h(x) = f(x)g(x)$, where g is a differentiable function in $(-\infty, \infty)$ and
    $f(a)g'(a) \neq 0$. Show that the tangent line to the graph of $y = h(x)$ at $(a, h(a))$
    and the tangent line to the graph of $y = g(x)$ at $(a, g(a))$ intersect on the x axis.

67. Let $f : (0, \infty) \to \mathbb{R}$ satisfy $f(xy) = f(x) + f(y)$, for all $x, y \in (0, \infty)$. If f
    is differentiable at $x = 1$, show that f is differentiable at every $x \in (0, \infty)$ and
    $f'(x) = \dfrac{1}{x} f'(1)$, for all $x \in (0, \infty)$.

68. If $x^n y^m = (x + y)^{n+m}$, prove that $xy' = y$.

69. Show that the equation $x^8 + x - 1 = 0$ has two real roots, but no more than two.

70. Let $f : \mathbb{R} \to \mathbb{R}$ be a continuous function.

    (a) Suppose that f attains each of its values exactly two times. Let $f(x_1) = f(x_2) = c$ for some $c \in \mathbb{R}$ and $f(x) > c$ for some $x \in [x_1, x_2]$. Show that
    f attains its maximum in $[x_1, x_2]$ exactly at one point.

    (b) Using (a) show that f can not attain each of its values exactly two times.

71. Let $a_0, a_1, \ldots, a_n$ be real numbers with the property that

$$a_0 + \frac{a_1}{2} + \frac{a_2}{3} + \cdots + \frac{a_n}{n+1} = 0.$$

Prove that the equation

$$a_0 + a_1 x + a_2 x^2 + \cdots a_n x^n = 0$$

has at least one solution in the interval $(0, 1)$.

72. Show that $ex \leq e^x$, for all $x \in \mathbb{R}$.

73. Prove that $|\cos x - \cos y| \leq |x - y|$, for all $x, y \in \mathbb{R}$.

74. For any positive integer n and real numbers x, y, find all solutions of equation
    $(x^n + y^n) = (x + y)^n$.

75. Prove that $\dfrac{1}{9} < \sqrt{66} - 8 < \dfrac{1}{8}$.

76. Let $0 \leq a < b < \dfrac{\pi}{2}$. Prove that

$$\frac{b - a}{\cos^2 a} < \tan b - \tan a < \frac{b - a}{\cos^2 b}.$$

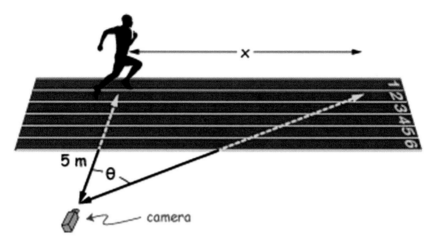

Fig. 3.3 The runner is moving

77. If $0 < x < 1$, prove that

$$\sqrt{\frac{1-x}{1+x}} < \frac{\ln(1+x)}{\sin^{-1} x}.$$

78. Suppose that f is a differentiable function on $[0, 1]$ with $f(0) = 0$ and $f(1) = 1$. For each positive integer n, show that there exist distinct numbers $a_1, a_2, \ldots, a_n$ in $[0, 1]$ such that

$$\sum_{i=1}^{n} \frac{1}{f'(a_i)} = n.$$

79. A man 6 ft tall walks at a speed of 4 ft/s toward a street lamp which is 14 ft above the ground. How fast is the length of the man's shadow decreasing?

80. Compute the following limit

$$\lim_{x \to 0} \frac{e^x - \sum_{k=0}^{n} k! x^k}{x^n}, \quad (n \in \mathbb{N}).$$

81. Compute the following limit

$$\lim_{x \to 0} \frac{\sin x - \sum_{k=0}^{n} (-1)^k \frac{x^{2k+1}}{(2k+1)!}}{x^{2k+1}}, \quad (n \ge 0).$$

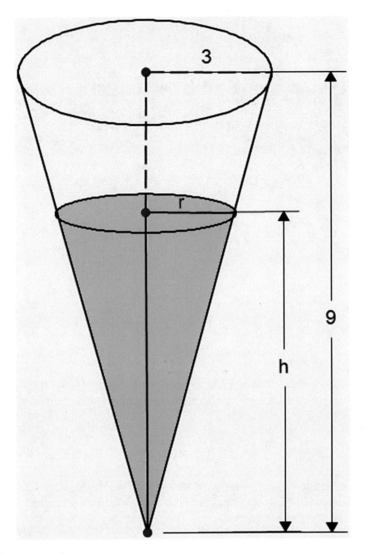

Fig. 3.4 A water tank

82. Let f be a differentiable function on some deleted neighborhood N of a and suppose that f and f' have no zeros in N. Find

(a) $\lim\limits_{x \to a} |f(x)|^{f(x)}$ if $\lim\limits_{x \to a} f(x) = 0$,

(b) $\lim\limits_{x \to a} |f(x)|^{\frac{1}{f(x)-1}}$ if $\lim\limits_{x \to a} f(x) = 1$,

(c) $\lim\limits_{x \to a} |f(x)|^{\frac{1}{f(x)}}$ if $\lim\limits_{x \to a} f(x) = \infty$.

Fig. 3.5 A laser pointer is placed on a platform

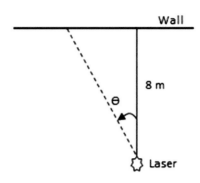

83. Suppose that f and g are differentiable functions and g' has no zeros in (a, b). Suppose that $\lim\limits_{x \to b^-} \dfrac{f'(x)}{g'(x)} = L$ and either

$$\lim_{x \to b^-} f(x) = \lim_{x \to b^-} g(x) = 0$$

or

$$\lim_{x \to b^-} f(x) = \infty \text{ and } \lim_{x \to b^-} g(x) = \pm \infty.$$

Find $\lim\limits_{x \to b^-} \left(1 + f(x)\right)^{\frac{1}{g(x)}}$.

84. A television camera is positioned 5 meters from the edge of a track, as shown in Fig. 3.3, in order to follow runners as they pass and approach the finish line. If the runner is moving at 5 m/s as he passes the finish line 10 meters away (where the lane numbers are), how fast (in degrees per second) must the camera be rotating just as it pans across the line?

85. A water tank has the shape of an inverted circular cone with base radius 3 meters and height 9 meters (see Fig. 3.4). If water is being pumped into the tank at a rate of $2 \, \text{m}^3/\text{min}$, find the rate at which the water level is rising when the water is 4 meters deep

86. A laser pointer is placed on a platform that rotates at a rate of 20 revolutions per minute. The beam hits a wall 8 m away, producing a dot of light that moves horizontally along the wall. Let θ be the angle between the beam and the line through the searchlight perpendicular to the wall (see Fig. 3.5). How fast is this dot moving when $\theta = \dfrac{\pi}{6}$?

87. The volume of a balloon is decreasing at a rate proportional to its surface area. Show that the radius of the balloon shrinks at a constant rate.

88. The measure of one of the acute angles of a right triangle is decreasing at the rate of $\dfrac{\pi}{36}$ rad/s. If the length of the hypotenuse is constant and 40 cm, find how fast the area is changing when the measure of the acute angle is $\dfrac{\pi}{6}$.

Chapter 4
Optimization Problems

4.1 Basic Concepts and Theorems

One of the most common applications of calculus involves the determination of minimum and maximum values. Consider how frequently you hear or read terms such as greatest profit, least cost, shortest time, greatest voltage, optimum size, greatest revenue, least size, greatest strength, greatest distance, and so on. Problems of this type ask for the best value of some variable quantity and hence are called *optimization problems*. If the reader is interested to see more discussion about these concepts, we refer him/her to [1–16].

Problem-Solving Strategy

(1) Read problem carefully, until you understand it.
(2) Draw a picture if relevant.
(3) Introduce variables. List every relation in the picture and in the problem as an equation or algebraic expression, and identify the unknown variable.
(4) Express quantity to be optimized in terms of one variable.
(5) Find the domain of variable.
(6) Test the critical points and endpoints in the domain of the unknown variable. Use what you know about the shape of the function's graph. Use the first and second derivatives to identify and classify the function's critical points.
(7) Find the absolute maximum or minimum and answer the question.

4.2 Problems

225. Let $f(x) = (\ln x)^{\ln x}$.

(1) Find the domain of f.
(2) Find the local extremum points of f.

Solution. (1) We must have $\ln x > 0$. So, the domain of f is $(1, \infty)$.
   (2) We can write $f(x) = e^{\ln x \cdot \ln(\ln x)}$. So, we get

$$f'(x) = \left(\frac{1}{x}\ln(\ln x) + \ln x \cdot \frac{\frac{1}{x}}{\ln x}\right)e^{\ln x \cdot \ln(\ln x)}$$
$$= \frac{1}{x}\Big(\ln(\ln x) + 1\Big)e^{\ln x \cdot \ln(\ln x)}.$$

Note that $\dfrac{1}{x}e^{\ln x \cdot \ln(\ln x)} > 0$. Moreover,

$$\ln(\ln x) + 1 > 0 \Leftrightarrow \ln(\ln x) > -1$$
$$\Leftrightarrow \ln x > \frac{1}{e}$$
$$\Leftrightarrow x > e^{\frac{1}{e}}.$$

We have $f'(e^{\frac{1}{e}}) = 0$. If $x < e^{\frac{1}{e}}$, then $f'(x) < 0$ and if $x > e^{\frac{1}{e}}$, then $f'(x) > 0$. So, f has a local minimum at $x = e^{\frac{1}{e}}$.

226. *Suppose that the function f is defined on $[-1, e]$ as follows:*

$$f(x)\begin{cases} x\ln|x| & if\ x \neq 0 \\ 0 & if\ x = 0. \end{cases}$$

Determine the points of local maximum and minimum.

Solution. The function is continuous and has a finite derivative throughout $[-1, e]$ expect at 0, where

$$f'(0) = \lim_{x \to 0}\frac{f(x) - f(0)}{x - 0} = \lim_{x \to 0}\ln|x| = -\infty. \tag{4.1}$$

It follows from (4.1) that f decreases at the point $x = 0$. The equation $f'(x) = 1 + \ln|x| = 0$ has two roots: $x_1 = -\dfrac{1}{e}$ and $x_2 = \dfrac{1}{e}$. We also have $f''(x) = \dfrac{1}{x}$ $(x \neq 0)$, and so $f''(-\dfrac{1}{e}) < 0$ and $f''(\dfrac{1}{e}) > 0$. Consequently, $-\dfrac{1}{e}$ is a point of local maximum and $\dfrac{1}{e}$ is a point of local minimum.

227. *Find the absolute maximum value of the following function:*

$$f(x) = \frac{1}{1 + |x|} + \frac{1}{1 + |x - 2|}.$$

Solution. It is easy to see that

$$
f(x) = \begin{cases}
\dfrac{1}{1-x} + \dfrac{1}{1-(x-2)} & \text{if } x < 0 \\[2mm]
\dfrac{1}{1+x} + \dfrac{1}{1-(x-2)} & \text{if } 0 \le x < 2 \\[2mm]
\dfrac{1}{1+x} + \dfrac{1}{1+(x-2)} & \text{if } x \ge 2.
\end{cases}
$$

Then, we obtain

$$
f'(x) = \begin{cases}
\dfrac{1}{(1-x)^2} + \dfrac{1}{(3-x)^2} & \text{if } x < 0 \\[2mm]
\dfrac{-1}{(1+x)^2} + \dfrac{1}{(3-x)^2} = \dfrac{8(x-1)}{(3-x)^2(x+1)^2} & \text{if } 0 < x < 2 \\[2mm]
\dfrac{-1}{(1+x)^2} - \dfrac{1}{(x-1)^2} & \text{if } x > 2.
\end{cases}
$$

Consequently, we have

$$
\begin{aligned}
&f'(x) > 0 \quad \text{if } x < 0, \\
&f'(x) < 0 \quad \text{if } 0 < x < 1, \\
&f'(1) = 0, \\
&f'(x) > 0 \quad \text{if } 1 < x < 2, \\
&f'(x) < 0 \quad \text{if } x > 2.
\end{aligned}
$$

So, by the first derivative test, f has local maximum at points $x = 0$ and $x = 2$. Since $f(0) = f(2) = \dfrac{4}{3}$, $f(1) = 1$ and $\lim\limits_{x \to \pm\infty} f(x) = 0$, it follows that $\dfrac{4}{3}$ is the absolute maximum value of f.

228. *Let $a_1, a_2, \ldots, a_n \in \mathbb{R}$ and for every $x \in \mathbb{R}$*

$$
f(x) = \sqrt{(x-a_1)^2 + (x-a_2)^2 + \cdots + (x-a_n)^2}.
$$

Find the point of absolute minimum of the function f.

Solution. We consider $g(x) = (x-a_1)^2 + (x-a_2)^2 + \cdots + (x-a_n)^2$. Clearly, the point of minimum of f and g are same. We have

$$
\begin{aligned}
g'(x) &= 2(x-a_1) + 2(x-a_2) + \cdots + 2(x-a_n) \\
&= 2nx - 2(a_1 + \cdots + a_n)
\end{aligned}
$$

and $g''(x) = 2n$. Now, if $c = \dfrac{a_1 + \cdots + a_n}{n}$, then $g'(c) = 0$ and $g''(c) > 0$. Consequently, c is the point of minimum of f.

229. *Find the absolute minimum and the absolute maximum values of the following function:*

$$f(x) = x^{\frac{1}{x}}$$

Solution. Note that the domain of f is equal to $(0, \infty)$. Now, on the domain, we have

$$\ln(f(x)) = \frac{1}{x}\ln x. \tag{4.2}$$

We take derivative from both sides of (4.2). Then, we get

$$\frac{f'(x)}{f(x)} = \frac{-1}{x^2}\ln x + \frac{1}{x^2},$$

or equivalently,

$$f'(x) = x^{\frac{1}{x}}\left(\frac{1 - \ln x}{x^2}\right).$$

If $0 < x < e$, then $1 - \ln x > 0$, and so $f'(x) > 0$.
If $e < x$, then $1 - \ln x < 0$, which implies that $f'(x) < 0$.

Consequently, f has the absolute maximum value at $x = e$ and $f(e) = e^{\frac{1}{e}}$ is the absolute maximum. Moreover, we observe that f has no absolute minimum value.

230. *A given rectangular area is to be fenced off in a field that lies along a straight river. If no fencing is needed along the river, show that the least amount of fencing will be required when the length of the field is twice its width.*

Solution. Suppose that x and y are the length and width of the field, respectively. Then the area of the field is $A = xy$. The fencing required is $F(x) = x + 2y = x + \frac{2A}{x}$. Taking derivative implies that $F'(x) = 1 - \frac{2A}{x^2}$. Solving $F'(x) = 0$ we get $x = \sqrt{2A}$, and hence $y = \frac{1}{2}\sqrt{2A}$ as required. In order to investigate that $F(x)$ has minimum at $x = \sqrt{2A}$ we observe that $F''(\sqrt{2A}) > 0$.

231. *A rectangular area of A m^2 is to be fenced off. Two opposite sides will use fencing costing a per meter and the remaining sides will use fencing costing b per meter. Find the dimensions of the rectangle of least cost.*

Solution. Suppose that x and y are the length and width of the rectangle, respectively, where the sides that are x meter long cost a per meter and the sides that are y meter long cost b per meter. The area of rectangle is $A = xy$. So, we conclude that the total cost is $C(x) = 2ax + 2by = 2ax + \frac{2bA}{x}$. We determine the minimum of $C(x)$ for $x > 0$. We have $C'(x) = 2a - \frac{2bA}{x^2} = \frac{2ax^2 - 2bA}{x^2}$ and $C''(x) = \frac{4bA}{x^3}$. We observe that $\sqrt{\dfrac{b}{a}A}$ is the only value for x such that $C'\left(\sqrt{\dfrac{b}{a}A}\right) = 0$. Since $C''\left(\sqrt{\dfrac{b}{a}A}\right) > 0$, it

follows that the minimum cost is $C\left(\sqrt{\dfrac{b}{a}A}\right)$.

232. *A closed rectangular box with a square base is to have a surface area of S cm². What is the maximum possible volume such a box can contain?*

Solution. Let x be the length and y be the width of the box. The volume of the box is $V(x) = x^2 y$. The surface area of the box is $S = 2x^2 + 4xy$. This implies that $y = \dfrac{S - 2x^2}{4x}$. Consequently, we obtain

$$V(x) = x^2 \cdot \frac{S - 2x^2}{4x} = \frac{S}{4}x - \frac{1}{2}x^3.$$

Then, we have

$$V'(x) = \frac{S}{4} - \frac{3}{2}x^2.$$

If $V'(x) = 0$, then $x = \sqrt{\dfrac{S}{6}}$. Clearly, $V''\left(\sqrt{\dfrac{S}{6}}\right) < 0$. Therefore, V has a local maxi-

mum at $x = \sqrt{\dfrac{S}{6}}$. Since it is the only local maximum for positive x, it follows that

the absolute maximum volume is $V\left(\sqrt{\dfrac{S}{6}}\right)$.

233. *What is the largest possible area of a triangle, two of whose sides are a and b?*

Solution. Suppose that θ is the angle between the two sides. The area of the triangle is

$$S(\theta) = \frac{1}{2}ab \sin \theta,$$

where $0 < \theta < \pi$. Note that a and b are constants. Thus, we have

$$S'(\theta) = \frac{1}{2}ab \cos \theta,$$

Setting the derivative equal to 0 and solving, we obtain $\theta = \dfrac{\pi}{2}$. Since $S\left(\dfrac{\pi}{2}\right) = \dfrac{1}{2}ab$,

it follows that the maximum area occurs when $\theta = \dfrac{\pi}{2}$.

234. *Of all isosceles triangles with a fixed perimeter, which one has the maximum area?*

Fig. 4.1 Isosceles triangle

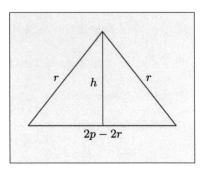

Solution. Suppose that r is the length of the equal sides, and h is the height to the base. Assume that the fixed perimeter is equal to $2p$. So, the base is $2p - 2r$, look at Fig. 4.1.

We have $r^2 = h^2 + (p - r)^2$. This implies that $h = \sqrt{r^2 - (p - r)^2} = \sqrt{2pr - p^2}$. We denote $S(r)$ the area of triangle, i.e.,

$$S(r) = (p - r)h = (p - r)\sqrt{2pr - p^2}.$$

Then, we obtain

$$S'(r) = -\sqrt{2pr - p^2} + (p - r)\frac{p}{\sqrt{2pr - p^2}}$$
$$= \frac{2p^2 - 3pr}{\sqrt{2pr - p^2}}.$$

If $S'(r) = 0$, then $2p - 3r = 0$, and so $r = \dfrac{2}{3}p$. Now it is easy to see that the maximum area occurs when the length of equal sides is $\dfrac{2}{3}p$ and the base is $\dfrac{2}{3}p$ too. Therefore, the triangle that maximizes the area is equilateral.

235. *Find a point on the altitude of an isosceles triangle such that the sum of its distance from the vertices is the smallest possible.*

Solution. See Fig. 4.2. Suppose that $\overline{BH} = a$, $\overline{AH} = h$ and θ is the triangle between BH and BO. If D is the required sum of distances, then

$$D(\theta) = \overline{OB} + \overline{OC} + \overline{OA} = 2\overline{OB} + \overline{OA}$$
$$= 2\frac{a}{\cos \theta} + h - a \tan \theta.$$

Taking derivative from D gives

$$D'(\theta) = 2a\frac{\sin \theta}{\cos^2 \theta} - a\frac{1}{\cos^2 \theta} = \frac{a(2\sin \theta - 1)}{\cos^2 \theta}.$$

Fig. 4.2 A point on the altitude of an isosceles triangle

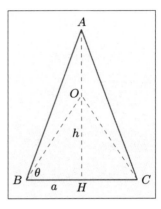

If $D'(\theta) = 0$, then $2 \sin \theta - 1 = 0$. This yields $\sin \theta = \dfrac{1}{2}$, and consequently $\theta = \dfrac{\pi}{6}$. Moreover,

$$D''(\theta) = \frac{2a \cos^2 \theta + 2a \sin \theta (2 \sin \theta - 1)}{\cos^3 \theta}.$$

Since $D''(\dfrac{\pi}{6}) > 0$, it follows that the minimum occurs at $\theta = \dfrac{\pi}{6}$, and the minimum of D is equal to $h + a\sqrt{3}$.

236. *Given a circle, prove that the square is the rectangle of maximal area that can be inscribed in the circle.*

Solution. Let the positive numbers x and y denote the lengths of the sides of the inscribed rectangle, and r denote the radius of the circle. Then, $x^2 + y^2 = 4r^2$ which implies that $y = \sqrt{4r^2 - x^2}$. Therefore,

$$\text{Area of rectangle} = f(x) = xy = x\sqrt{4r^2 - x^2}.$$

Then, we have

$$f'(x) = 2\left(\frac{2r^2 - x^2}{\sqrt{4r^2 - x^2}}\right).$$

Now, $\sqrt{2}r$ is the root of $f'(x) = 0$. If $0 < x < \sqrt{2}r$, then $f'(x) > 0$. If $x > \sqrt{2}r$, then $f'(x) < 0$. Thus, $f(x)$ has a maximum at $x = \sqrt{2}r$. Since $y = \sqrt{4r^2 - x^2}$, it follows that $y = \sqrt{2}r$. Therefore, $x = y$ and so the rectangle is a square.

237. *Let a cylindrical has a constant surface area S. Determine the dimensions that produce the maximum volume.*

Solution. Let x be the radius and h be the height of the cylinder having its total surface area S. We have $S = 2\pi x h + 2\pi x^2$. So, we obtain

$$h = \frac{S - 2\pi x^2}{2\pi x} = \frac{S}{2\pi x} - x. \tag{4.3}$$

Now the volume of the cylinder is

$$V = \pi x^2 h = \pi x^2 \left(\frac{S}{2\pi x} - x \right) = \frac{S}{2}x - \pi x^3.$$

Hence, $\dfrac{dV}{dx} = \dfrac{S}{2} - 3\pi x^2$. If $\dfrac{dV}{dx} = 0$, then we obtain $x = \sqrt{\dfrac{S}{6\pi}}$. Since $\dfrac{d^2V}{dx^2} = -6\pi x < 0$ for all $x > 0$, it follows that the volume is maximum when $x = \sqrt{\dfrac{S}{6\pi}}$.

Setting $x = \sqrt{\dfrac{S}{6\pi}}$ in (4.3), we obtain

$$h = \frac{S}{2\pi\sqrt{\dfrac{S}{6\pi}}} - \sqrt{\frac{S}{6\pi}} = \sqrt{\frac{2S}{3\pi}}.$$

Therefore, the ratio of hight to the radius is given as $\dfrac{h}{x} = 2$, or equivalently, the height of the cylinder is equal to the diameter of its base.

238. *The sum of the squares of two non-negative numbers is equal to* 4. *How should they be chosen such that the product of their cubes is maximum?*

Solution. Suppose that x and y are non-negative numbers such that $x^2 + y^2 = 4$ or $y = \sqrt{4 - x^2}$. Then, we conclude that $0 \le x \le 2$ and $2x + 2y\dfrac{dy}{dx} = 0$. So, $\dfrac{dy}{dx} = -\dfrac{x}{y}$. Now, if $F = x^3 y^3$, then by implicit differentiation we get

$$\frac{dF}{dx} = 3x^2 y^3 + x^3 \left(3y^2 \frac{dy}{dx} \right) = 3x^2 y^3 + 3x^3 y^2 \left(-\frac{x}{y} \right)$$
$$= 3x^2 y^3 - 3x^4 y = 3x^2 y(y^2 - x^2).$$

If $\dfrac{dF}{dx} = 0$, then $xy = 0$ or $y^2 - x^2 = 0$. Consequently, we obtain three critical points $x = 0$, $x = 2$ and $x = \sqrt{2}$. Since F is equal to 0, 8 and 0 when $x = 0$, $x = \sqrt{2}$ and $x = 2$, respectively, we conclude that the maximum value of F is at $x = \sqrt{2}$ and $y = \sqrt{2}$.

239. *Find the points on the graph* $3x^2 + 10xy + 3y^2 = 9$ *closest to the origin.*

Solution. The distance of each point (x, y) on the graph from origin is $\sqrt{x^2 + y^2}$. So, it is enough to minimize $F = x^2 + y^2$ on the graph. By implicit differentiation, we have

$$\frac{dF}{dx} = 2x + 2y\frac{dy}{dx} \tag{4.4}$$

and

$$6x + 10y + 10x\frac{dy}{dx} + 6y\frac{dy}{dx} = 0,$$

which implies that

$$\frac{dy}{dx} = -\frac{3x + 5y}{5x + 3y}. \tag{4.5}$$

If we substitute (4.5) into (4.4), then we get

$$\begin{aligned}
\frac{dF}{dx} &= 2x - 2y\frac{3x + 5y}{5x + 3y} \\
&= \frac{10x^2 + 6xy - 6xy - 10y^2}{5x + 3y} \\
&= \frac{10x^2 - 10y^2}{5x + 3y}.
\end{aligned}$$

If $\dfrac{dF}{dx} = 0$, then $x^2 = y^2$, or equivalently $y = \pm x$. But $y = -x$ is impossible. So, if $y = x$, then we have $16x^2 = 9$, which implies that $x = \pm\dfrac{3}{4}$. Therefore, $(\dfrac{3}{4}, \dfrac{3}{4})$ and $(-\dfrac{3}{4}, -\dfrac{3}{4})$ are two points closest to the origin.

240. *Two vertices of a rectangle are on the positive x-axis. The other two vertices are on the lines $y = 2x - 1$ and $y = -\dfrac{5}{4}x + 5$. What is the maximum area of rectangle?*

Solution. See Fig. 4.3. Suppose that t is the x-coordinate of the vertex A. Then, the y-coordinate of the vertices C and D is $2t - 1$. The x-coordinate of vertex C is obtained by solving the equation $y = -\dfrac{5}{4}x + 5$ for x when $y = 2t - 1$. Consequently, we get $x = \dfrac{-8t + 24}{5}$. This is also x-coordinate of B. So, the base of rectangle is

$$\frac{-8t + 24}{5} - t = \frac{-13t + 24}{5}.$$

Therefore, the area of rectangle is

$$A(t) = (2t - 1)\left(\frac{-13t + 24}{5}\right) = \frac{1}{5}(-26t^2 + 61t - 24).$$

Taking the first and second derivative we get

$$A'(t) = \frac{1}{5}(-52t + 61) \text{ and } A''(t) = -\frac{52}{5} < 0.$$

Fig. 4.3 Rectangle of
Problem 240

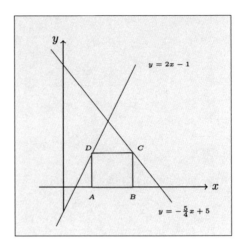

If $A'(t) = 0$, then $t = \dfrac{61}{52}$. Since the second derivative is negative, it follows that the

rectangle has the maximum area when $t = \dfrac{61}{52}$, and the maximum area of rectangle

is equal to $A\left(\dfrac{61}{52}\right)$.

241. *A piece of wire L cm long is cut two pieces. One piece is bent into the shape of
a circle and the other into the shape of a square. How should the wire be cut so that*

(1)   the combined area of the two figures is as small as possible.
(2)   the combined area of the two figures is as large as possible.

Solution. Let x be the length used to form the square. This means that the rest of
the wire $L - x$ is the circumference of the circle. Since x is the amount of wire cut
for the square, it follows that each side has length of $\dfrac{x}{4}$. The circumference of the

circle is $2\pi r = L - x$. Solving for r we get $r = \dfrac{L - x}{2\pi}$. Now, the area enclosed by
the wire is

$$f(x) = \left(\frac{x}{4}\right)^2 + \pi\left(\frac{L - x}{2\pi}\right)^2$$
$$= \left(\frac{1}{16} + \frac{1}{4\pi}\right)x^2 - \frac{L}{2\pi}x + \frac{L^2}{4\pi},$$

where $x \in [0, L]$. We take derivative of f, then we get

$$f'(x) = 2\left(\frac{1}{16} + \frac{1}{4\pi}\right)x - \frac{L}{2\pi}.$$

If $f'(x) = 0$, then we get $x = \dfrac{4L}{\pi + 4}$, the critical point. Since f is a continuous function on closed interval $[0, L]$, it has absolute maximum and absolute

minimum at critical point or at the ends of the interval. So, the minimum of f is equal to $\min\{f(0), f(\frac{4L}{\pi + 4}), f(L)\}$ and the maximum of f is equal to $\max\{f(0), f(\frac{4L}{\pi + 4}), f(L)\}$. Therefore, after computations we get

(1) f has minimum if the radius of circle is $\dfrac{L}{2(\pi + 4)}$ cm and the length of side of square is $\dfrac{L}{\pi + 4}$;

(2) f has maximum if the radius of circle is $\dfrac{L}{2\pi}$ cm and there is no square.

242. *Solve Problem (241) if one piece of wire is bent into the shape of an equilateral triangle and the other piece is bent into the shape of a square.*

Solution. Let x be the length that is cut for the wire to form the square. Then the rest of the wire $L - x$ is the circumference of the equilateral triangle. So, each side of square has length of $\dfrac{x}{4}$.

- The area of the square is $\left(\dfrac{x}{4}\right)^2 = \dfrac{x^2}{16}$.
- The area of the triangle is $\dfrac{\sqrt{3}}{4}\left(\dfrac{L - x}{3}\right)^2$.
- Total area of these figures is

$$g(x) = \frac{x^2}{16} + \frac{\sqrt{3}}{4}\left(\frac{L - x}{3}\right)^2 = \frac{x^2}{16} + \frac{L^2 + x^2 - 2Lx}{12\sqrt{3}}.$$

where $x \in [0, L]$. If we take derivative, then we have

$$g'(x) = \frac{x}{8} + \frac{L^2 + x^2 - 2Lx}{12\sqrt{3}} = \left(\frac{1}{8} + \frac{1}{6\sqrt{3}}\right)x - \frac{L}{6\sqrt{3}}.$$

If $g'(x) = 0$, then we obtain $x = \dfrac{4L}{3\sqrt{3} + 4}$, the critical point. Now, we observe that $g'(\dfrac{4L}{3\sqrt{3} + 4}) = \dfrac{1}{8} + \dfrac{1}{6\sqrt{3}} > 0$. So, by the second derivative test, g has strict local minimum at $x = \dfrac{4L}{3\sqrt{3} + 4}$. It is easy to see that this point is absolute minimum too. Moreover, we have $g(0) = \dfrac{L^2}{12\sqrt{3}} < g(L) = \dfrac{L^2}{16}$. So, we have maximum at $x = L$. So, for obtaining the maximum area, we must don't cut the wire and use it to form a square.

243. *A mosque window is to be in the shape of a rectangle surmounted by a semicircle. If the perimeter of the window is L cm, express its area as a function of its semicircular radius r. What is its maximum area?*

Fig. 4.4 A rectangle
surmounted by a semicircle

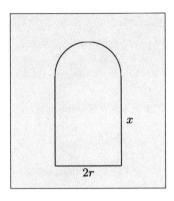

Solution. Since the radius of the semicircle is r, it follows that the base of the window is $2r$. Suppose that x is the height of the rectangle (see Fig. 4.4). We denote $S(x)$ the area of semicircle and rectangle, i.e.,

$$S(r) = \frac{1}{2}\pi r^2 + 2rx. \tag{4.6}$$

On the other hand, we have $\pi r + 2x + 2r = L$, or equivalently

$$x = \frac{L - \pi r}{2} - r. \tag{4.7}$$

Substituting (4.7) in (4.6) we get

$$S(r) = \frac{1}{2}\pi r^2 + 2r\left(\frac{L - \pi r}{2} - r\right).$$

Now, we compute the derivation of $S(r)$. We obtain

$$S'(r) = \pi r + 2\left(\frac{L - \pi r}{2} - r\right) + 2r\left(\frac{-\pi}{2} - 1\right).$$

If $S'(r) = 0$, then

$$r = \frac{L}{\pi + 4}.$$

Obviously, we observe that the maximum area occurs at $r = \dfrac{L}{\pi + 4}$. So, the maximum area is equal to $S\left(\dfrac{L}{\pi + 4}\right)$.

244. *A window is to be constructed in the shape of an equilateral triangle on top of a rectangle. If its perimeter is to be L cm, what is the maximum possible area of the window?*

Fig. 4.5 An equilateral
triangle on top of a rectangle

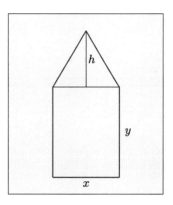

Solution. Consider Fig. 4.5. Since $h^2 + \left(\frac{x}{2}\right)^2 = x^2$, it follows that $h = \frac{\sqrt{3}}{2}x$. Since
the perimeter of the window is L, it follows that $3x + 2y = L$, or equivalently $y = \frac{L - 3x}{2}$. We denote $S(x)$ the area of the window. Then, $S(x)$ is the sum of areas of
the triangular top and rectangular bottom.

Hence, we have

$$S(x) = \frac{\sqrt{3}}{4}x^2 + x\left(\frac{L - 3x}{2}\right)$$
$$= \left(\frac{\sqrt{3}}{4} - \frac{3}{2}\right)x^2 + \frac{L}{2}x,$$

where $x \in [0, \frac{L}{3}]$. So, we get

$$S'(x) = \left(\frac{\sqrt{3}}{2} - 3\right)x + \frac{L}{2}.$$

If $S'(x) = 0$, then

$$x = \frac{L}{6 - \sqrt{3}}.$$

Obviously, we observe that the maximum area occurs at $x = \frac{L}{6 - \sqrt{3}}$. So, the maximum area is equal to $S\left(\frac{L}{6 - \sqrt{3}}\right)$.

245. *Find the dimensions of the rectangle of largest area that can be inscribed in
an equilateral triangle of side L if one side of the rectangle lies on the base of the
triangle.*

Fig. 4.6 Rectangle inscribed in an equilateral triangle

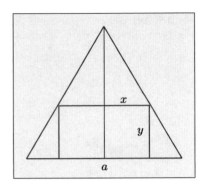

Solution. Consider Fig. 4.6. If h is the height of the equilateral triangle, then $a^2 = \left(\frac{a}{2}\right)^2 + h^2$. This implies that $h = \frac{\sqrt{3}}{2}a$. Using similarity triangles we have

$$\frac{y}{\frac{a}{2} - x} = \frac{h}{\frac{a}{2}} = \sqrt{3}.$$

This implies that $y = \sqrt{3}\left(\frac{a}{2} - x\right)$, where $x \in [0, \frac{a}{2}]$. Suppose that $S(x)$ is the area of the rectangle. Then, we have

$$S(x) = (2x)y = 2\sqrt{3}x\left(\frac{a}{2} - x\right) = \sqrt{3}ax - 2\sqrt{3}x^2.$$

Thus, $S'(x) = \sqrt{3}a - 4\sqrt{3}x$. If $S'(x) = 0$, then $x = \frac{a}{4}$. Now, we observe that the maximum area is $S\left(\frac{a}{4}\right) = \frac{\sqrt{3}}{8}a$. Therefore, the width of the rectangle is $2x = \frac{a}{2}$ and the height of the rectangle is $y = \frac{\sqrt{3}}{4}a$.

246. *Find the dimensions of the rectangle of largest area whose base is on the x-axis and whose upper two vertices lie on the parabola $y = 12 - x^2$.*

Solution. Consider Fig. 4.7.

Since P lies on the parabola, its coordinates satisfy the equation $y = 12 - x^2$. Suppose that $S(x)$ is the area of the rectangle. Then, we have

$$S(x) = (2x)y = 2x(12 - x^2) = 24x - 2x^3,$$

where $x \in [0, \sqrt{12}]$. Hence, $S'(x) = 24 - 6x^2$. If $S'(x) = 0$, then $x = 2$. Now, we observe that $S(2) = 32$ is the maximum of the area. Consequently, the width of the rectangle is $2x = 4$ and the height of the rectangle is $y = 8$.

Fig. 4.7 Rectangle of
Problem 246

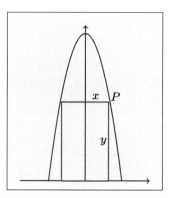

Fig. 4.8 A right circular
cylinder is inscribed in a
right circular cone

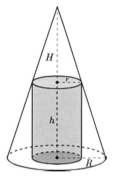

247. *A right circular cylinder is inscribed in a right circular cone with radius R and height H. Find the maximum volume of the cylinder.*

Solution. Let r be the radius of the base of a cylinder inscribed in the cone and let h be its height. According to the similar triangles in Fig. 4.8, we have

$$\frac{H}{R} = \frac{h}{R - r}$$

and so we conclude that $h = \dfrac{H(R - r)}{R}$. Hence, we can obtain a function that gives the volume of cylinder in terms of r, i.e.,

$$V = V(r) = (\text{area of base of cylinder}) \cdot (\text{height of cylinder})$$
$$= \pi r^2 \cdot \frac{H(R - r)}{R},$$

where $r \in (0, R)$. Now, we compute the first and second derivative and find the critical points. We obtain

Fig. 4.9 A right circular
cylinder is inscribed in a
sphere

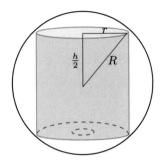

$$V'(r) = \frac{\pi H}{R} r(2R - 3r),$$
$$V''(r) = \frac{2\pi H}{R}(R - 3r).$$

If $V'(r) = 0$, then $r = 0$ or $r = \frac{2}{3}R$. Obviously $r = 0$ does not give a maximum
value for V. So, we examine $r = \frac{2}{3}r$. Since $V''(\frac{2}{3}R) = -2\pi H < 0$, by the second
derivative test, the critical value $r = \frac{2}{3}R$ gives a maximum value

$$V(\frac{2}{3}R) = \frac{4}{27}\pi H R^2.$$

248. *Find the height h and radius r of a right circular cylinder of maximum volume
that can be cut within a sphere of radius R.*

Solution. The axis of the cylinder must lie on a diameter of the sphere. From Fig. 4.9,
we obtain $R^2 = r^2 + \left(\frac{h}{2}\right)^2$ and hence $r = \sqrt{R^2 - \left(\frac{h}{2}\right)^2}$. The volume of the cylinder
in terms of h is equal to

$$V = V(h) = \pi r^2 h = \pi \left(R^2 - \left(\frac{h}{2}\right)^2\right)h = \pi R^2 h - \frac{\pi}{4}h^3,$$

where $0 \le h \le 2R$.

Now, we compute the first and the second derivative and determine the critical
points. We get

$$V'(h) = \pi R^2 - \frac{3}{4}\pi h^2,$$
$$V''(h) = -\frac{3}{2}\pi h.$$

Fig. 4.10 A right circular
cone is circumscribed about
a sphere

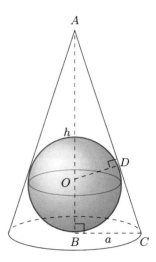

So, the critical number is $h = \dfrac{2}{\sqrt{3}} R$. Since the second derivative is negative, it follows

that there is a local maximum at $h = \dfrac{2}{\sqrt{3}} R$. Since $V(0) = 0$ and $V(2R) = 0$, we

deduce that the absolute maximum occurs when $h = \dfrac{2}{\sqrt{3}} R$ and $r = \sqrt{\dfrac{2}{3}} R$.

249. *Find the dimensions of the right circular cone of minimum volume which can be circumscribed about a sphere of radius r.*

Solution. Suppose that a is the radius of the base of cone and $h + r$ is the height of the cone, see Fig. 4.10. Since the triangles ABC and ADO are similar, it follows that

$$\frac{\overline{BC}}{\overline{DO}} = \frac{\overline{AB}}{\overline{AD}},$$

and so

$$\frac{a}{r} = \frac{h + r}{\sqrt{h^2 - r^2}}.$$

The volume of the cone is

$$V(h) = \frac{\pi}{3} a^2 (h + r) = \frac{\pi r^2 (h + r)^2}{3(h - r)},$$

and hence we get

$$V'(h) = \frac{\pi r^2 (h + r)(h - 3r)}{3(h - r)^2}.$$

The critical point greater than r is $h = 3r$, and by the first derivation test, this point is a local minimum. Finally, we observe that this point is absolute minimum too.

4.3 Exercises

Easier Exercises

1. If a and b are positive numbers, find the x-coordinate which gives the absolute maximum value of $f(x) = x^a(1 - x)^b$ on the interval $[0, 1]$.
2. Find the rectangle of area A with the smallest perimeter.
3. What is the maximum value of the sum of two numbers (not necessary positive) whose product is a given number a? What is the minimum value?
4. What is the maximum value of the product of two numbers whose difference is a given number a? What is the minimum value?
5. A straight piece of wire of L cm is cut into two pieces. One piece is bent into a square and the another piece is bent into a rectangle with aspect ratio 3. What are dimensions, in centimeters of the square and the rectangle such that the sum of their area is minimized?
6. A L cm piece of wire is cut into two pieces and once piece is bent into a square and the another is bent into an equilateral triangle. Where should the wire cut so that the total area enclosed by both is

   (a) minimum?
   (b) maximum?

7. A rectangle is inscribed in a semicircle with radius r with one of its sides at the diameter of the semicircle. Find the dimensions of the rectangle so that its area is a maximum see Fig. 4.11.
8. Find the tangent to the graph of function $f(x) = x^2 - 1$ which cuts off the triangle of minimum area from the fourth quadrant. What is the minimum area?
9. If $P(a, b)$ is a point in the first quadrant, find the line through P which cuts off the triangle of minimum area from the first quadrant. What is the minimum area?
10. Find the shortest distance between the point $(4, 1)$ and the parabola $f(x) = \dfrac{1}{x^2}$.
11. Find the largest vertical distance between the graphs of $f(x) = \sqrt{x}$ and $g(x) = \sqrt[3]{x}$ on the interval $[0, 1]$.
12. Given two points $P_1(0, 4)$ and $P_2(3, 5)$. Find the point Q on the x-axis for which the sum of the distance $\overline{P_1 Q}$ and $\overline{Q P_2}$ is the smallest.
13. A closed box with a square base (see Fig. 4.12) is to have a volume of 2000 cubic inches. The material for the top and bottom of the box is to cost 3 dollars per square inch, and the material for the sides is to cost 1.5 dollars per square inch. If the cost of the material is to be the least, find the dimensions of the box.

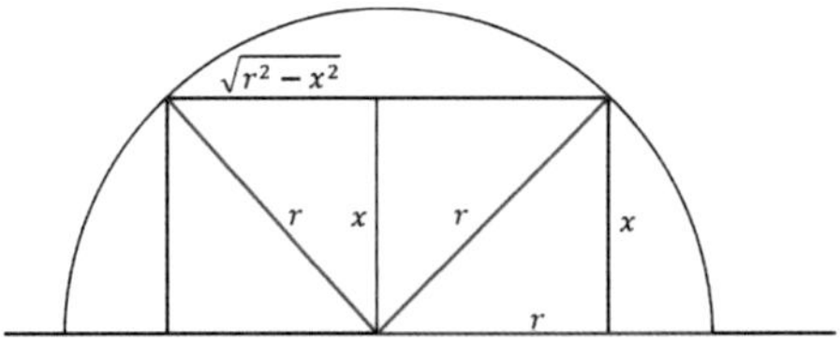

Fig. 4.11 A rectangle is inscribed in a semicircle with radius r

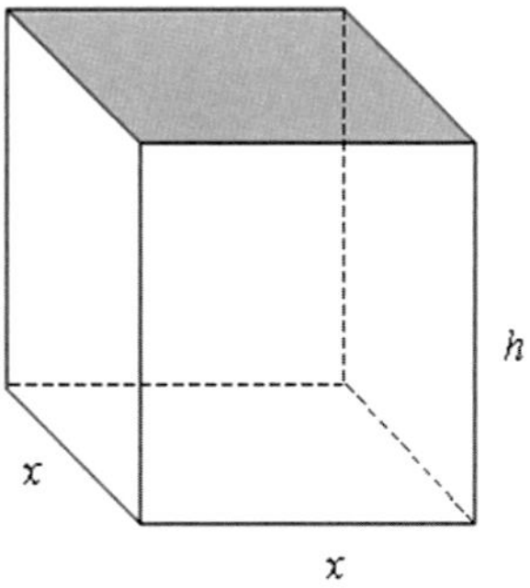

Fig. 4.12 A closed box with a square base

14. A box manufacturer is to produce a closed box of specific volume whose base is a rectangle having a length that is three times its width. Find the most economical dimensions.
15. Each rectangular page of a book must contain 30 cm^2 of printed text, and each page must have 2 cm margins at top and bottom, and 1 cm margin at each side. What is the minimum possible area of such a page?
16. A direct current generator has an electromotive force of E volts and an internal resistance of r ohms, where E and r are constants. If R ohms is the external resistance, the total resistance is $(r + R)$ ohms, and if P wats is the power, then

$$P = \frac{E^2 R}{(r + R)^2}.$$

What external resistance will consume the most power?
17. Two posts, one is 12 meters high and the other is 28 meters high are placed 30 meters apart. They are to be stayed by 2 wires attached to a single stake running from ground level to the top of each post. Where should the stake be placed to use the least wire?

Harder Exercises

18. A snowman is an anthropomorphic sculpture made from snow as well as some pieces of coal, a carrot, a hat and a scarf (see Fig. 4.13). For the purposes of this

Fig. 4.13 A snowman

question, we consider a snowman to consist of a spherical head of radius a and a spherical body of radius b, and we also assume that the snow does not melt and its density does not change while it is being sculpted. The research shows that the cuteness F of a snowman is given by

$$F = \begin{cases} \left(\frac{a}{b}\right)^2\left(1 - \frac{a}{b}\right)(a^2 + ab + b^2) & \text{if } 0 \le a < b \\ 0 & \text{if } 0 < b \le a. \end{cases}$$

Find the dimensions of the cutest snowman that can be built with $\dfrac{4\pi}{3}$ m^3 of snow.

19. A 1125 ft^3 open-top rectangular tank with a square base x ft on a side and y ft deep is to be built with its top flush with the ground to catch runoff water. The costs associated with the tank involve not only the material form which the tank is made but also an excavation charge proportional to the product xy.

 (a) If the total cost is
$$c = 5(x^2 + 4xy) + 10xy,$$

 what values of x and y will minimize it?

 (b) Give a possible scenario for the cost function in part (a).

20. In an elliptical sport field we want to design a rectangular soccer field with the maximum possible area. The sport field is given by the graph of $\dfrac{x^2}{a^2} + \dfrac{y^2}{b^2} = 1$. Find the length $2x$ and width $2y$ of the pitch (in terms of a and b) that maximize the area of the pitch.

21. Find an equation of the tangent line to the graph of function $f(x) = x^3 - 3x^2 + 5x$ that has the least slop.

22. Find the point on the line $3x + y = 6$ that is closest to the origin.

23. Find the dimensions of the rectangle of largest area that has its base on the x-axis and its other two vertices above the x-axis and lying on the parabola $f(x) = 12 - x^2$.

24. Find the smallest volume of a right circular cone circumscribed about a hemisphere of radius r.

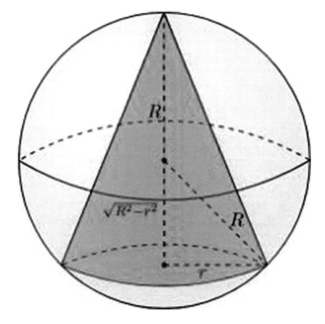

Fig. 4.14 A cone inscribed inside a sphere

Fig. 4.15 Constructing cone

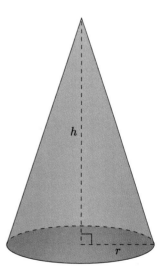

25. Prove by the method of this chapter that the shortest distance from the point $P(x_0, y_0)$ to the line L having the equation $ax + by + c = 0$ is

$$\frac{|ax_0 + by_0 + c|}{\sqrt{a^2 + b^2}}.$$

26. Show that the volume of the largest cone that can be inscribed inside a sphere of radius r is $\frac{32\pi r^3}{81}$.

27. Show that the volume of the largest cone that can be inscribed inside a sphere of radius R is $\frac{32\pi R^3}{81}$ (see Fig. 4.14).

28. A rectangle is to inscribed under the graph of function $f(x) = 4\cos(\frac{x}{2})$ from $x = -\pi$ to π. What are the dimensions of the rectangle with largest area, and what is the largest area?

29. A right triangle whose hypotenuse is $\sqrt{3}$ meters long is revolved about one of its legs to generate a right circular cone, see Fig. 4.15. Find the radius (r), height (h) and the volume of the greatest volume that can be made in this way.

30. A fence, 8 feet high, is parallel to the wall of a building and 1 foot from the building. What is the shortest plank that can go over the fence, from the level ground, to prop the wall?

Chapter 5
Integrals

5.1 Basic Concepts and Theorems

Antiderivative and the Indefinite Integral

A function F is called an *antiderivative* of a function f on an interval I if $F'(x) = f(x)$ for all $x \in I$. For instance, see Fig. 5.1.

Let $F(x)$ be any antiderivative of $f(x)$ on an interval I. Then, every other antiderivative of $f(x)$ on I is of the form $F(x) + C$, where C is an arbitrary constant. The expression $F(x) + C$ is called the *indefinite integral* of $f(x)$, denoted by

$$\int f(x)dx. \tag{5.1}$$

In the notation of indefinite integral (5.1), $f(x)$ is called *integrand*. If f and g have indefinite integral on the same interval and k is any constant, then it is not difficult to see that

$$\int kf(x)dx = k \int f(x)dx,$$
$$\int \big(f(x) + g(x)\big)dx = \int f(x)dx + \int g(x)dx.$$

The Definite Integral

Let f be a function defined on $[a, b]$. We divide $[a, b]$ into n subintervals by choosing $n - 1$ intermediate points between a and b. Let $x_0 = a$, $x_n = b$ and $x_0 < x_1 < \cdots < x_{n-1} < x_n$. Let $\Delta_i x$ be the length of the subinterval $[x_{i-1}, x_i]$. A set of such points

© The Author(s), under exclusive license to Springer Nature Singapore Pte Ltd. 2020
B. Davvaz, *Examples and Problems in Advanced Calculus: Real-Valued Functions*,
https://doi.org/10.1007/978-981-15-9569-1_5

Fig. 5.1 Example of antiderivative

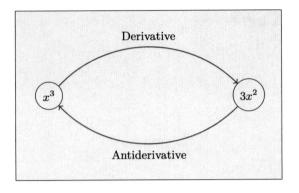

is called a *partition* of $[a, b]$. Let Δ be a such partition. The length of the longest subinterval of the partition Δ called the *norm of the partition* is denoted by $\|\Delta\|$. Now, choose a point ξ_i in each subinterval $[x_{i-1}, x_i]$, where $1 \leq i \leq n$. Form the sum

$$\sum_{i=1}^{n} f(\xi_i)\Delta_i x.$$

This sum is called *Riemann sum* of f on $[a, b]$.

If f is a function defined on $[a, b]$, then the *definite integral* of f from a to b is defined by

$$\int_a^b f(x)dx = \lim_{\|\Delta\| \to 0} \sum_{i=1}^{n} f(\xi_i)\Delta_i x,$$

if the limit exists. In this case, we say that f is *integrable* on $[a, b]$.

Integrability of continuous functions: If f is a continuous function on $[a, b]$, or if f has at most finitely many jump discontinuities on $[a, b]$, then f is integrable on $[a, b]$.

When f and g are integrable on $[a, b]$, the definite integral satisfies the following rules:

(1) $\displaystyle\int_b^a f(x)dx = -\int_a^b f(x)dx,$

(2) $\displaystyle\int_a^a f(x)dx = 0,$

(3) $\displaystyle\int_a^b kf(x)dx = k\int_a^b f(x)dx,$

(4) $\displaystyle\int_a^b \left(f(x) \pm g(x)\right)dx = \int_a^b f(x)dx \pm \int_a^b g(x)dx,$

(5) $\displaystyle\int_a^b f(x)dx = \int_a^c f(x)dx + \int_c^b f(x)dx,$ for all c between a and b.

(6) If $f(x) \geq 0$ for all $x \in [a, b]$, then $\displaystyle\int_a^b f(x)dx \geq 0$.

Mean value theorem for integral: If f is a continuous function on $[a, b]$, then there exists $c \in [a, b]$ such that

$$\frac{1}{b-a}\int_a^b f(x)dx = f(c).$$

The first fundamental theorem of calculus: Let f be a continuous function on $[a, b]$ and let x be any number in $[a, b]$. If F is the function defined by

$$F(x) = \int_a^x f(t)dt,$$

then $F'(x) = f(x)$, for all $x \in [a, b]$.

The (second) fundamental theorem of calculus: If f is a continuous function on $[a, b]$, then

$$\int_a^b f(x)dx = F(b) - F(a) = F(x)\big|_a^b,$$

where F is any antiderivative of f on $[a, b]$.

Method of Integration

The fundamental theorem of calculus tells us how to evaluate a definite integral once we have an antiderivative for the integrand function. Furthermore, there exist a number of techniques for finding indefinite integrals for many combinations of functions. The chain rule for differentiating a composite function leads to an important method of integration, known as integration by substitution.

Substitution rule: If $u = g(x)$ is a differentiable function whose range is contained in an interval I and f is continuous on I, then

$$\int f\big(g(x)\big)g'(x)dx = \int f(u)du, \tag{5.2}$$

where the substitution $u = g(x)$ is made in the right integral of (5.2) after its evaluation.

Another important method of integration that is quite useful is integration by parts, which is a consequence of the product rule for differentiation.

Integration by parts: If f and g are differentiable functions, then

$$\int f(x)g'(x)dx = f(x)g(x) - \int f'(x)g(x)dx. \tag{5.3}$$

Sometimes it is easier to remember the formula if we write it in differential form. Let $u = f(x)$ and $v = g(x)$. Then, $du = f'(x)dx$ and $dv = g'(x)dx$. Using the substitution rule, the integration by parts formula (5.3) becomes

$$\int u\,dv = uv - \int v\,du.$$

Integrals with integrands that are combinations of the trigonometric functions are called *trigonometric integrals*. We consider the case where the integrand is a product of trigonometric functions. We begin with the integral

$$\int \sin^m x \cos^n x\,dx$$

involving two exponents m and n. We can divide the appropriate substitution into two cases according to m and n being odd or even.

(1) If one of the exponents m, n is a positive odd integer, we use one of the identities $\sin^2 x = 1 - \cos^2 x$ and $\cos^2 x = 1 - \sin^2 x$ to bring the integrand into the form $f(\sin x)\cos x$ or $f(\cos x)\sin x$;
(2) If both exponents are positive even integers, we make repeated use of the identities $\sin^2 x = \dfrac{1 - \cos 2x}{2}$ and $\cos^2 x = \dfrac{1 + \cos 2x}{2}$ to reduce the integrand to one in lower power of $\cos 2x$.

Now, consider the following integral

$$\int \tan^m x \sec^n x\,dx$$

involving two exponents m and n.

(1) If m is a positive odd integer, we use the identity $\tan^2 x = \sec^2 x - 1$ to bring the integrand into the form $f(\sec x)\sec x \tan x$;
(2) If n is a positive even integer, we use the identity $\sec^2 x = \tan^2 x + 1$ to bring the integrand into the form $f(\tan x)\sec^2 x$.

We can use a similar method to evaluate

$$\int \cot^m x \csc^n x\,dx,$$

with the help of identities $\cot^2 x = \csc^2 x - 1$ and $csc^2 x = \cot^2 x + 1$.

Trigonometric substitution: If the integrand contains an expression of the form $\sqrt{a^2 + x^2}$, $\sqrt{a^2 - x^2}$ or $\sqrt{x^2 - a^2}$, where $a > 0$, it is often possible to perform the integration by making a trigonometric substitution that results in an integral involving trigonometric functions. We consider each form as a separate case.

(1) In an integral involving $\sqrt{a^2 + x^2}$ make the substitution $x = a \tan u$, where $-\dfrac{\pi}{2} < u < \dfrac{\pi}{2}$; or $x = a \cot u$, where $0 < u < \pi$.

(2) In an integral involving $\sqrt{a^2 - x^2}$ make the substitution $x = a \sin u$, where $-\dfrac{\pi}{2} \leq u \leq \dfrac{\pi}{2}$; or $x = a \cos u$, where $0 \leq u \leq \pi$.

(3) In an integral involving $\sqrt{x^2 - a^2}$ make the substitution $x = a \sec u$, where $0 \leq u < \dfrac{\pi}{2}$.

Note that in all above cases we can use *hyperbolic substitutions* too.

Integration of rational fractions: We consider the integration of expressions of the form

$$\int \frac{P(x)}{Q(x)} dx = \int \frac{a_0 + a_1 x + \cdots + a_n x^n}{b_0 + b_1 x + \cdots + b_m x^m} dx$$

where $a_n \neq 0$, $b_m \neq 0$ and $n < m$. To do this it is often necessary to write $\dfrac{P(x)}{Q(x)}$ as the sum of *partial fractions*. After $Q(x)$ has been factored into products of linear and quadratic factors, the method of determining the partial factors depends on the nature of these factors. A quadratic polynomial (or factor) is irreducible if it cannot be written as the product of two linear factors with real coefficients. That is, the polynomial has no real roots. We consider the various cases separately.

(1) Let $x - r$ be a linear factor of $Q(x)$. Assume that $(x - r)^k$ is the highest power of $x - r$ that divides $Q(x)$. Then, to this factor, assign a sum of k partial fractions:

$$\frac{A_1}{(x - r)} + \frac{A_2}{(x - r)^2} + \cdots + \frac{A_k}{(x - r)^k}.$$

Do this for each distinct linear factor of $Q(x)$.

(2) Let $x^2 + px + q$ be an irreducible quadratic factor of $Q(x)$ so that $x^2 + px + q$ has no real roots. Suppose that $(x^2 + px + q)^t$ is the highest power of this factor that divides $Q(x)$. Then, to this factor, assign a sum of t partial fractions:

$$\frac{B_1 x + C_1}{(x^2 + px + q)} + \frac{B_2 x + C_2}{(x^2 + px + q)^2} + \cdots + \frac{B_t x + C_t}{(x^2 + px + q)^t}.$$

Do this for each distinct quadratic factor of $Q(x)$.

(3) Set the original fraction $\dfrac{P(x)}{Q(x)}$ equal to the sum of all these partial fractions. Clear the resulting equation of fractions and arrange the terms in decreasing powers of x.

(4) Equate the coefficients of corresponding powers of x and solve the resulting equations for the undetermined coefficients.

Integration of rational function of sine and cosine: If an integrand is a rational function of $\sin x$ and $\cos x$, it can be reduced to a rational function of u by the substitution

$$u = \tan \frac{1}{2}x,$$

where $-\pi < x < \pi$. Then, we obtain

$$\cos x = \frac{1 - u^2}{1 + u^2}, \quad \sin x = \frac{2u}{1 + u^2} \quad \text{and} \quad dx = \frac{2}{1 + u^2}du.$$

Improper Integrals

Integrals with infinite limits of integration are *improper integrals of the first type*.

(1) If f is a continuous function on $[a, \infty)$, then

$$\int_a^\infty f(x)dx = \lim_{b \to \infty} \int_a^b f(x)dx.$$

(2) If f is a continuous function on $(-\infty, b]$, then

$$\int_{-\infty}^b f(x)dx = \lim_{a \to -\infty} \int_a^b f(x)dx.$$

(3) If f is a continuous function on $(-\infty, \infty)$, then

$$\int_{-\infty}^\infty f(x)dx = \int_{-\infty}^c f(x)dx + \int_c^\infty f(x)dx,$$

where c is any real number. In each case, if the limit exists and finite we say that the *improper integral converges* and that the limit is the value of the improper integral. If the limit fails to exist, the *improper integral diverges*.

Integrals of functions that become infinite at a point within the interval of integration are *improper integrals of the second type*.

(1) If f is a continuous function on $(a, b]$ and $\lim_{x \to a^+} f(x) = \pm\infty$, then

$$\int_a^b f(x)dx = \lim_{c \to a^+} \int_c^b f(x)dx.$$

(2) If f is a continuous function on $[a, b)$ and $\lim_{x \to b^-} f(x) = \pm\infty$, then

$$\int_a^b f(x)dx = \lim_{c \to b^-} \int_a^c f(x)dx.$$

(3) If f is unbounded at c, where $a < c < b$, and continuous on $[a, c) \cup (c, b]$, then

$$\int_a^b f(x)dx = \int_a^c f(x)dx + \int_c^b f(x)dx,$$

In each case, if the limit exists and finite we say the *improper integral converges* and that the limit is the value of the improper integral. If the limit does not exist, the *integral diverges*.

Comparison test for improper integral: Let f and g be two continuous functions such that $0 \le f(x) \le g(x)$ for all $x \ge a$. Then

(1) If $\displaystyle\int_a^\infty g(x)dx$ converges, so does $\displaystyle\int_a^\infty f(x)dx$;

(2) If $\displaystyle\int_a^\infty f(x)dx$ diverges, so does $\displaystyle\int_a^\infty g(x)dx$.

Limit comparison test for improper integral: If the positive functions f and g are continuous on $[a, \infty)$ and

$$\lim_{x \to \infty} \frac{f(x)}{g(x)} = L, \ 0 < L < \infty,$$

then

$$\int_a^\infty f(x)dx \text{ and } \int_a^\infty g(x)dx$$

both converge or both diverge.

There are analogs of the (limit) comparison test for other kinds of improper integrals. Although the improper integrals of two functions from a to ∞ may both converge, this does not mean that their integrals necessarily have the same value.

If the reader is interested to see the proof of theorems that are presented in this chapter, we refer him/her to [1–16].

5.2 Problems

250. *Find*
$$I = \lim_{n \to \infty} \left(\frac{1}{n+1} + \frac{1}{n+2} + \cdots + \frac{1}{2n} \right).$$

Solution. We can write

$$\frac{1}{n+1} + \frac{1}{n+2} + \cdots + \frac{1}{2n} = \frac{1}{n}\left(\frac{1}{1 + \frac{1}{n}} + \frac{1}{1 + \frac{2}{n}} + \cdots + \frac{1}{1 + \frac{n}{n}} \right)$$

$$= \sum_{i=1}^n \frac{1}{n}\left(\frac{1}{1 + \frac{i}{n}} \right).$$

If we consider $f(x) = \dfrac{1}{x}$ on $[1, 2]$, then

$$
\begin{aligned}
I &= \lim_{n \to \infty} \sum_{i=1}^{n} \frac{1}{n}\left(\frac{1}{1 + \dfrac{i}{n}}\right) \\
&= \int_{1}^{2} f(x)\,dx \\
&= \int_{1}^{2} \frac{1}{x}\,dx \\
&= \ln 2.
\end{aligned}
$$

251. *Find*

$$
I = \lim_{m \to \infty}\left(\lim_{n \to \infty} \sum_{i=1}^{n} \frac{mn}{(mi)^2 + n^2}\right).
$$

Solution. We have

$$
\begin{aligned}
I &= \lim_{m \to \infty}\left(\lim_{n \to \infty} \frac{1}{n} \sum_{i=1}^{n} \frac{m}{m^2(\dfrac{i}{n})^2 + 1}\right). \\
&= \lim_{m \to \infty}\left(\int_{0}^{1} \frac{m}{m^2 x^2 + 1}\,dx\right) \\
&= \lim_{m \to \infty}\left(\tan^{-1} mx \Big|_{0}^{1}\right) \\
&= \lim_{m \to \infty}\left(\tan^{-1} m - \tan^{-1} 0\right) \\
&= \frac{\pi}{2}.
\end{aligned}
$$

252. *Prove that*

$$
\lim_{n \to \infty}\left(\frac{1}{n} + \frac{n^2}{(n+1)^3} + \frac{n^2}{(n+2)^3} + \cdots + \frac{1}{8n}\right) = \frac{3}{8},
$$

by applying the definition of integral.

Solution. The $(i+1)$-th term of this series is

$$
\frac{n^2}{(n+i)^3} = \frac{1}{n} \frac{1}{(1 + \dfrac{i}{n})^3},
$$

If we consider $f(x) = \dfrac{1}{(1+x)^3}$, then f is an integrable function on $[0, 1]$. Thus, we have

$$\lim_{n\to\infty}\left(\frac{1}{n}+\frac{n^2}{(n+1)^3}+\frac{n^2}{(n+2)^3}+\cdots+\frac{1}{8n}\right)=\lim_{n\to\infty}\frac{1}{n}\sum_{i=0}^{n}\frac{1}{(1+\frac{i}{n})^3}$$

$$=\int_0^1\frac{1}{(1+x)^3}dx=\frac{3}{8}.$$

253. *Prove that*

$$\lim_{n\to\infty}\left((1+\frac{1}{n})(1+\frac{2}{n})\cdots(1+\frac{4n}{n})\right)^{\frac{1}{n}}=\frac{5^5}{e^4},$$

by applying the definition of integral.

Solution. Suppose that

$$y_n=\left((1+\frac{1}{n})(1+\frac{2}{n})\cdots(1+\frac{4n}{n})\right)^{\frac{1}{n}}.$$

Then, we obtain

$$\ln y_n=\frac{1}{n}\left(\ln(1+\frac{1}{n})+\ln(1+\frac{2}{n})+\cdots+\ln(1+\frac{4n}{n})\right).$$

Here, the i-th term is $\frac{1}{n}\ln(1+\frac{i}{n})$ and the number of terms is $4n$. If we consider $f(x)=\ln(1+x)$, then f is an integrable function on $[0,4]$. Thus, we have

$$\lim_{n\to\infty}\ln y_n=\int_0^4\ln(1+x)dx$$

$$=(1+x)\ln(1+x)-x\Big|_0^4$$

$$=5\ln 5-4=\ln\frac{5^5}{e^4}.$$

Since $\ln$ is a continuous function, it follows that $\lim\limits_{n\to\infty}y_n=\frac{5^5}{e^4}$.

254. *Show that if f is continuous on $[a,b]$, then*

$$\left|\int_a^b f(x)dx\right|\le\int_a^b|f(x)|dx.$$

Solution. Clearly, we have

$$- |f(x)| \le f(x) \le |f(x)|. \tag{5.4}$$

Since f is continuous on $[a, b]$, it follows that $|f|$ is continuous on $[a, b]$ too. So, f and $|f|$ are integrable on $[a, b]$. By integrating from (5.4), we get

$$- \int_a^b |f(x)| dx \le \int_a^b f(x) dx \le \int_a^b |f(x)| dx.$$

This completes the proof of desired inequality.

255. *Let* $P = \{a_0, , a_1, \dots, a_n\}$ *be any partition of the interval* $[1, x]$, *where* $x > 1$.

(1) Integrate suitable step functions that are constant on the open subintervals of P to derive the following inequalities

$$\sum_{k=1}^n \left(\frac{a_k - a_{k-1}}{a_k} \right) < \ln x < \sum_{k=1}^n \left(\frac{a_k - a_{k-1}}{a_{k-1}} \right).$$

(2) Interpret the inequalities of part (1) geometrically in terms of areas.
(3) Specialize the partition to show that for every integer $n > 1$,

$$\sum_{k=2}^n \frac{1}{k} < \ln n < \sum_{k=1}^{n-1} \frac{1}{k}.$$

Solution. (1) Suppose that $1 = a_0 < a_1 < \cdots < a_n = x$. If $t \in [a_{k-1}, a_k]$, then $\dfrac{1}{a_k} < \dfrac{1}{t} < \dfrac{1}{a_{k-1}}$. Hence, we have

$$\int_{a_{k-1}}^{a_k} \frac{1}{a_k} dt < \int_{a_{k-1}}^{a_k} \frac{1}{t} dt < \int_{a_{k-1}}^{a_k} \frac{1}{a_{k-1}} dt,$$

or equivalently

$$\frac{a_k - a_{k-1}}{a_k} < \int_{a_{k-1}}^{a_k} \frac{1}{t} dt < \frac{a_k - a_{k-1}}{a_{k-1}}.$$

This implies that

$$\sum_{k=1}^n \frac{a_k - a_{k-1}}{a_k} < \sum_{k=1}^n \int_{a_{k-1}}^{a_k} \frac{1}{t} dt < \sum_{k=1}^n \frac{a_k - a_{k-1}}{a_{k-1}}.$$

Therefore, we deduce that

Fig. 5.2 Geometrically interpretation of Problem (5.20)

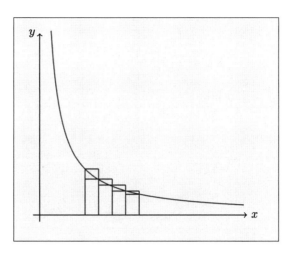

$$\sum_{k=1}^{n} \frac{a_k - a_{k-1}}{a_k} < \int_0^x \frac{1}{t} dt < \sum_{k=1}^{n} \frac{a_k - a_{k-1}}{a_{k-1}}.$$

This completes the proof of part (1).

(2) $\ln x$ is greater than the step function integral consisting of rectangular strips within $\frac{1}{x}$ and less than rectangular strips covering over $\frac{1}{x}$ (see Fig. 5.2).

(3) Taking $a_k = 1 + k$ implies that $a_n = 1 + n = x$. Now, applying part (1) gives

$$\sum_{k=1}^{n} \frac{1}{1+k} < \ln(1+n) < \sum_{k=1}^{n} \frac{1}{k}.$$

Since $\sum_{k=1}^{n} \frac{1}{1+k} = \sum_{k=2}^{n+1} \frac{1}{k}$, the proof of part (3) completes.

256. *Let f be a positive, continuous and increasing function on $[1, \infty)$.*

(1) Show that

$$\sum_{k=1}^{n-1} f(k) \le \int_1^n f(x)dx \le \sum_{k=2}^{n} f(k).$$

(2) By using (1), prove that

$$n^n e^{-n+1} \le n! \le (n+1)^{n+1} e^{-n}.$$

Solution. (1) Since f is increasing, it follows that

$$f(k) \le f(x) \le f(k+1), \quad \text{for all } x \in [k, k+1] \text{ and } k \in \mathbb{N}.$$

Hence, we get

$$\int_k^{k+1} f(k)dx \le \int_k^{k+1} f(x)dx \le \int_k^{k+1} f(k+1)dx,$$

or equivalently

$$f(k) \le \int_k^{k+1} f(x)dx \le f(k+1).$$

Therefore, we have

$$\sum_{k=1}^{n-1} f(k) \le \sum_{k=1}^{n-1} \int_k^{k+1} f(x)dx \le \sum_{k=1}^{n-1} f(k+1),$$

which implies that

$$\sum_{k=1}^{n-1} f(k) \le \int_1^n f(x)dx \le \sum_{k=2}^{n} f(k).$$

(2) Let $f(x) = \ln x$. It is easy to see that $\int \ln x dx = x \ln x - x + C$, and so $\int_1^n \ln x dx = n \ln n - n + 1$. Now, by using (1), we obtain

$$\sum_{k=1}^{n-1} \ln k \le n \ln n - n + 1 \le \sum_{k=2}^{n} \ln k. \tag{5.5}$$

From the left inequality in (5.5), we get

$$\begin{aligned}
\ln(n!) &= \sum_{k=1}^{n} \ln k \\
&\le (n+1) \ln(n+1) - (n+1) + 1 \\
&= \ln(n+1)^{n+1} - n \\
&= \ln(n+1)^{n+1} - \ln e^n \\
&= \ln \left(\frac{(n+1)^{n+1}}{e^n} \right),
\end{aligned}$$

which implies that

$$n! \le (n+1)^{n+1} e^{-n}.$$

On the other hand, from the right inequality in (5.5), we have

$$n \ln n - n + 1 \le \ln 2 + \cdots + \ln n,$$

or equivalently,

$$\ln(n^n) + \ln(e^{-n+1}) \le \ln(n!).$$

This implies that

$$\ln\left(n^n \cdot e^{-n+1}\right) \le \ln(n!),$$

and consequently we get

$$n^n e^{-n+1} \le n!.$$

257. *Without computing the integral, prove that*

$$\frac{1}{3\sqrt{2}} \le \int_0^1 \frac{x^2}{\sqrt{1+x}} dx \le \frac{1}{3}.$$

Solution. For every $0 \le x \le 1$ we have

$$\frac{x^2}{\sqrt{2}} \le \frac{x^2}{\sqrt{1+x}} \le x^2. \tag{5.6}$$

Consequently, by integrating from (5.6), we obtain

$$\int_0^1 \frac{x^2}{\sqrt{2}} dx \le \int_0^1 \frac{x^2}{\sqrt{1+x}} dx \le \int_0^1 x^2 dx,$$

and so

$$\frac{x^3}{3\sqrt{2}}\Big|_0^1 \le \int_0^1 \frac{x^2}{\sqrt{1+x}} dx \le \frac{1}{3}x^3\Big|_0^1.$$

This completes the proof of desired inequality.

258. *Let a be a positive real number, $x_1 = a$, $x_2 = 2a$ and $x_3 = 3a$. Compute the following integral:*

$$I = \int_{-a}^a |x(x - x_1)(x - x_2)(x - x_3)| dx.$$

Solution. Obviously, $x - x_i \le 0$ for $i = 1, 2, 3$, where $x \in [-a, a]$. So, we obtain

$$I = \int_{-a}^{0} x(x - x_1)(x - x_2)(x - x_3)dx - \int_{0}^{a} x(x - x_1)(x - x_2)(x - x_3)dx$$

$$= \int_{-a}^{0} \left(x^4 - 6ax^3 + 11a^2x^2 - 6a^3x\right)dx - \int_{0}^{a} \left(x^4 - 6ax^3 + 11a^2x^2 - 6a^3x\right)dx$$

$$= \left(\frac{1}{5}x^5 - \frac{3a}{2}x^4 + \frac{11a^2}{3}x^3 - 3a^3x^2\right)\Big|_{-a}^{0} - \left(\frac{1}{5}x^5 - \frac{3a}{2}x^4 + \frac{11a^2}{3}x^3 - 3a^3x^2\right)\Big|_{0}^{a}$$

$$= 9a^5.$$

259. *Evaluate*

(1) $\int_{0}^{n} [x]dx$, *where n is a positive integer;*

(2) $\int_{a}^{b} [x]dx$, *where a and b are real numbers with $0 \le a < b$.*

Solution. (1) We find that

$$\int_{0}^{n} [x]dx = \sum_{i=1}^{n} \left(\int_{i-1}^{i} [x]dx\right)$$

$$= \sum_{i=1}^{n} (i - 1)$$

$$= \frac{(n - 1)n}{2}.$$

(2) We have

$$\int_{a}^{b} [x]dx = \int_{0}^{b} [x]dx - \int_{0}^{a} [x]dx$$

$$= \int_{0}^{[b]} [x]dx + \int_{[b]}^{b} [x]dx - \int_{0}^{[a]} [x]dx - \int_{[a]}^{a} [x]dx$$

$$= \frac{([b] - 1)[b]}{2} + \int_{[b]}^{b} [b]dx - \frac{([a] - 1)[a]}{2} - \int_{[a]}^{a} [a]dx$$

$$= \frac{([b] - 1)[b]}{2} + [b](b - [b]) - \frac{([a] - 1)[a]}{2} - [a](a - [a]).$$

260. *Compute the following integral:*

$$\int_{0}^{2} [x^2]dx.$$

Solution. We consider the following cases:

(1) $0 \le x < 1$ implies $0 \le x^2 < 1$,
(2) $1 \le x < \sqrt{2}$ implies $1 \le x^2 < 2$,

(3) $\sqrt{2} \leq x < \sqrt{3}$ implies $2 \leq x^2 < 3$,
(4) $\sqrt{3} \leq x < 2$ implies $3 \leq x^2 < 4$.

Therefore, we have

$$
\begin{aligned}
\int_0^2 [x^2]dx &= \int_0^1 [x^2]dx + \int_1^{\sqrt{2}} [x^2]dx + \int_{\sqrt{2}}^{\sqrt{3}} [x^2]dx + \int_{\sqrt{3}}^2 [x^2]dx \\
&= \int_1^{\sqrt{2}} dx + \int_{\sqrt{2}}^{\sqrt{3}} 2dx + \int_{\sqrt{3}}^2 3dx \\
&= (\sqrt{2} - 1) + 2(\sqrt{3} - \sqrt{2}) + 3(2 - \sqrt{3}) \\
&= 5 - \sqrt{2} - \sqrt{3}.
\end{aligned}
$$

261. *If n is a positive integer, prove that*

$$
\int_0^n [x]^2 dx = \frac{n(n-1)(2n-1)}{6}.
$$

Solution. If $k - 1 \leq x < k$, then $[x]^2 = (k-1)^2$, for all $1 \leq k \leq n$. So,

$$
\begin{aligned}
\int_0^n [x]^2 dx &= \int_0^1 [x]^2 dx + \int_1^2 [x]^2 dx + \int_2^3 [x]^2 dx + \cdots + \int_{n-1}^n [x]^2 dx \\
&= \int_1^2 1^2 dx + \int_2^3 2^2 dx + \cdots + \int_{n-1}^n (n-1)^2 dx \\
&= 1^2 + 2^2 + \cdots + (n-1)^2 \\
&= \frac{n(n-1)(2n-1)}{6}.
\end{aligned}
$$

262. *Show that*

(1) $\displaystyle \int_0^{\ln n} [e^x]dx = n \ln n - \ln(n!),$

(2) $n! \geq n^n e^{1-n}.$

Solution. (1) For every positive integer k, if $\ln k \leq x < \ln(k+1)$, then $[e^x] = k$.
Hence, we have

$$
\begin{aligned}
\int_0^{\ln n} [e^x]dx &= \int_{\ln 1}^{\ln 2} [e^x]dx + \int_{\ln 2}^{\ln 3} [e^x]dx + \int_{\ln 3}^{\ln 4} [e^x]dx + \cdots + \int_{\ln(n-1)}^{\ln n} [e^x]dx \\
&= 1(\ln 2 - \ln 1) + 2(\ln 3 - \ln 2) + 3(\ln 4 - \ln 3) + \cdots + (n-1)\big(\ln n - \ln(n-1)\big) \\
&= n \ln n - \ln(n!).
\end{aligned}
$$

(2) Since $[e^x] \leq e^x$, it follows that

$$n \ln n - \ln(n!) = \int_0^{\ln n} [e^x] dx \le \int_0^{\ln n} e^x dx = n - 1.$$

Consequently, we have $\ln \dfrac{n^n}{n1} \le n - 1$. This implies that $n! \ge n^n e^{1-n}$.

263. *Prove that*

$$\int_x^1 \frac{1}{1+t^2} dt = \int_1^{\frac{1}{x}} \frac{1}{1+t^2} dt.$$

Solution. For the left integral, we consider the substitution $u = \dfrac{1}{t}$. Then, we have $t = \dfrac{1}{u}$ and $dt = -\dfrac{1}{u^2} du$. Therefore, we get

$$\int_x^1 \frac{1}{1+t^2} dt = \int_{\frac{1}{x}}^1 \frac{1}{1+\frac{1}{u^2}} \cdot \frac{-1}{u^2} du = \int_{\frac{1}{x}}^1 \frac{-1}{1+u^2} du$$

$$= \int_1^{\frac{1}{x}} \frac{1}{1+u^2} du = \int_1^{\frac{1}{x}} \frac{1}{1+t^2} dt.$$

264. *Prove that*

$$\int_{c\alpha}^{c\beta} f(t) dt = c \int_\alpha^\beta f(ct) dt.$$

Solution. Applying the substitution $t = cx$ for the left integral, then $dt = cdx$. So, we get

$$\int_{c\alpha}^{c\beta} f(t) dt = \int_\alpha^\beta cf(cx) dx = c \int_\alpha^\beta f(ct) dt.$$

265. *Prove that*

$$\sum_{k=0}^n (-1)^k \binom{n}{k} \frac{1}{k+m+1} = \sum_{k=0}^m (-1)^k \binom{m}{k} \frac{1}{k+n+1}.$$

Solution. It is easy to see that

$$\frac{1}{k+m+1} = \int_0^1 t^{k+m} dt.$$

So, we have

$$\sum_{k=0}^{n}(-1)^k \binom{n}{k}\frac{1}{k+m+1} = \sum_{k=0}^{n}(-1)^k \binom{n}{k}\int_0^1 t^{k+m}dt$$

$$= \int_0^1 \left(\sum_{k=0}^{n}(-1)^k \binom{n}{k}t^{k+m}\right)dt$$

$$= \int_0^1 t^m \left(\sum_{k=0}^{n}\binom{n}{k}(-t)^k\right)dt$$

$$= \int_0^1 t^m (1-t)^n dt.$$

Note that

$$\int_0^1 t^m (1-t)^n dt = \int_0^1 t^n (1-t)^m dt.$$

This completes the proof of our equality.

266. *Suppose that f is a continuous function. If*

$$g(x) = \int_{-1}^{1} f(t)|x-t|dt,$$

find $g''(x)$.

Solution. We have

$$g(x) = \int_{-1}^{x} f(t)|x-t|dt + \int_{x}^{1} f(t)|x-t|dt$$

$$= \int_{-1}^{x} f(t)(x-t)dt + \int_{x}^{1} f(t)(t-x)dt$$

$$= x\int_{-1}^{x} f(t)dt - \int_{-1}^{x} tf(t)dt + \int_{x}^{1} tf(t)dt - x\int_{x}^{1} f(t)dt.$$

So, the first derivative of g is equal to

$$g'(x) = \int_{-1}^{x} f(t)dt + \int_{1}^{x} f(t).$$

Consequently, we obtain $g''(x) = f(x) + f(x) = 2f(x)$.

267. *If f is a positive and non-increasing function on $[0, 1]$, prove that*

$$\frac{\int_0^1 xf^2(x)dx}{\int_0^1 xf(x)dx} \le \frac{\int_0^1 f^2(x)dx}{\int_0^1 f(x)dx}.$$

Solution. Assume that

$$g(x) = \int_0^x f^2(t)dt \int_0^x tf(t)dt - \int_0^x tf^2(t)dt \int_0^x f(t)dt.$$

For every $x \in [0, 1]$ we have

$$
\begin{aligned}
g'(x) &= f^2(x) \int_0^x tf(t)dt + xf(x) \int_0^x f^2(t)dt \\
&\quad -xf^2(x) \int_0^x f(t)dt - f(x) \int_0^x tf^2(t)dt \\
&= f(x) \int_0^x (t - x)\big(f(x) - f(t)\big)f(t)dt \ge 0.
\end{aligned}
$$

Hence, g is an increasing function. Since $g(0) = 0$, it follows that $g(1) \ge 0$. This completes the proof.

268. *If f is a continuous function, prove that*

$$\int_0^x f(u)(x - u)du = \int_0^x \left(\int_0^u f(t)dt \right)du.$$

Solution. We have

$$
\begin{aligned}
\frac{d}{dx}\left(\int_0^x f(u)(x - u)du \right) &= \frac{d}{dx}\left(x \int_0^x f(u)du - \int_0^x uf(u)du \right) \\
&= \int_0^x f(u)du + xf(x) - xf(x) \qquad\qquad (5.7) \\
&= \int_0^x f(u)du,
\end{aligned}
$$

and

$$\frac{d}{dx}\left(\int_0^x \left(\int_0^u f(t)dt \right)du \right) = \int_0^x f(t)dt. \qquad\qquad (5.8)$$

In Eqs. (5.7) and (5.8), we have two antiderivatives of $\int_0^x f(t)dt$. Since two functions with the same derivative on an interval can differ only by a constant, it follows that

$$\int_0^x f(u)(x - u)du = \int_0^x \left(\int_0^u f(t)dt \right)du + C.$$

If $x = 0$, then $C = 0$. This completes the proof.

269. *Find*

$$\lim_{h \to 0} \left(\frac{1}{h} \int_1^{1+h} \sqrt{t^4 + 1}\, dt \right).$$

Solution. Suppose that $f(x) = \int_0^x \sqrt{t^4 + 1}\, dt$. Then, we have

$$\frac{1}{h} \int_1^{1+h} \sqrt{t^4 + 1}\, dt = \frac{f(1+h) - f(1)}{h}.$$

Therefore, the desired limit is

$$\lim_{h \to 0} \frac{f(1+h) - f(1)}{h} = f'(1) = \sqrt{1^4 + 1} = \sqrt{2}.$$

270. *Let f be a continuous function on a neighborhood of $x = 1$.*

(1) Find $f(1)$ if $\displaystyle\int_0^{x^3} f(t)\, dt = x^2 \sin \pi x$.

(2) Find $f(1)$ if $\displaystyle\int_0^{f(x)} t^3\, dt = x^2 \sin \pi x$.

Solution. We take derivative from both sides of the equality. Then, we obtain

$$3x^2 f(x^3) = 2x \sin \pi x + \pi x^2 \cos \pi x.$$

Now, assume that $x = 1$. Then, we get $3f(1) = -\pi$, and so $f(1) = -\dfrac{\pi}{3}$.

(2) We compute the integral in the left side of the equality. Then, we obtain

$$\frac{1}{4} t^4 \Big|_0^{f(x)} = \frac{1}{4} f^4(x) = x^2 \sin \pi x.$$

Now, letting $x = 1$, we get $\dfrac{1}{4} f^4(1) = 0$. This implies that $f(1) = 0$.

271. *Suppose that f is defined on $[0, \dfrac{\pi}{3}]$ as follows:*

$$f(x) = \int_{\frac{x}{2}}^x \frac{\sin t}{t}\, dt.$$

Determine the point that f reaches to its maximum on $[0, \dfrac{\pi}{3}]$.

Solution. First, we evaluate the derivation of f. We obtain

$$f'(x) = \frac{\sin x}{x} - \frac{1}{2} \cdot \frac{\sin(\frac{x}{2})}{\frac{x}{2}} = \frac{\sin x - \sin(\frac{x}{2})}{x}.$$

Since sine function is increasing on $[0, \frac{\pi}{3}]$, it follows that $\sin \frac{x}{2} < \sin x$. So, we conclude that $f'(x) > 0$, which implies that f is increasing on $[0, \frac{\pi}{3}]$. Therefore, f reaches to its maximum at $x = \frac{\pi}{3}$.

272. *Find the extremum of the following function:*

$$F(x) = \int_0^x \frac{\sin t}{t} dt, \quad (for \ x > 0).$$

Solution. We have $F'(x) = \frac{\sin x}{x}$. Critical points are $x = n\pi$, for all $n \in \mathbb{N}$. If we evaluate the second derivative, we get $F''(x) = \frac{x \cos x - \sin x}{x^2}$. Since $F''(n\pi) = \frac{1}{n\pi}(-1)^n \neq 0$, it follows that $x = n\pi$ is an extremum point, for all $n \in \mathbb{N}$. It is maximum, if n is odd, and it is minimum, if n is even.

273. *A function f, continuous on the positive real axis, has the property that for all choices of $x > 0$ and $y > 0$,*

$$\int_x^{xy} f(t)dt \tag{5.9}$$

is independent of x (and therefore depends only on y). If $f(2) = 2$, compute the value of the integral $A(x) = \int_1^x f(t)dt$, for all $x > 0$.

Solution. We take derivative of (5.9) relative to x. Then, by assumption we obtain $0 = -f(x) + yf(xy)$. If $x = 2$, then $2 = yf(2y)$. Put $2y = t$, then $f(t) = \frac{4}{t}$. Therefore, $A(x) = \int_1^x \frac{4}{t} dt = 4 \ln x$.

274. *A function f, continuous on the positive real axis, has the property that*

$$\int_1^{xy} f(t)dt = y \int_1^x f(t)dt + x \int_1^y f(t)dt, \tag{5.10}$$

for all $x > 0$ and all $y > 0$. If $f(1) = 3$, compute $f(x)$ for each $x > 0$.

Solution. Let x be fixed and y be variable. We take derivative of both sides of (5.10) relative to y. We obtain

$$xf(xy) = \int_1^x f(t)dt + xf(y). \tag{5.11}$$

In (5.11), let $y = 1$. Then, we observe that

$$xf(x) = \int_1^x f(t)dt + 3x. \tag{5.12}$$

Now, we take derivative of both sides of (5.12) relative to x. Then, we have $f(x) + xf'(x) = f(x) + 3$, which implies that $f'(x) = \dfrac{3}{x}$. Thus, $f(x) = 3\ln x + C$. Since $f(1) = 3$, it follows that $C = 3$. Therefore, $f(x) = 3\ln x + 3$.

275. *Let $f : [0, 1] \to [0, \infty)$ be a continuous function such that*

$$f^2(x) \le 1 + 2\int_0^x f(t)dt, \ \ for \ all \ x \in [0, 1].$$

Prove that $f(x) \le 1 + x$, for every $x \in [0, 1]$.

Solution. Suppose that

$$F(x) = 1 + 2\int_0^x f(t)dt.$$

Then, we get

$$F'(x) = 2f(x) \le 2\sqrt{F(x)}. \tag{5.13}$$

On the other hand, we have

$$\int_0^x \frac{F'(t)}{2\sqrt{F(t)}}dt = \sqrt{F(x)} - \sqrt{F(0)} = \sqrt{F(x)} - 1. \tag{5.14}$$

From (5.13) and (5.14) we conclude that

$$\sqrt{F(x)} - 1 = \int_0^x \frac{F'(t)}{2\sqrt{F(t)}}dt \le \int_0^x dt = x. \tag{5.15}$$

Now, from (5.13) and (5.15) we obtain $f(x) \le \sqrt{F(x)} \le x + 1$. This completes the proof of desired inequality.

276. *Let f be a real function with the following conditions:*

(1) $f(1) = 1$;

(2) $f'(x) = \dfrac{1}{x^2 + (f(x))^2}$, *for all* $x \geq 1$.

Prove that $\lim\limits_{x \to \infty} f(x)$ *exists, and it is less than* $1 + \dfrac{\pi}{4}$.

Solution. From (2) we have $f'(x) > 0$ for $x \geq 1$. So, f is monotone increasing. If $t > 1$, then $f(t) > f(1) = 1$. Hence,

$$f'(t) = \frac{1}{t^2 + (f(t))^2} < \frac{1}{t^2 + 1}.$$

Since $f(x) - f(1) = \displaystyle\int_1^x f'(t)dt$, it follows that

$$f(x) = 1 + \int_1^x f'(t)dt < 1 + \int_1^x \frac{1}{t^2+1}dt < 1 + \int_1^\infty \frac{1}{t^2+1}dt.$$

Thus, $f(x) < 1 + \dfrac{\pi}{4}$, i.e., f is bounded from above. Moreover, since f is increasing, it follows that $\lim\limits_{x \to \infty} f(x)$ exists. Also, we have

$$\lim_{x \to \infty} f(x) = 1 + \int_1^\infty f'(t)dt \leq 1 + \int_1^\infty \frac{1}{t^2+1}dt = 1 + \frac{\pi}{4}.$$

277. *(1) If f is a monotone continuous function on $[a, b]$ such that $af(a) = bf(b)$, show that*

$$\int_{f(b)}^{f(a)} f^{-1}(x)dx = \int_a^b f(x)dx.$$

(2) If p and q are positive real numbers, show that

$$\int_0^1 (1 - x^p)^{\frac{1}{q}}dx = \int_0^1 (1 - x^q)^{\frac{1}{p}}dx.$$

Solution. (1) Since f is a decreasing continuous function on $[a, b]$, it follows that f^{-1} exists in $[f(b), f(a)]$. Moreover, f^{-1} is decreasing and continuous too. So, by substitution $x = f(u)$ we have

$$\int_{f(b)}^{f(a)} f^{-1}(x)dx = \int_{b}^{a} f^{-1}(f(u))f'(u)du$$

$$= -\int_{a}^{b} uf'(u)du$$

$$= af(a) - bf(b) + \int_{a}^{b} f(u)du$$

$$= \int_{a}^{b} f(x)dx.$$

(2) If $f(x) = (1 - x^p)^{\frac{1}{q}}$, then $f^{-1}(x) = (1 - x^q)^{\frac{1}{p}}$. Now, it is enough to apply part (1) for f and f^{-1} on $[0, 1]$.

278. *Let p and q be positive real numbers such that $\dfrac{1}{p} + \dfrac{1}{q} = 1$. If f and g are integrable, $f \geq 0$, $g \geq 0$ and*

$$\int_{a}^{b} f^p(x)dx = 1 = \int_{a}^{b} g^q(x)dx,$$

prove that

$$\int_{a}^{b} f(x)g(x)dx \leq 1.$$

Solution. By Problem (214), we have

$$f(x)g(x) \leq \frac{f^p(x)}{p} + \frac{g^q(x)}{q},$$

for all $a \leq x \leq b$. So, we have

$$\int_{a}^{b} f(x)g(x)dx \leq \int_{a}^{b} \frac{f^p(x)}{p}dx + \int_{a}^{b} \frac{g^q(x)}{q}dx$$

$$= \frac{1}{p}\int_{a}^{b} f^p(x)dx + \frac{1}{q}\int_{a}^{b} g^q(x)dx$$

$$= \frac{1}{p} + \frac{1}{q} = 1.$$

279. *Let p and q be positive real numbers such that $\dfrac{1}{p} + \dfrac{1}{q} = 1$. If f and g are integrable, prove that*

$$\left| \int_{a}^{b} f(x)g(x)dx \right| \leq \left(\int_{a}^{b} |f(x)|^p dx \right)^{\frac{1}{p}} \cdot \left(\int_{a}^{b} |g(x)|^q dx \right)^{\frac{1}{q}}.$$

This is Hölder's inequality. When $p = q = 2$, it is usually called Schwarz inequality.

Solution. We put

$$F(x) = \frac{|f(x)|}{\left(\int_a^b |f(x)|^p dx\right)^{\frac{1}{p}}} \text{ and } G(x) = \frac{|g(x)|}{\left(\int_a^b |g(x)|^q dx\right)^{\frac{1}{q}}}.$$

Then, we obtain

$$\int_a^b F^p(x)dx = \int_a^b \frac{|f(x)|^p}{\int_a^b |f(x)|^p dx} dx = 1$$

and

$$\int_a^b G^q(x)dx = \int_a^b \frac{|g(x)|^q}{\int_a^b |g(x)|^q dx} dx = 1.$$

So, by Problem (278), we get

$$\int_a^b F(x)G(x)dx \le 1$$

or

$$\int_a^b \frac{|f(x)|}{\left(\int_a^b |f(x)|^p dx\right)^{\frac{1}{p}}} \cdot \frac{|g(x)|}{\left(\int_a^b |g(x)|^q dx\right)^{\frac{1}{q}}} dx \le 1,$$

which implies that

$$\int_a^b |f(x)| \cdot |g(x)|dx \le \left(\int_a^b |f(x)|^p dx\right)^{\frac{1}{p}} \cdot \left(\int_a^b |g(x)|^q dx\right)^{\frac{1}{q}}.$$

This completes the proof.

280. *Using Hölder's inequality, establish Minkowski inequality:*

$$\left(\int_a^b (|f(x)| + |g(x)|)^p dx\right)^{\frac{1}{p}} \le \left(\int_a^b |f(x)|^p dx\right)^{\frac{1}{p}} + \left(\int_a^b |g(x)|^p dx\right)^{\frac{1}{p}}.$$

Solution. We suppose that f and g are non-zero. Obviously, we have

$$(|f(x)| + |g(x)|)^p = |f(x)|(|f(x)| + |g(x)|)^{p-1} + |g(x)|(|f(x)| + |g(x)|)^{p-1}.$$

So, we have

$$\int_a^b (|f(x)| + |g(x)|)^p dx$$
$$= \int_a^b |f(x)|(|f(x)| + |g(x)|)^{p-1} dx + \int_a^b |g(x)|(|f(x)| + |g(x)|)^{p-1} dx.$$

(5.16)

On the other hand, by Hölder's inequality,

$$\int_a^b |f(x)|(|f(x)| + |g(x)|)^{p-1} dx$$
$$\leq \left(\int_a^b |f(x)|^p dx \right)^{\frac{1}{p}} \cdot \left(\int_a^b (|f(x)| + |g(x)|)^{q(p-1)} dx \right)^{\frac{1}{q}} \qquad (5.17)$$
$$= \left(\int_a^b |f(x)|^p dx \right)^{\frac{1}{p}} \cdot \left(\int_a^b (|f(x)| + |g(x)|)^p dx \right)^{\frac{1}{q}}.$$

Similarly, we have

$$\int_a^b |g(x)|(|f(x)| + |g(x)|)^{p-1} dx$$
$$\leq \left(\int_a^b (|f(x)| + |g(x)|)^p dx \right)^{\frac{1}{q}} \cdot \left(\int_a^b |g(x)|^p dx \right)^{\frac{1}{p}}. \qquad (5.18)$$

Now, we apply (5.17) and (5.18) into (5.16). Then,

$$\int_a^b (|f(x)| + |g(x)|)^p dx$$
$$\leq \left(\int_a^b (|f(x)| + |g(x)|)^p dx \right)^{\frac{1}{q}} \left(\left(\int_a^b |f(x)|^p dx \right)^{\frac{1}{p}} + \left(\int_a^b |g(x)|^p dx \right)^{\frac{1}{p}} \right).$$

Therefore, we observe that

$$\left(\int_a^b (|f(x)| + |g(x)|)^p dx \right)^{1-\frac{1}{q}} \leq \left(\int_a^b |f(x)|^p dx \right)^{\frac{1}{p}} + \left(\int_a^b |g(x)|^p dx \right)^{\frac{1}{p}}.$$

This completes the proof.

281. *Let f and g be two positive continuous functions on $[a, b]$ such that for all $a \leq x \leq b$, $f(x)g(x) \geq 1$. Prove that for all $p > 0$,*

$$\int_a^b f(x)^p dx \cdot \int_a^b g(x)^p dx \geq (b - a)^2.$$

Solution. By assumption, we have

$$f(x)^{\frac{p}{2}} g(x)^{\frac{p}{2}} \geq 1,$$

for all $a \leq x \leq b$ and all $p > 0$. Now, Hölder's inequality implies that

$$b - a = \int_a^b dx \leq \int_a^b f(x)^{\frac{p}{2}} g(x)^{\frac{p}{2}} dx$$

$$\leq \left(\int_a^b f(x)^p dx \right)^{\frac{1}{2}} \cdot \left(\int_a^b g(x)^p dx \right)^{\frac{1}{2}}.$$

This completes the proof.

282. *Let f be a non-negative continuous function on $[a, b]$ and M be the maximum value of f on $[a, b]$. Prove that*

$$M = \lim_{n \to \infty} \left(\int_a^b f(x)^n dx \right)^{\frac{1}{n}}.$$

Solution. Since $0 \leq f(x) \leq M$, for all $x \in [a, b]$, it follows that $f(x)^n \leq M^n$, and so

$$\int_a^b f(x)^n dx \leq \int_a^b M^n dx = M^n (b - a),$$

for all $n \in \mathbb{N}$. Therefore, we conclude that

$$\lim_{n \to \infty} \left(\int_a^b f(x)^n dx \right)^{\frac{1}{n}} \leq \lim_{n \to \infty} M (b - a)^{\frac{1}{n}} = M. \tag{5.19}$$

For the converse, suppose that $\epsilon > 0$ is given. Then, by the definition of M, there exists $x_0 \in [a, b]$ such that $M - \epsilon < f(x_0)$. Without loss of generality, we may assume that $a < x_0 < b$ and $\epsilon < M$. Since f is continuous at x_0, it follows that there exists $\delta > 0$ such that

$$[x_0 - \delta, x_0 + \delta] \subseteq [a, b] \text{ and } 0 < M - \epsilon < f(x),$$

for all $x \in (x_0 - \delta, x_0 + \delta)$. Thus, we have

$$\int_{x_0 - \delta}^{x_0 + \delta} (M - \epsilon)^n dx \leq \int_{x_0 - \delta}^{x_0 + \delta} f(x)^n dx \leq \int_a^b f(x)^n dx.$$

Therefore, we obtain

$$M - \epsilon = \lim_{n \to \infty} (M - \epsilon)(2\delta)^{\frac{1}{n}} = \lim_{n \to \infty} \left(\int_{x_0 - \delta}^{x_0 + \delta} (M - \epsilon)^n dx \right)^{\frac{1}{n}}$$

$$\leq \lim_{n \to \infty} \left(\int_a^b f(x)^n dx \right)^{\frac{1}{n}}.$$

Since $\epsilon > 0$ is arbitrary, it follows that

$$M \leq \lim_{n\to\infty} \left(\int_a^b f(x)^n dx \right)^{\frac{1}{n}}. \tag{5.20}$$

By (7.22) and (5.20), our proof completes.

283. *If f, g, h and k are four polynomials, show that*

$$P(x) = \left(\int_1^x f(x)h(x)dx \right) \cdot \left(\int_1^x g(x)k(x)dx \right)$$
$$- \left(\int_1^x f(x)k(x)dx \right) \cdot \left(\int_1^x g(x)h(x)dx \right)$$

is divided by $(x-1)^4$.

Solution. Obviously, P is a polynomial. It is easy to check that

$$(x-1)^4 \text{ divided } P \iff P'''(1) = 0.$$

We check this statement by differentiation. We have

$$P'(x) = f(x)h(x)\left(\int_1^x g(x)k(x)dx \right) + g(x)k(x)\left(\int_1^x f(x)h(x)dx \right)$$
$$- f(x)k(x)\left(\int_1^x g(x)h(x)dx \right) - g(x)h(x)\left(\int_1^x f(x)k(x)dx \right).$$

Then, we get

$$P''(x) = \left(f'(x)h(x) + f(x)h'(x) \right)\left(\int_1^x g(x)k(x)dx \right) + f(x)g(x)h(x)k(x)$$
$$+ \left(g'(x)k(x) + g(x)k'(x) \right)\left(\int_1^x f(x)h(x)dx \right) + f(x)g(x)h(x)k(x)$$
$$- \left(f'(x)k(x) + f(x)k'(x) \right)\left(\int_1^x g(x)h(x)dx \right) - f(x)g(x)h(x)k(x)$$
$$- \left(g'(x)h(x) + g(x)h'(x) \right)\left(\int_1^x f(x)k(x)dx \right) - f(x)g(x)h(x)k(x).$$

Finally, after simplification, we have

$$P'''(x) = \Big(f''(x)h(x) + 2f'(x)h'(x) + f(x)h''(x)\Big)\Big(\int_1^x g(x)k(x)dx\Big)$$
$$+\Big(g''(x)k(x) + 2g'(x)k'(x) + g(x)k''(x)\Big)\Big(\int_1^x f(x)h(x)dx\Big)$$
$$-\Big(f''(x)k(x) + 2f'(x)k'(x) + f(x)k''(x)\Big)\Big(\int_1^x g(x)h(x)dx\Big)$$
$$-\Big(g''(x)h(x) + 2g'(x)h'(x) + g(x)h''(x)\Big)\Big(\int_1^x f(x)k(x)dx\Big).$$

Now, clearly, we observe that $P'''(1) = 0$.

284. *A function f, defined for all positive real numbers, satisfies the following two conditions: $f(1) = 1$ and $f'(x^2) = x^3$, for all $x > 0$. Compute $f(4)$.*

Solution. If $x = \sqrt{t}$, then $f'(t) = t\sqrt{t}$. Hence, $f(t) = \int t\sqrt{t}dt = \frac{2}{5}t^2\sqrt{t} + C$.
Since $f(1) = 1$, it follows that $f(t) = \frac{2}{5}t^2\sqrt{t} + \frac{3}{5}$. This implies that $f(4) = \frac{67}{5}$.

285. *Let f be a continuous function on $[a, b]$ and $\int_a^b f(t)dt \neq 0$. Show that for any number $k \in (0, 1)$ there is a number $c \in (a, b)$ such that*

$$\int_a^c f(t)dt = k \int_a^b f(t)dt.$$

Solution. We consider the function g for which

$$g(x) = \frac{\int_a^x f(t)dt}{\int_a^b f(t)dt}.$$

Then, g is a continuous function and $g(a) = 0$ and $g(b) = 1$. So, $g(a) < k < g(b)$.
Now, by the intermediate value theorem, there exists $c \in (a, b)$ such that $g(c) = k$.
This completes the proof.

286. *Find the function f such that $f(0) = 0$ and*

$$f'(\ln x) = \begin{cases} 1 & \text{if } 0 < x \leq 1 \\ x & \text{if } x > 1. \end{cases}$$

Solution. If $\ln x = t$, then

$$0 < x \le 1 \Leftrightarrow t \le 0,$$
$$x > 1 \Leftrightarrow t > 0.$$

Therefore, we have

$$f(x) = \int_0^x f'(t)dt = \begin{cases} \int_0^x dt & \text{if } x \le 0 \\ \int_0^x e^t dt & \text{if } x > 0 \end{cases} = \begin{cases} x & \text{if } x \le 0 \\ e^x - 1 & \text{if } x < 0. \end{cases}$$

287. *Show that*

$$e^{-\frac{1}{e}} \le \int_0^1 x^x dx \le 1.$$

Solution. Note that

$$\lim_{x \to 0^+} x^x = \lim_{x \to 0^+} e^{x \ln x} = e^{\lim_{x \to 0^+} x \ln x} = e^0 = 1.$$

So, the function f defined by

$$f(x) = \begin{cases} x^x & \text{if } 0 < x \le 1 \\ 1 & \text{if } x = 0 \end{cases}$$

is continuous on $[0, 1]$. Let $x \neq 0$. Then, we have

$$\ln(f(x)) = x \ln x. \tag{5.21}$$

We take derivative of both sides of (5.21), we find that

$$\frac{f'(x)}{f(x)} = \ln x + 1 \text{ or } f'(x) = x^x(\ln x + 1).$$

If $f'(x) = 0$, then $\ln x + 1 = 0$. This implies that $x = \dfrac{1}{e}$ is a critical point. Now, it is easy to check that f attains its minimum at $x = \dfrac{1}{e}$ and attains its maximum at $x = 0$ and $x = 1$. The minimum value of f is

$$f\left(\frac{1}{e}\right) = \left(\frac{1}{e}\right)^{\frac{1}{e}} = e^{-\frac{1}{e}}.$$

This completes the proof of desired inequalities.

288. *Prove that*

$$\int_0^{\frac{\pi}{2}} e^{-a\sin x}dx < \frac{\pi}{2a}(1 - e^{-a}), \text{ where } a > 0.$$

Solution. We consider $f(x) = \dfrac{\sin x}{x}$, for all $x \in (0, \dfrac{\pi}{2})$. Then, $f'(x) = \dfrac{x\cos x - \sin x}{x^2}$.

By Problem (175), we have $\sin x \leq x$, and by Problem (178), we get $\dfrac{x\cos x}{\sin x} < \dfrac{\sin x}{x}$.

So, we conclude that $\cos x < \dfrac{\sin x}{x}$. Consequently, $f'(x) < 0$, which implies that f is decreasing. Since $0 < x < \dfrac{\pi}{2}$, it follows that $f(\dfrac{\pi}{2}) < f(x)$ or $\dfrac{2}{\pi} < \dfrac{\sin x}{x}$. Now, we can write $-a\sin x < -\dfrac{2a}{\pi}x$, and so

$$e^{-a\sin x} < e^{-\frac{2a}{\pi}x},$$

for all $x \in (0, \dfrac{\pi}{2})$. Hence, we get

$$\int_0^{\frac{\pi}{2}} e^{-a\sin x}dx < \int_0^{\frac{\pi}{2}} e^{-\frac{2a}{\pi}x}dx$$

$$= -\frac{\pi}{2a}e^{-\frac{2a}{\pi}x}\Big|_0^{\frac{\pi}{2}}$$

$$= \frac{\pi}{2a}(1 - e^{-a}).$$

289. *Let f be a real function such that $|f''(x)| \leq m$, for all $x \in [0, a]$ and f has local extremum in $(0, a)$. Show that $|f'(0)| + |f'(a)| \leq am$.*

Solution. If f has local extremum at $c \in (0, a)$, then $f'(c) = 0$. Clearly, f is integrable on $[0, c]$ and $[c, a]$. Since $-m \leq f''(x) \leq m$, it follows that

$$-\int_0^c mdx \leq \int_0^c f''(x)dx \leq \int_0^c mdx, \tag{5.22}$$

$$-\int_c^a mdx \leq \int_c^a f''(x)dx \leq \int_c^a mdx. \tag{5.23}$$

By (5.22) we obtain $-mc \leq f'(c) - f'(0) \leq mc$. This implies that

$$|f'(0)| \leq mc. \tag{5.24}$$

Similarly, by (5.23) we obtain $-m(a - c) \leq f'(a) - f'(c) \leq m(a - c)$. This implies that

$$|f'(a)| \leq ma - mc. \tag{5.25}$$

Now, from (5.24) and (5.25) we conclude that $|f'(0)| + |f'(a)| \leq am$.

290. *Let $f \geq 0$, f be continuous on $[a, b]$, and $\int_a^b f(x)dx = 0$. Prove that $f(x) = 0$, for all $x \in [a, b]$.*

Solution. Suppose that there exists $c \in [a, b]$ such that $f(c) \neq 0$. Since $f \geq 0$, it follows that $f(c) > 0$. Assume that $f(c) = \lambda > 0$.

Let $\epsilon = \dfrac{\lambda}{2}$. Since f is continuous at c, it follows that there exists $\delta > 0$ such that $|f(x) - f(c)| < \dfrac{\lambda}{2}$, whenever $x \in (c - \delta, c + \delta)$.

If $c - \delta < x < c + \delta$, then $f(c) - \dfrac{\lambda}{2} < f(x) < f(c) + \dfrac{\lambda}{2}$. Since $f(c) = \lambda$, it follows that $0 < \dfrac{\lambda}{2} < f(x)$. Now, we can write

$$\int_a^b f(x)dx = \int_a^{c-\frac{\delta}{2}} f(x)dx + \int_{c-\frac{\delta}{2}}^{c+\frac{\delta}{2}} f(x)dx + \int_{c+\frac{\delta}{2}}^b f(x)dx$$

$$\geq \int_{c-\frac{\delta}{2}}^{c+\frac{\delta}{2}} f(x)dx$$

$$\geq \frac{\lambda}{2}\left(c + \frac{\delta}{2} - c + \frac{\delta}{2}\right)$$

$$= \frac{\lambda\delta}{2}.$$

So, $\int_a^b f(x)dx > 0$, and this is a contradiction.

291. *If f is a real continuous function such that for every $x \in \mathbb{R}$*

$$\int_0^1 f(xt)dt = 0,$$

prove that f is the zero function.

Solution. Suppose that $x \neq 0$. We introduce a new variable u by letting $u = xt$. Then, $du = xdt$. So, we get

$$\int_0^1 f(xt)dt = \int_0^x f(u)\frac{du}{x} = \frac{1}{x}\int_0^x f(u)du.$$

Hence, for every $x \neq 0$ we have

$$\int_0^x f(u)du = 0. \tag{5.26}$$

Clearly, the equality (5.26) holds for $x = 0$ too. Therefore, we have

$$\frac{d}{dx}\int_0^x f(u)du = 0.$$

This implies that $f(x) = 0$ for all $x \in \mathbb{R}$.

292. *If*

$$f(x) = \int_1^x \frac{1}{1+t^2}dt,$$

show that $f(4) - f(2) < \dfrac{2}{5}$.

Solution. Since f is continuous on $[2, 4]$ and differentiable in $(2, 4)$, by the mean value theorem there is $\xi \in (2, 4)$ such that

$$f(4) - f(2) = f'(\xi)(4 - 2).$$

On the other hand, we have

$$f'(\xi) = \frac{1}{1+\xi^2} < \frac{1}{1+2^2} = \frac{1}{5}.$$

Consequently, we have $f(4) - f(2) < \dfrac{2}{5}$.

293. *If $n \in \mathbb{N}$ is fixed and f is a continuous function on $[0, 1]$ such that $\int_0^1 f(x)dx = 1$, show that $f(\alpha) = n\alpha^{n-1}$, for some $\alpha \in (0, 1)$.*

Solution. We define

$$g(x) = \int_0^x f(t)dt - x^n.$$

Then, g is continuous on $[0, 1]$ and is differentiable in $(0, 1)$. Moreover, $g(1) = g(0) = 0$. So, by Rolle's theorem, there is $\alpha \in (0, 1)$ such that $g'(\alpha) = 0$. This completes the proof.

294. *Let f be a differentiable function such that $f(3) = 9$ and $f'(3) = -2$, and let*

$$g(x) = \frac{1}{x}\int_0^x f(t)dt$$

has a critical point at $x = 3$. *Determine whether the critical point of g at $x = 3$ is a local maximum, minimum or not.*

Solution. First, we calculate the first and second derivation of g. We obtain

$$g'(x) = \frac{xf(x) - \int_0^x f(t)dt}{x^2} = \frac{f(x)}{x} - \frac{1}{x^2}\int_0^x f(t)dt$$

and

$$g''(x) = \frac{xf'(x) - f(x)}{x^2} - \frac{x^2 f(x) - 2x\int_0^x f(t)dt}{x^4}$$

$$= \frac{1}{x}f'(x) - \frac{2}{x^2}f(x) + \frac{2}{x^3}\int_0^x f(t)dt.$$

Since $g'(3) = 0$, it follows that $0 = \frac{f(3)}{3} - \frac{1}{9}\int_0^3 f(t)dt$. This implies that

$$\int_0^3 f(t)dt = 27.$$

On the other hand, we get

$$g''(3) = \frac{1}{3}f'(3) - \frac{2}{9}f(3) + \frac{2}{27}\int_0^3 f(t)dt$$

$$= -\frac{2}{3} < 0.$$

Consequently, g has a local maximum at $x = 3$.

295. *Let $f(x) = e^{g(x)}$ such that*

$$g(x) = \int_2^x \frac{t}{1 + t^4}dt.$$

Find $f'(2)$.

Solution. By the chain rule and the fundamental theorem of calculus, we have

$$f'(x) = g'(x)e^{g(x)} = \frac{x}{1 + x^4}e^{g(x)}.$$

Consequently, we get $f'(2) = \frac{2}{1 + 2^4}e^0 = \frac{2}{17}.$

296. *Let f be a continuous function on $[0, \frac{\pi}{4}]$. Show that there exists $\alpha \in [0, \frac{\pi}{4}]$* such that

$$2\cos(2\alpha) \int_0^{\frac{\pi}{4}} f(t)dt = f(\alpha).$$

Solution. We consider $F(x) = \int_0^x f(t)dt$ and $G(x) = \sin(2x)$. Then, F and G are continuous on $[0, \frac{\pi}{4}]$ and differentiable in $(0, \frac{\pi}{4})$. So, by Cauchy mean value theorem, there exists $\alpha \in (0, \frac{\pi}{4})$ such that

$$\left(F(\frac{\pi}{4}) - F(0)\right)G'(\alpha) = \left(G(\frac{\pi}{4}) - G(0)\right)F'(\alpha)$$

This implies that $2\cos(2\alpha) \int_0^{\frac{\pi}{4}} f(t)dt = f(\alpha)$.

297. *Suppose that f and g are continuous functions on $[a, b]$ such that $\int_a^b f(x)dx = \int_a^b g(x)dx$. Prove that for some $x \in [a, b]$, $f(x) = g(x)$.*

Solution. Since $f - g$ is a continuous function on $[a, b]$, by the mean value theorem for integrals, there is $c \in [a, b]$ such that

$$(f - g)(c) = \frac{1}{b - a} \int_a^b (f(x) - g(x))dx = 0.$$

So, $f(c) = g(c)$.

298. *Suppose that $f : [0, \infty) \to \mathbb{R}$ is defined by*

$$f(x) = \int_0^x \frac{\sin^2 t}{1 + t^2} dt.$$

Show that $0 \le f(x) \le x$, for all $x \ge 0$.

Solution. By applying the mean value theorem on $[0, x]$, there exists $\xi \in (0, x)$ such that $f'(\xi) = \frac{f(x) - f(0)}{x - 0} = \frac{f(x)}{x}$. Since $0 \le f'(\xi) = \frac{\sin^2 \xi}{1 + \xi^2} \le 1$, it follows that $0 \le \frac{f(x)}{x} \le 1$. This implies that $0 \le f(x) \le x$.

299. *Let f be a differentiable function on $[a, b]$ and $f(a) = f(b) = 0$. Prove that there exists $\xi \in (a, b)$ such that*

$$|f'(\xi)| \geq \frac{2}{(b-a)^2} \int_a^b f(x)dx.$$

Solution. Since f is continuous on $[a, b]$, by the mean value theorem for integral, there exists $c \in [a, b]$ such that

$$\int_a^b f(x)dx = f(c)(b - a). \tag{5.27}$$

Now, we apply the mean value theorem for intervals $[a, c]$ and $[c, b]$. Then, there exist $c_1 \in (a, c)$ and $c_2 \in (c, b)$ such that

$$f(c) - f(a) = f'(c_1)(c - a),$$
$$f(b) - f(c) = f'(c_2)(b - c).$$

Thus, we obtain $f(c) \leq |f'(c_1)|(c - a)$ and $f(c) \leq |f'(c_2)|(b - c)$. We set $|f'(\xi)| = \max\{|f'(c_1)|, |f'(c_2)|\}$. Then, we have $f(c) \leq |f'(\xi)|(c - a)$ and $f(c) \leq |f'(\xi)|(b - c)$. So, $2f(c) \leq |f'(\xi)|(b - a)$. This implies that $\dfrac{2f(c)}{b - a} \leq |f'(\xi)|$. Now, by using (5.27) we obtain $|f'(\xi)| \geq \dfrac{2}{(b-a)^2} \int_a^b f(x)dx$.

300. *Let f be a differentiable function with continuous derivative such that $|f'(x)| \leq M$, for all $x \in \mathbb{R}$. Prove that*

(1) If $f(c) = 0$ for some $c \in \mathbb{R}$, then

$$\left| \int_c^t f(x)dx \right| \leq \frac{1}{2}M(t - c)^2, \ \text{for all } t \in \mathbb{R}.$$

(2) If $f(a) = f(b) = 0$ for some $a < b$, then

$$\left| \int_a^b f(x)dx \right| \leq \frac{1}{4}M(b - a)^2.$$

Solution. (1) We define

$$g(t) = \int_c^t f(x)dx.$$

By Taylor's theorem, we can write

$$g(t) = g(c) + g'(c)(t - c) + \frac{g''(\xi)}{2!}(t - c)^2,$$

where ξ is a number between c and t. On the other hand, $g(c) = 0$ and $g'(c) = f(c) = 0$. Thus, we have

$$g(t) = \frac{g''(\xi)}{2!}(t-c)^2 = \frac{f'(\xi)}{2}(t-c)^2.$$

Therefore,

$$\left| \int_c^t f(x)dx \right| = |g(t)| \le \frac{1}{2}M(t-c)^2.$$

(2) Let $t = \dfrac{a+b}{2}$. Then, we get

$$\begin{aligned}
\left| \int_a^b f(x)dx \right| &\le \left| \int_a^t f(x)dx + \int_t^b f(x)dx \right| \\
&\le \left| \int_a^t f(x)dx \right| + \left| \int_b^t f(x)dx \right| \\
&\le \frac{M}{2}(t-a)^2 + \frac{M}{2}(t-b)^2 \\
&= \frac{M}{2}\left(\frac{b-a}{2}\right)^2 + \frac{M}{2}\left(\frac{a-b}{2}\right)^2 \\
&= \frac{M}{4}(b-a)^2.
\end{aligned}$$

301. *Let f be continuously differentiable function on $[a, x]$. By the mean value theorem for integrals, for each x, there is $\theta_x \in (a, x)$ such that*

$$\int_a^x f(t)dt = f(\theta_x)(x-a).$$

If $f'(a) \ne 0$, prove that

$$\lim_{x \to a} \frac{\theta_x - a}{x - a} = \frac{1}{2}.$$

Solution. Suppose that $F(x) = \int_a^x f(t)dt$. By Taylor's theorem, we have

$$F(x) = F(a) + (x-a)F'(a) + \frac{(x-a)^2}{2}F''(\xi_x),$$

where $\xi_x \in (a, x)$. Moreover, $F(a) = 0$, $F'(x) = f(x)$ and $F''(x) = f'(x)$. So, we get

$$F(x) = (x-a)f(a) + \frac{(x-a)^2}{2}f'(\xi_x).$$

On the other hand, since $(x-a)f(\theta_x) = F(x)$, it follows that

$$f(\theta_x) = f(a) + \frac{x-a}{2}f'(\xi_x),$$

or equivalently,

$$\frac{f(\theta_x) - f(a)}{x - a} = \frac{1}{2} f'(\xi_x).$$

Therefore, we have

$$
\begin{aligned}
\lim_{x \to a} \frac{1}{2} f'(\xi_x) &= \lim_{x \to a} \frac{f(\theta_x) - f(a)}{x - a} \\
&= \lim_{x \to a} \left(\frac{f(\theta_x) - f(a)}{\theta_x - a} \cdot \frac{\theta_x - a}{x - a} \right) \\
&= \lim_{\theta_x \to a} \frac{f(\theta_x) - f(a)}{\theta_x - a} \cdot \lim_{x \to a} \frac{\theta_x - a}{x - a} \\
&= f'(a) \lim_{x \to a} \frac{\theta_x - a}{x - a},
\end{aligned}
$$

which implies that

$$\frac{1}{2} f'(a) = f'(a) \lim_{x \to a} \frac{\theta_x - a}{x - a}.$$

Since $f'(a) \neq 0$, it follows that $\displaystyle \lim_{x \to a} \frac{\theta_x - a}{x - a} = \frac{1}{2}$.

302. *Find*

$$\lim_{x \to 0^+} \frac{\displaystyle \int_0^{\sin x} \sqrt{\tan t} \, dt}{\displaystyle \int_0^{\tan x} \sqrt{\sin t} \, dt}.$$

Solution. Since

$$L = \lim_{x \to 0^+} \frac{\displaystyle \int_0^{\sin x} \sqrt{\tan t} \, dt}{\displaystyle \int_0^{\tan x} \sqrt{\sin t} \, dt} = \frac{0}{0},$$

we apply L'Hospital's rule. Hence, we get

$$L = \lim_{x \to 0^+} \frac{\cos x \sqrt{\tan(\sin x)}}{(1 + \tan^2 x) \sqrt{\sin(\tan x)}} = \frac{0}{0}.$$

In continue, using L'Hospital's rule will make our calculations very complicated. So, we do not use L'Hospital's rule again. But, we have

$$L = \lim_{x \to 0^+} \frac{\cos^3 x \sqrt{\tan(\sin x)}}{\sqrt{\sin(\tan x)}}.$$

Now, we compute

$$\lim_{x \to 0^+} \frac{\sqrt{\tan(\sin x)}}{\sqrt{\sin(\tan x)}} = \frac{0}{0}.$$

Indeed, we have

$$\lim_{x \to 0^+} \frac{\sqrt{\tan(\sin x)}}{\sqrt{\sin(\tan x)}} = \lim_{x \to 0^+} \frac{\frac{\sqrt{\sin x}}{\sqrt{x}} \cdot \frac{\sqrt{\tan(\sin x)}}{\sqrt{\sin x}}}{\frac{\sqrt{\tan x}}{\sqrt{x}} \cdot \frac{\sqrt{\sin(\tan x)}}{\sqrt{\tan x}}} = 1.$$

Therefore, we conclude that $L = 1$.

303. *Find constants a and b such that*

$$\lim_{x \to 0} \frac{1}{bx - \sin x} \int_0^x \frac{t^2}{\sqrt{a + t}} dt = 1. \tag{5.28}$$

Solution. To determine the limit and using L'Hospital's rule, we let

$$f(x) = \int_0^x \frac{t^2}{\sqrt{a + t}} dt \text{ and } g(x) = bx - \sin x$$

and we find that

$$\frac{f'(x)}{g'(x)} = \frac{\frac{x^2}{\sqrt{a + x}}}{b - \cos x} = \frac{x^2}{(b - \cos x)\sqrt{a + x}}. \tag{5.29}$$

If $b \neq 1$, then the limit of (5.29) is equal to 0, and so the left side of (5.28) will be equal to 0 too, and this is impossible. Thus, we deduce that $b = 1$. In this case, we may remove the indeterminacy at this stage by L'Hospital's rule again. We find that

$$\frac{f''(x)}{g''(x)} = \frac{2x}{\sqrt{a + x}\sin x + \frac{1 - \cos x}{2\sqrt{a + x}}} = \frac{4x\sqrt{a + x}}{2(a + x)\sin x + 1 - \cos x},$$

and this is indeterminate as $x \to 0$. Again, we apply L'Hospital's rule, then

$$\frac{f'''(x)}{g'''(x)} = \frac{4\sqrt{a + x} + \frac{2x}{\sqrt{a + x}}}{2(a + x)\cos x + 2\sin x + \sin x} = \frac{4a + 6x}{\sqrt{a + x}\left(2(a + x)\cos x + 3\sin x\right)}.$$

This quotient is not indeterminate as $x \to 0$. The correct limit, $\dfrac{2}{\sqrt{a}}$, is obtained by substituting 0 for x. Now, we must have $\dfrac{2}{\sqrt{a}} = 1$, and this yields $a = 4$.

304. *Suppose that* $f(x) = \displaystyle\int_{1}^{x} e^{3t}\sqrt{9t^4 + 1}\,dt$ *and* $g(x) = x^n e^{3x}$. *If* $\displaystyle\lim_{x\to\infty} \dfrac{f'(x)}{g'(x)} = 1$, *find* n.

Solution. We have
$$f'(x) = e^{3x}\sqrt{9x^4 + 1},$$
$$g'(x) = nx^{n-1}e^{3x} + 3x^n e^{3x}.$$

So, we get
$$\lim_{x\to\infty} \frac{f'(x)}{g'(x)} = \lim_{x\to\infty} \frac{e^{3x}\sqrt{9x^4 + 1}}{nx^{n-1}e^{3x} + 3x^n e^{3x}} = \lim_{x\to\infty} \frac{\sqrt{9x^4 + 1}}{nx^{n-1} + 3x^n}.$$

The above limit will be equal to 1 if $n = 2$.

305. *Let m and n be arbitrary integers. Verify the following formulas, of great important in applied mathematics:*

(1) $\displaystyle\int_{0}^{2\pi} \sin mx \cos nx\,dx = 0,$

(2) $\displaystyle\int_{0}^{2\pi} \sin mx \sin nx\,dx = \int_{0}^{2\pi} \cos mx \cos nx\,dx = \begin{cases} 0 & if\ m \neq n \\ \pi & m = n. \end{cases}$

Solutions. Without loss of generality, we assume that m and n are positive. Let $m \neq n$. Then, we have

$$\int_{0}^{2\pi} \sin mx \cos nx\,dx = \frac{1}{2}\int_{0}^{2\pi} \big(\sin(m+n)x + \sin(m-n)x\big)dx$$
$$= \frac{1}{2}\Big(-\frac{1}{m+n}\cos(m+n)x - \frac{1}{m-n}\cos(m-n)x\Big)\Big|_{0}^{2\pi} = 0,$$
$$\int_{0}^{2\pi} \sin mx \sin nx\,dx = \frac{1}{2}\int_{0}^{2\pi} \big(\cos(m-n)x - \cos(m+n)x\big)dx$$
$$= \frac{1}{2}\Big(\frac{1}{m-n}\sin(m-n)x - \frac{1}{m+n}\sin(m+n)x\Big)\Big|_{0}^{2\pi} = 0,$$

$$\int_0^{2\pi} \cos mx \cos nx dx = \frac{1}{2} \int_0^{2\pi} \big(\cos(m+n)x + \cos(m-n)x\big)dx$$

$$= \frac{1}{2}\Big(\frac{1}{m+n}\sin(m+n)x + \frac{1}{m-n}\sin(m-n)x\Big)\Big|_0^{2\pi} = 0.$$

Now, let $m = n$. If $m \neq 0$, then we have

$$\int_0^{2\pi} \sin^2 mx dx = \frac{1}{2}\int_0^{2\pi}(1 - \cos 2mx)dx = \frac{1}{2}\Big(x - \frac{1}{2m}\sin mx\Big)\Big|_0^{2\pi} = \pi,$$

$$\int_0^{2\pi} \cos^2 mx dx = \frac{1}{2}\int_0^{2\pi}(1 + \cos 2mx)dx = \frac{1}{2}\Big(x + \frac{1}{2m}\sin mx\Big)\Big|_0^{2\pi} = \pi.$$

306. *Let f be a real continuous function and periodic with period p. If*

$$F(x) = \int_x^{x+p} f(t)dt,$$

show that F is a constant function.

Solution. For each $x \in \mathbb{R}$, we have $f(x) = f(x + p)$. Clearly, we have

$$F(x) = \int_x^0 f(t)dt + \int_0^p f(t)dt + \int_p^{x+p} f(t)dt. \qquad (5.30)$$

Substituting $t = u + p$ we get

$$\int_p^{x+p} f(t)dt = \int_0^x f(u+p)du = \int_0^x f(u)du = \int_0^x f(t)dt. \qquad (5.31)$$

If we put (5.31) into (5.30), we obtain

$$F(x) = \int_0^p f(t)dt,$$

that is F is a constant function.

307. *Suppose that f is continuous on $\mathbb{R}$ and periodic with period p. Verify the formula*

$$\int_a^{a+p} f(x)dx = \int_0^p f(x)dx \quad (a \ arbitrary)$$

which shows that f has the same integral over every integral of length p.

Solution. Let

$$F(x) = \int_x^{x+p} f(x)dx.$$

Then, we get $F'(x) = f(x + p) - f(x) = 0$. So, we conclude that F is a constant function. In particular, $F(a) = F(0)$.

308. *If f is continuous on $\mathbb{R}$ and periodic with period p, show that*

$$\int_a^{a+np} f(x)dx = n \int_0^p f(x)dx \quad (a \ is \ arbitrary)$$

for each $n \in \mathbb{N}$.

Solution. Using Problem (307), we obtain

$$\int_a^{a+np} f(x)dx = \int_a^{a+p} f(x)dx + \int_{a+p}^{a+2p} f(x)dx + \cdots + \int_{a+(n-1)p}^{a+np} f(x)dx$$

$$= \underbrace{\int_0^p f(x)dx + \int_0^p f(x)dx + \cdots + \int_0^p f(x)dx}_{n \text{ times}}$$

$$= n \int_0^p f(x)dx.$$

309. *Let f be continuous on $[-a, a]$. Show that*

$$\int_{-a}^a f(x)dx = 2 \int_0^a f(x)dx$$

if f is even, while

$$\int_{-a}^a f(x)dx = 0$$

if f is odd.

Solution. Suppose that f is even. Then, $f(x) = f(-x)$. We use integration by substitution $t = -x$. Then, we get

$$\int_{-a}^a f(x)dx = \int_{-a}^0 f(x)dx + \int_0^a f(x)dx$$

$$= \int_a^0 f(t)(-dt) + \int_0^a f(x)dx$$

$$= 2 \int_0^a f(x)dx.$$

Now, let f be odd. Then, $f(x) = -f(-x)$. Again, we use integration by substitution $t = -x$. We get

$$
\begin{aligned}
\int_{-a}^{a} f(x)dx &= \int_{-a}^{0} f(x)dx + \int_{0}^{a} f(x)dx \\
&= \int_{-a}^{0} -f(-x)dx + \int_{0}^{a} f(x)dx \\
&= \int_{a}^{0} -f(t)(-dt) + \int_{0}^{a} f(x)dx \\
&= -\int_{0}^{a} f(t)dt + \int_{0}^{a} f(x)dx \\
&= 0.
\end{aligned}
$$

310. *Compute the following integral:*

$$
I = \int_{-\frac{1}{2}}^{\frac{1}{2}} \cos x \cdot \ln\left(\frac{1+x}{1-x}\right) dx.
$$

Solution. We know that $f(x) = \cos x$ is an even function. Suppose that $g(x) = \ln\left(\dfrac{1+x}{1-x}\right)$. Then, we get

$$
g(-x) = \ln\left(\frac{1-x}{1+x}\right) = \ln\left(\frac{1+x}{1-x}\right)^{-1} = -\ln\left(\frac{1+x}{1-x}\right) = -g(x).
$$

So, $g(x)$ is odd. Consequently, $f(x)g(x)$ is an odd function. Therefore, by Problem (309), we conclude that $I = 0$.

311. *If $f : [a, b] \rightarrow \mathbb{R}$ is monotone increasing on $[a, b]$, show that*

$$
F(x) = \int_{a}^{x} f(t)dt
$$

is a convex function.

Solution. Suppose that $a \leq y < x \leq b$. Then, we have

$$
\begin{aligned}
&\lambda F(x) + (1 - \lambda)F(y) - F(\lambda x + (1 - \lambda)y) \\
&= \lambda \int_{a}^{x} f(t)dt + (1 - \lambda) \int_{a}^{y} f(t)dt - \int_{a}^{\lambda x + (1-\lambda)y} f(t)dt \\
&= \lambda \int_{a}^{x} f(t)dt - \lambda \int_{a}^{y} f(t)dt - \int_{y}^{\lambda x + (1-\lambda)y} f(t)dt
\end{aligned}
$$

$$= \lambda \int_y^x f(t)dt - \int_y^{\lambda x+(1-\lambda)y} f(t)dt$$

$$= \int_{\lambda x+(1-\lambda)y}^x \lambda f(t) + \int_y^{\lambda x+(1-\lambda)y} (\lambda - 1)f(t)dt$$

$$\geq \lambda f\Big(\lambda x + (1-\lambda)y\Big)\Big(x - \lambda x - (1-\lambda)y\Big)$$

$$+ (\lambda - 1)f\Big(\lambda x + (1-\lambda)y\Big)\Big(\lambda x + (1-\lambda)y - y\Big)$$

$$= \lambda(1-\lambda)f(\lambda x + (1-\lambda)y)(x-y) + (\lambda-1)\lambda f(\lambda x + (1-\lambda)y)(x-y) = 0.$$

Thus, F is convex on $[a, b]$.

312. *Let $g : \mathbb{R} \to \mathbb{R}$ be a continuous function such that for all integrable function $f : [0, 1] \to \mathbb{R}$,*

$$g\Big(\int_0^1 f(t)dt\Big) \leq \int_0^1 g\big(f(t)\big)dt. \tag{5.32}$$

Prove that g is a convex function.

Solution. Assume that $x, y \in \mathbb{R}$ and $0 < \lambda < 1$ are arbitrary. We define $f : [0, 1] \to \mathbb{R}$ by

$$f(t) = \begin{cases} x & \text{if } 0 \leq t \leq \lambda \\ y & \text{if } \lambda < t \leq 1. \end{cases}$$

Then, f is integrable and

$$\int_0^1 f(t)dt = \int_0^\lambda f(t)dt + \int_\lambda^1 f(t)dt$$

$$= \int_0^\lambda xdt + \int_\lambda^1 ydt \tag{5.33}$$

$$= \lambda x + (1-\lambda)y.$$

Similarly, we have

$$\int_0^1 g\big(f(t)\big)dt = \lambda g(x) + (1-\lambda)g(y). \tag{5.34}$$

Now, by (5.32), (5.33) and (5.34) we conclude that $g\big(\lambda x + (1-\lambda)y\big) \leq \lambda g(x) + (1-\lambda)g(y)$.

313. *Let f and g be two linear independent solutions of the equation*

$$y'' + P(x)y' + Q(x)y = 0,$$

and let $x_0 \in [a, b]$ be fixed and $W(f, g)$ be the Wronkian of f and g. Prove that

(1) $W(f, g)(x) = W(f, g)(x_0)e^{-\int_{x_0}^x P(t)dt}$.

(2) $W(f, g)(x_0) \neq 0$ *if and only if* $W(f, g)(x) \neq 0$, *for all* $x \in [a, b]$.

Solution. (1) We have $W(f, g)(x) = f(x)g'(x) - f'(x)g(x)$. So, we obtain

$$
\begin{aligned}
W'(f, g)(x) &= f'(x)g'(x) + f(x)g''(x) - f''(x)g(x) - f'(x)g'(x) \\
&= f(x)g''(x) - f''(x)g(x) \\
&= f(x)\Big(- P(x)g'(x) - Q(x)g(x)\Big) - \Big(- P(x)f'(x) - Q(x)f(x)\Big)g(x) \\
&= -P(x)\Big(f(x)g'(x) - f'(x)g(x)\Big) \\
&= -P(x)W(f, g)(x).
\end{aligned}
$$

So, we have

$$
\frac{W'(f, g)(x)}{W(f, g)(x)} = -P(x).
$$

Therefore,

$$
\int_{x_0}^x \frac{W'(f, g)(t)}{W(f, g)(t)}dt = -\int_{x_0}^x P(t)dt,
$$

which implies that

$$
\ln\Big(W(f, g)(x)\Big) - \ln\Big(W(f, g)(x_0)\Big) = -\int_{x_0}^x P(t)dt.
$$

This completes the proof of desired equality.

(2) Since the exponential function does not vanish, it follows from (1).

314. *Prove the following generalized mean value theorem for integrals: Let f and g be two continuous functions on $[a, b]$ and suppose that g does not change sign on $[a, b]$. Then, there is a point $\xi \in [a, b]$ such that*

$$
\int_a^b f(x)g(x)dx = f(\xi) \int_a^b g(x)dx.
$$

(If $g(x) = 1$, this reduces to the ordinary mean value theorem for integrals).

Solution. Without loss of generality, assume that $g(x) \geq 0$, for all $x \in [a, b]$. Since g is continuous, it follows that $\int_a^b g(x)dx \geq 0$. If $\int_a^b g(x)dx = 0$, then $g(x) = 0$, for all $x \in [a, b]$ and the statement of the problem is valid. Now, assume that $\int_a^b g(x)dx > 0$. Suppose that $f(x)$ attains its minimum m at x_1 and maximum M at x_2. Then, we have

$$
m = f(x_1) \leq f(x) \leq f(x_2) = M, \text{ for all } x \in [a, b].
$$

This implies that

$$\int_a^b f(x_1)g(x)dx \leq \int_a^b f(x)g(x)dx \leq \int_a^b f(x_2)g(x)dx.$$

Thus, we obtain

$$f(x_1)\int_a^b g(x)dx \leq \int_a^b f(x)g(x)dx \leq f(x_2)\int_a^b g(x)dx,$$

or equivalently,

$$f(x_1) \leq \frac{\int_a^b f(x)g(x)dx}{\int_a^b g(x)dx} \leq f(x_2).$$

Since f is continuous, by the intermediate value theorem, there is $\xi \in [a, b]$ such that

$$f(\xi) = \frac{\int_a^b f(x)g(x)dx}{\int_a^b g(x)dx}.$$

This completes the proof of our statement. If $g(x) < 0$, for all $x \in [a, b]$, the proof is similar.

315. *If f is a continuous function on $[0, 1]$, show that*

$$\lim_{n \to \infty} \int_0^1 x^n f(x)dx = 0.$$

Solution. We apply Problem (314) for $f(x)$ and $g_n(x) = x^n$. Then, there exists $\xi_n \in (0, 1)$ such that

$$\int_0^1 x^n f(x)dx = f(\xi_n) \int_0^1 x^n dx = \frac{f(\xi_n)}{n+1}.$$

Since f is continuous on $[0, 1]$, we conclude that it is bounded and so $\lim_{n \to \infty} \frac{f(\xi_n)}{n+1} = 0$, the proof is done.

316. *Establish the following inequalities:*

$$\frac{1}{10\sqrt{2}} \leq \int_0^1 \frac{x^9}{\sqrt{1+x}}dx \leq \frac{1}{10}.$$

Solution. Let $f(x) = \dfrac{1}{\sqrt{1+x}}$ and $g(x) = x^9$, where $0 \leq x \leq 1$. Then, by the mean value theorem for integrals of continuous functions, there exists $c \in [0, 1]$ such that

$$\int_0^1 \frac{x^9}{\sqrt{1+x}}\,dx = \int_0^1 f(x)g(x)\,dx = f(c)\int_0^1 g(x)\,dx$$

$$= \frac{1}{\sqrt{1+c^2}}\int_0^1 x^9\,dx = \frac{1}{\sqrt{1+c^2}}\cdot\frac{1}{10}.$$

Now, the proof is completed, since $0 \leq c \leq 1$.

317. *Show that*

$$\frac{\pi^2}{9} \leq \int_{\frac{\pi}{6}}^{\frac{\pi}{2}} \frac{x}{\sin x}\,dx \leq \frac{2\pi^2}{9}.$$

Solution. Suppose that $f(x) = \dfrac{1}{\sin x}$ and $g(x) = x$. Clearly, f and g are continuous on $[\frac{\pi}{6}, \frac{\pi}{2}]$ and $g(x) > 0$, for all $x \in [\frac{\pi}{6}, \frac{\pi}{2}]$. Then, by Problem (314), there is $\alpha \in [\frac{\pi}{6}, \frac{\pi}{2}]$ such that

$$\int_{\frac{\pi}{6}}^{\frac{\pi}{2}} \frac{x}{\sin x}\,dx = \int_{\frac{\pi}{6}}^{\frac{\pi}{2}} f(x)g(x)\,dx = f(\alpha)\int_{\frac{\pi}{6}}^{\frac{\pi}{2}} g(x)\,dx = \frac{1}{\sin\alpha}\int_{\frac{\pi}{6}}^{\frac{\pi}{2}} x\,dx$$

$$= \frac{1}{\sin\alpha}\cdot\left(\frac{1}{2}x^2\right)\Big|_{\frac{\pi}{6}}^{\frac{\pi}{2}} = \frac{1}{\sin\alpha}\cdot\frac{\pi^2}{9}.$$

Since $\dfrac{\pi}{6} \leq \alpha \leq \dfrac{\pi}{2}$, it follows that $\dfrac{1}{2} \leq \sin\alpha \leq 1$. So, we get $1 \leq \dfrac{1}{\sin\alpha} \leq 2$. This completes the proof.

318. *Evaluate the integral*

$$I = \int \sqrt{\frac{e^x - 1}{e^x + 1}}\,dx, \quad for\ x > 0.$$

Solution. If we use substitution $u = \sqrt{\dfrac{e^x - 1}{e^x + 1}}$ with $0 < u < 1$, then we have $u^2 = \dfrac{e^x - 1}{e^x + 1}$. This implies that $e^x = \dfrac{1 + u^2}{1 - u^2}$, or equivalently $x = \ln(1 + u^2) - \ln(1 - u^2)$.

Hence, $dx = \left(\dfrac{2u}{1 + u^2} + \dfrac{2u}{1 - u^2}\right)du$. Therefore, we get

$$I = \int u \left(\frac{2u}{u^2 + 1} + \frac{2u}{1 - u^2} \right) du = \int \left(\frac{2u^2}{u^2 + 1} + \frac{2u^2}{1 - u^2} \right) du$$

$$= \int \left(\frac{2u^2 + 2 - 2}{u^2 + 1} + \frac{2u^2 - 2 + 2}{1 - u^2} \right) du = \int \left(-\frac{2}{u^2 + 1} + \frac{2}{1 - u^2} \right) du$$

$$= \int \left(-\frac{2}{u^2 + 1} + \frac{1}{1 + u} + \frac{1}{1 - u} \right) du$$

$$= -2 \tan^{-1} u + \ln(1 + u) + \ln(1 - u) + C$$

$$= -2 \tan^{-1} \sqrt{\frac{e^x - 1}{e^x + 1}} + \ln \left(1 + \sqrt{\frac{e^x - 1}{e^x + 1}} \right) + \ln \left(1 - \sqrt{\frac{e^x - 1}{e^x + 1}} \right) + C.$$

319. *Find*

$$I = \int \frac{1}{\sqrt[3]{x + 1} - 1} dx.$$

Solution. We make the substitution $t = \sqrt[3]{x + 1}$. Then, we have $dt = \frac{1}{3}(x + 1)^{-\frac{2}{3}} dx$. This implies that $3(x + 1)^{\frac{2}{3}} dt = dx$ or equivalently $3t^2 dt = dx$. So, we get

$$I = 3 \int \frac{t^2}{t - 1} dt.$$

Now, if we make the substitution $u = t - 1$, then $du = dt$. Therefore, we have

$$I = 3 \int \frac{(u + 1)^2}{u} du = 3 \int (u^2 + 2u + 1) u^{-1} du$$

$$= 3 \int (u + 2 + \frac{1}{u}) du = \frac{3}{2} u^2 + 6u + 3 \ln |u| + C$$

$$= \frac{3}{2}(t - 1)^2 + 6(t - 1) + 3 \ln |t - 1| + C$$

$$= \frac{3}{2}(\sqrt[3]{x + 1} - 1)^2 + 6(\sqrt[3]{x + 1} - 1) + 3 \ln(\sqrt[3]{x + 1} - 1) + C.$$

320. *Evaluate the integral*

$$I = \int x \sqrt{3 - 2x - x^2} dx.$$

Solution. Since $3 - 2x - x^2 = 4 - (x + 1)^2$, we use substitution $t = x + 1$. Then, we get

$$I = \int (t - 1) \sqrt{4 - t^2} dt = \int t \sqrt{4 - t^2} dt - \int \sqrt{4 - t^2} dt.$$

For the first integral, if we use substitution $u = 4 - t^2$, then $du = -2t dt$. Consequently, we have

$$\int t\sqrt{4-t^2}dt = -\frac{1}{2}\int \sqrt{u}\,du = -\frac{1}{3}u^{\frac{3}{2}} + C_1$$
$$= -\frac{1}{3}(4-t^2)^{\frac{3}{2}} + C_1.$$

For the second integral, let $t = 2\sin\theta$, where $-\frac{\pi}{2} \leq \theta \leq \frac{\pi}{2}$. Then, we have $\theta = \sin^{-1}\left(\frac{t}{2}\right)$, $dt = 2\cos\theta d\theta$ and

$$\sqrt{4-t^2} = \sqrt{4 - 4\sin^2\theta} = 2\cos\theta.$$

Therefore, we get

$$\int \sqrt{4-t^2}dt = \int 4\cos^2\theta d\theta = \int (2\cos 2\theta + 2)d\theta$$
$$= \sin 2\theta + 2\theta + C_2 = 2\sin\theta\cos\theta + 2\theta + C_2.$$

Note that

$$\cos\theta = \sqrt{1 - \sin^2\theta} = \sqrt{1 - \frac{t^2}{4}} = \frac{1}{2}\sqrt{4-t^2}.$$

Thus, we have

$$\int \sqrt{4-t^2}dt = \frac{1}{2}t\sqrt{4-t^2} + 2\sin^{-1}\left(\frac{t}{2}\right) + C_2.$$

Finally, we get

$$I = -\frac{1}{3}(4-t^2)^{\frac{3}{2}} - \frac{1}{2}t\sqrt{4-t^2} - 2\sin^{-1}\left(\frac{t}{2}\right) + C$$
$$= -\frac{1}{3}(3 - 2x - x^2)^{\frac{3}{2}} - \frac{1}{2}(x+1)\sqrt{3 - 2x - x^2} - 2\sin^{-1}\left(\frac{x+1}{2}\right) + C.$$

321. *Let p be a positive real number. Evaluate the integral*

$$I = \int \frac{1}{x\sqrt{x^{2p} + x^p + 1}}dx, \quad for\ all\ x > 0.$$

Solution. We observe that

$$I = \int \frac{1}{x^{p+1}\sqrt{1 + \dfrac{1}{x^p} + \dfrac{1}{x^{2p}}}}dx = \int \frac{1}{x^{p+1}\sqrt{\left(\dfrac{1}{x^p} + \dfrac{1}{2}\right)^2 + \dfrac{3}{4}}}dx.$$

Now, we use substitution $u = \dfrac{1}{x^p} + \dfrac{1}{2}$. Then, $du = \dfrac{-p}{x^{p+1}}dx$. Thus, we obtain

$$I = -\frac{1}{p} \int \frac{1}{\sqrt{u^2 + \frac{3}{4}}} du = -\frac{1}{p} \ln\left(u + \sqrt{u^2 + \frac{3}{4}}\right) + C$$

$$= -\frac{1}{p} \ln\left(\frac{1}{x^p} + \frac{1}{2} + \sqrt{1 + \frac{1}{x^p} + \frac{1}{x^{2p}}}\right) + C.$$

322. *Suppose that f is defined on $[1, \infty)$ by*

$$f(x) = \int_1^x \frac{\ln t}{1+t} dt.$$

Solve the equation $f(x) + f(\frac{1}{x}) = \frac{1}{2}$.

Solution. We have $f(\frac{1}{x}) = \int_1^{\frac{1}{x}} \frac{\ln t}{1+t} dt$. If we substitute $t = \frac{1}{u}$, then $dt = -\frac{1}{u^2} du$.
So, we get

$$f(\frac{1}{x}) = \int_1^x \frac{-\ln u}{1 + \frac{1}{u}} \cdot (\frac{-1}{u^2} du) = \int_1^x \frac{\ln u}{u(u+1)} du.$$

Consequently, we obtain

$$f(x) + f(\frac{1}{x}) = \int_1^x \frac{\ln t}{1+t} dt + \int_1^x \frac{\ln t}{t(t+1)} dt$$

$$= \int_1^x \frac{\ln t}{1+t} dt + \int_1^x \left(\frac{\ln t}{t} - \frac{\ln t}{1+t}\right) dt$$

$$= \int_1^x \frac{\ln t}{t} dt = \frac{1}{2}(\ln x)^2.$$

So, we must have $(\ln x)^2 = 1$. Since $x > 1$, it follows that $\ln x = 1$. Hence, $x = e$.

323. *Prove that for each continuous function f on $[0, a]$,*

$$\int_0^a \frac{f(x)}{f(x) + f(a - x)} dx = \frac{a}{2}.$$

Solution. We use substitution $t = a - x$. Then, we get

$$\int_0^a \frac{f(x)}{f(x)+f(a-x)}dx = \int_a^0 \frac{f(a-t)}{f(a-t)+f(t)}(-dt)$$
$$= \int_0^a \frac{f(a-t)}{f(a-t)+f(t)}dt$$
$$= \int_0^a \frac{f(a-t)+f(t)-f(t)}{f(a-t)+f(t)}dt$$
$$= \int_0^a \left(1 - \frac{f(t)}{f(t)+f(a-t)}\right)dt$$
$$= a - \int_0^a \frac{f(t)}{f(t)+f(a-t)}dt$$
$$= a - \int_0^a \frac{f(x)}{f(x)+f(a-x)}dx$$

Therefore, $2\int_0^a \frac{f(x)}{f(x)+f(a-x)}dx = a$. This completes the proof.

324. *Show that*

$$\int_0^\pi xe^{\sin x}dx = \frac{\pi}{2}\int_0^\pi e^{\sin x}dx.$$

Solution. We use the substitution $t = \pi - x$. Then, we have

$$\int_0^\pi xe^{\sin x}dx = -\int_\pi^0 (\pi - t)e^{\sin(\pi-t)}dt = \int_0^\pi (\pi - t)e^{\sin t}dt$$
$$= \pi \int_0^\pi e^{\sin t}dt - \int_0^\pi te^{\sin t}dt$$
$$= \pi \int_0^\pi e^{\sin t}dt - \int_0^\pi xe^{\sin x}dx.$$

Therefore, we obtain

$$\int_0^\pi xe^{\sin x}dx = \frac{\pi}{2}\int_0^\pi e^{\sin x}dx.$$

325. *Find the following integral:*

$$I = \int \frac{\sqrt{a^2 - x^2}}{x^4}dx, \text{ where } a > 0.$$

Solution. Making the substitution $x = \frac{1}{t}$, we have $dx = -\frac{1}{t^2}dt$. It follows that

$$I = -\int \frac{\sqrt{a^2 - \frac{1}{t^2}}}{\frac{1}{t^4}} \cdot \frac{1}{t^2}dt = -\int t\sqrt{a^2t^2 - 1}dt.$$

Again, if we use the substitution $u = \sqrt{a^2t^2 - 1}$, then $du = \dfrac{2a^2t}{2\sqrt{a^2t^2 - 1}}dt$ or $udu = a^2tdt$. Hence, we obtain

$$I = -\frac{1}{a^2} \int u^2 du = -\frac{1}{3a^2}u^3 + C.$$

After replacing u by $\sqrt{a^2t^2 - 1}$ and replacing t by $\dfrac{1}{x}$, we get

$$I = -\frac{(a^2 - x^2)^{\frac{3}{2}}}{3a^2x^3} + C.$$

326. *Find the following integral:*

$$\int \frac{1}{x^4 + 1}dx.$$

Solution. We have

$$\int \frac{1}{x^4 + 1}dx = \frac{1}{2} \int \frac{(x^2 + 1) - (x^2 - 1)}{x^4 + 1}dx$$
$$= \frac{1}{2} \int \frac{x^2 + 1}{x^4 + 1}dx - \frac{1}{2} \int \frac{x^2 - 1}{x^4 + 1}dx$$
$$= \frac{1}{2}I_1 - \frac{1}{2}I_2.$$

Now, we compute I_1 and I_2. We can write

$$\int \frac{x^2 + 1}{x^4 + 1}dx = \int \frac{1 + \dfrac{1}{x^2}}{x^2 + \dfrac{1}{x^2}}dx.$$

We set $x - \dfrac{1}{x} = t$. Then, $x^2 + \dfrac{1}{x^2} = t^2 + 2$ and $(1 + \dfrac{1}{x^2})dx = dt$. Thus, we obtain

$$I_1 = \int \frac{1}{t^2 + 2}dt = \frac{1}{\sqrt{2}} \arctan \frac{t}{\sqrt{2}} + C_1 = \frac{1}{\sqrt{2}} \arctan \left(\frac{1}{\sqrt{2}}\left(x - \frac{1}{x}\right)\right) + C_1.$$

Also, we can write

$$\int \frac{x^2 - 1}{x^4 + 1}dx = \int \frac{1 - \dfrac{1}{x^2}}{x^2 + \dfrac{1}{x^2}}dx.$$

We set $x + \dfrac{1}{x} = u$. Then, $x^2 + \dfrac{1}{x^2} = u^2 - 2$ and $(1 - \dfrac{1}{x^2})dx = du$. Thus, we obtain

$$I_2 = \int \frac{1}{u^2 - 2}du = \frac{1}{2\sqrt{2}}\ln\left(\frac{u - \sqrt{2}}{u + \sqrt{2}}\right) + C_2$$

$$= \frac{1}{2\sqrt{2}}\ln\left(\frac{x + \dfrac{1}{x} - \sqrt{2}}{x + \dfrac{1}{x} + \sqrt{2}}\right) + C_2 = \frac{1}{2\sqrt{2}}\ln\left(\frac{x^2 - \sqrt{2}x + 1}{x^2 + \sqrt{2}x + 1}\right) + C_2.$$

327. *Compute the following integral:*

$$I = \int \sqrt{\frac{9 - x}{1 + x}}dx.$$

Solution. We use substitution $t = 4 - x$. Then, $dt = -dx$ and

$$I = -\int \sqrt{\frac{5 + t}{5 - t}}dt = -\int \frac{\sqrt{25 - t^2}}{5 - t}dt.$$

Now, we use substitution $t = 5\sin u$, where $-\dfrac{\pi}{2} \le u < \dfrac{\pi}{2}$. Then, we have $dt = 5\cos u$ and $\sqrt{25 - t^2} = 5\cos u$. So, we get

$$I = -\int \frac{5\cos u}{5 - 5\sin u} \cdot 5\cos u\, du = -5\int \frac{\cos^2 u}{1 - \sin u}du = -5\int \frac{1 - \sin^2 u}{1 - \sin u}du$$
$$= -5\int (1 + \sin u)du = -5(u - \cos u) + C$$
$$= -5\left(\sin^{-1}\frac{t}{5} - \frac{1}{5}\sqrt{25 - t^2}\right) + C$$
$$= -5\left(\sin^{-1}(\frac{4 - x}{5}) - \frac{1}{5}\sqrt{25 - (4 - x)^2}\right) + C.$$

328. *Find*

$$I = \int e^{3x}\tan^{-1}(e^x)dx.$$

Solution. If we put $t = e^x$, then $dt = e^x dx$. So, we have

$$I = \int t^2 \tan^{-1}t\, dt.$$

Now, we use integration by parts. If $u = \tan^{-1} t$ and $dv = t^2 dt$, then $du = \dfrac{1}{1+t^2} dt$
and $v = \dfrac{1}{3} t^3$. Hence, we get

$$
\begin{aligned}
I &= \frac{t^3}{3} \tan^{-1} t - \frac{1}{3} \int \frac{t^3}{1+t^2} dt \\
&= \frac{t^3}{3} \tan^{-1} t - \frac{1}{3} \int \left(t - \frac{t}{1+t^2} \right) dt \\
&= \frac{t^3}{3} \tan^{-1} t - \frac{1}{6} t^2 + \frac{1}{6} \ln(1+t^2) + C \\
&= \frac{1}{3} e^{3x} \tan^{-1}(e^x) - \frac{1}{6} e^{2x} + \frac{1}{6} \ln(1+e^{2x}) + C.
\end{aligned}
$$

329. *Let $x > 0$ and*

$$
f_n(x) = \int_1^x (\ln t)^n dt, \quad \text{for all } n \in \mathbb{N}.
$$

Show that $\dfrac{1}{n} f_{n+1}(e) = f_{n-1}(e) - f_n(e)$, *for all* $n \geq 2$.

Solution. We consider $f_{n+1}(x)$ and apply integration by parts. If $u = (\ln t)^n$ and $dv = \ln t \, dt$, then $du = \dfrac{n}{t}(\ln t)^{n-1} dt$ and $v = t \ln t - t$. Thus, we get

$$
\begin{aligned}
f_{n+1}(x) &= (\ln t)^n (t \ln t - t) \Big|_1^x - \int_1^x \frac{n}{t} (\ln t)^{n-1} (t \ln t - t) dt \\
&= (\ln x)^n (x \ln x - x) - n \int_1^x \left((\ln t)^n - (\ln t)^{n-1} \right) dt \\
&= (\ln x)^n (x \ln x - x) - n \Big(f_n(x) - f_{n-1}(x) \Big).
\end{aligned}
$$

Therefore, $f_{n+1}(e) = -n f_n(e) + n f_{n-1}(e)$. This completes the proof.

330. *Let f be a real function which is continuously differentiable and increasing on $[0, \infty)$ with $f(0) = 0$. For each $a \in [0, \infty)$ and $b \in f([0, \infty))$, define*

$$
g(x) = bx - \int_0^x f(t) dt.
$$

Show that

(1) *g attains its maximum value at $f^{-1}(b)$,*

(2) $\displaystyle \int_0^{f^{-1}(b)} x f'(x) dx = g\left(f^{-1}(b) \right),$

(3) $\displaystyle\int_0^{f^{-1}(b)} xf'(x)dx = \int_0^b f^{-1}(x)dx.$

Solution. (1) Since f is increasing and f' is continuous, it follows that $f'(x) \geq 0$. We have $g'(x) = b - f(x)$. So, we get

$$g'(x) = 0 \Leftrightarrow b - f(x) = 0$$
$$\Leftrightarrow f(x) = b$$
$$\Leftrightarrow x = f^{-1}(b).$$

Since $g''(x) = -f'(x) \leq 0$, it follows that g attains its maximum value at $f^{-1}(b)$.

(2) We use integration by parts. If $u = x$ and $dv = f'(x)dx$, then $du = dx$ and $v = f(x)$. So, we get

$$\int_0^{f^{-1}(b)} xf'(x)dx = xf(x)\Big|_0^{f^{-1}(b)} - \int_0^{f^{-1}(b)} f(x)dx$$
$$= bf^{-1}(b) - \int_0^{f^{-1}(b)} f(x)dx$$
$$= g\Big(f^{-1}(b)\Big).$$

(3) Put $y = f(x)$ or $x = f^{-1}(y)$. Then, $f'(x)dx = dy$. So, we obtain

$$\int_0^{f^{-1}(b)} xf'(x)dx = \int_0^b f^{-1}(y)dy = \int_0^b f^{-1}(x)dx.$$

331. *Let f be a real function which is continuously differentiable and increasing on $[0, \infty)$ with $f(0) = 0$. Prove Young's inequality:*

$$\int_0^a f(x)dx + \int_0^b f^{-1}(x)dx \geq ab,$$

where a and b are arbitrary positive numbers.

Solution. Suppose that g is the function defined in Problem (330). For each $x \in [0, \infty)$, we have $g(x) \leq g\Big(f^{-1}(b)\Big)$. In particular,

$$g(a) \leq g\Big(f^{-1}(b)\Big). \tag{5.35}$$

According to the definition of g, we have

$$g(a) = ba - \int_0^a f(x)dx. \tag{5.36}$$

Moreover, from parts (2) and (3) of Problem (330), we have

$$\int_0^b f^{-1}(x)dx = g\Big(f^{-1}(b)\Big). \tag{5.37}$$

So, by Eqs. (5.35), (5.36) and (5.37), we obtain

$$\begin{aligned}
\int_0^a f(x)dx + \int_0^b f^{-1}(x)dx &= \Big(ab - g(a)\Big) + g\Big(f^{-1}(b)\Big)\\
&\geq \Big(ab - g(a)\Big) + g(a)\\
&= ab.
\end{aligned}$$

332. *For every positive integers m, n, prove that*

$$\int_0^1 x^m (1-x)^n dx = \frac{(m!)(n!)}{(m+n+1)!}.$$

Solution. We denote the desired integral by $I_{m,n}$. Then, we have

$$\begin{aligned}
I_{m,n} &= \int_0^1 x^m (1-x)(1-x)^{n-1} dx\\
&= \int_0^1 (x^m - x^{m+1})(1-x)^{n-1} dx\\
&= \int_0^1 x^m (1-x)^{n-1} dx - \int_0^1 x^{m+1}(1-x)^{n-1} dx\\
&= I_{m,n-1} - \int_0^1 x^{m+1}(1-x)^{n-1} dx.
\end{aligned}$$

In order to compute the right integral we apply integration by parts. Assume that $u = x^{m+1}$ and $dv = (1-x)^{n-1} dx$. Then, $du = (m+1)x^m dx$ and $v = -\dfrac{1}{n}(1-x)^n$. So, we get

$$\begin{aligned}
\int_0^1 x^{m+1}(1-x)^{n-1} dx &= x^{m+1}\Big(-\frac{(1-x)^n}{n}\Big)\Big|_0^1 - \int_0^1 (m+1)x^m\Big(-\frac{(1-x)^n}{n}\Big)dx\\
&= \frac{m+1}{n}\int_0^1 x^m (1-x)^n dx\\
&= \frac{m+1}{n} I_{m,n}.
\end{aligned}$$

Therefore, we have $I_{m,n} = I_{m,n-1} - \dfrac{m+1}{n} I_{m,n}$. This implies that

$$I_{m,n} = \frac{n}{m+n+1} I_{m,n-1}.$$

Consequently, we obtain

$$I_{m,n} = \frac{n}{m+n+1} \cdot \frac{n-1}{m+n} \cdot \frac{n-2}{m+n-1} \cdots \frac{1}{m+2} \int_0^1 x^m dx$$

$$= \frac{n!}{(m+n+1)(m+n)\dots(m+1)}$$

$$= \frac{(m!)(n!)}{(m+n+1)!}.$$

333. *If*

$$I_n = \int_0^1 (1-x^2)^n dx,$$

show that $(2n+1)I_n = 2nI_{n-1}$.

Solution. We apply integration by parts. If $u = (1-x^2)^n$ and $dv = dx$, then $du = -2nx(1-x^2)^{n-1}dx$ and $v = x$. So, we have

$$I_n = x(1-x^2)^n \Big|_0^1 + 2n \int_0^1 x^2(1-x^2)^{n-1}dx$$

$$= -2n \int_0^1 (1-x^2-1)(1-x^2)^{n-1}dx$$

$$= -2n \int_0^1 \left((1-x^2)^n - (1-x^2)^{n-1}\right)dx$$

$$= -2nI_n + 2nI_{n-1}.$$

Therefore, we obtain $(2n+1)I_n = 2nI_{n-1}$.

334. *Suppose that f is a real, continuously differentiable function on $[a, b]$, $f(a) = f(b) = 0$ and*

$$\int_a^b f^2(x)dx = 1.$$

Prove that

$$\int_a^b x f(x) f'(x)dx = -\frac{1}{2}.$$

Solution. We apply integration by parts. If $u = f^2(x)$ and $dv = dx$, then $du = 2f(x)f'(x)dx$ and $v = x$. Hence, we have

$$\int_a^b f^2(x)dx = xf^2(x)\Big|_a^b - \int_a^b 2xf(x)f'(x)dx.$$

Thus, we obtain

$$1 = bf^2(b) - af^2(a) - 2\int_a^b xf(x)f'(x)dx.$$

This completes the proof.

335. *Show that if the second derivative f'' is continuous on $[a, b]$, then*

$$\int_a^b xf''(x)dx = bf'(b) - af'(a) + f(a) - f(b).$$

Solution. We apply integration by parts. If $u = x$ and $dv = f''(x)dx$, then $du = dx$ and $v = f'(x)$. So, we get

$$\begin{aligned}
\int_a^b xf''(x)dx &= xf'(x)\Big|_a^b - \int_a^b f'(x)dx \\
&= xf'(x)\Big|_a^b - f(x)\Big|_a^b \\
&= bf'(b) - af'(a) + f(a) - f(b).
\end{aligned}$$

336. *Define*

$$f(x) = \int_x^{x+1} \sin(e^t)dt.$$

Show that $e^x|f(x)| < 2$.

Solution. Let $e^t = y$. Then, $e^t dt = dy$. So, we obtain

$$f(x) = \int_{e^x}^{e^{x+1}} \frac{\sin y}{y}dy.$$

Now, we use integration by parts. If $u = \dfrac{1}{y}$ and $dv = \sin y dy$, then $du = -\dfrac{1}{y^2}dy$ and $v = -\cos y$. So, we have

$$\begin{aligned}
f(x) &= \frac{-\cos y}{y}\Big|_{e^x}^{e^{x+1}} - \int_{e^x}^{e^{x+1}} \frac{\cos y}{y^2}dy \\
&= \frac{\cos(e^x)}{e^x} - \frac{\cos(e^{x+1})}{e^{x+1}} - \int_{e^x}^{e^{x+1}} \frac{\cos y}{y^2}dy.
\end{aligned}$$

Hence, we obtain

$$e^x f(x) = \cos(e^x) - \frac{1}{e}\cos(e^{x+1}) - e^x \int_{e^x}^{e^{x+1}} \frac{\cos y}{y^2} dy.$$

Therefore,

$$e^x |f(x)| \leq |\cos(e^x)| + \frac{1}{e}|\cos(e^{x+1})| + e^x \int_{e^x}^{e^{x+1}} \frac{1}{y^2} dy$$

$$< 1 + \frac{1}{e} + e^x \left(\frac{-1}{e^{x+1}} + \frac{1}{e^x} \right)$$

$$= 1 + \frac{1}{e} - \frac{1}{e} + 1 = 2.$$

337. *Let f be a continuously differentiable function on $[a, b]$. Prove that*

$$\lim_{n \to \infty} \int_a^b f(x) \sin nx\, dx = 0.$$

Solution. First, we use integration by parts. Let $u = f(x)$ and $dv = \sin nx\, dx$. Then, $du = f'(x)dx$ and $v = -\frac{1}{n}\cos nx$. Thus, we have

$$\int_a^b f(x) \sin nx\, dx = \left(-\frac{1}{n}\cos nx f(x) \right)\Big|_a^b + \frac{1}{n}\int_a^b \cos nx f'(x)dx$$

$$= \frac{f(a)\cos na - f(b)\cos nb}{n} + \frac{1}{n}\int_a^b \cos nx f'(x)dx.$$

Let $C = \int_a^b |f'(x)|dx$. Then, we get

$$\left| \int_a^b f(x) \sin nx\, dx \right| \leq \left| \frac{f(a)\cos na - f(b)\cos nb}{n} \right| + \frac{1}{n}\left| \int_a^b \cos nx f'(x)dx \right|$$

$$\leq \frac{|f(a)| + |f(b)|}{n} + \frac{1}{n}\int_a^b |\cos nx| \cdot |f'(x)|dx$$

$$\leq \frac{|f(a)| + |f(b)|}{n} + \frac{1}{n}\int_a^b |f'(x)|dx$$

$$\leq \frac{|f(a)| + |f(b)|}{n} + \frac{C}{n}.$$

Now, we conclude that $\int_a^b f(x) \sin nx\, dx \to 0$ as $n \to \infty$.

338. *(1) If $\int_0^2 t f''(t)dt \neq 0$, find the integer n such that*

$$n \int_0^1 x f''(2x) dx = \int_0^2 t f''(t) dt.$$

(2) If $f(0) = 1$, $f(2) = 3$, $f'(2) = 5$, evaluate $\int_0^1 x f''(2x) dx$.

Solution. (1) We consider the change of variable $2x = t$. Then, we obtain

$$n \int_0^1 x f''(2x) dx = \frac{n}{4} \int_0^2 t f''(t) dt.$$

So, we must have

$$\frac{n}{4} \int_0^2 t f''(t) dt = \int_0^2 t f''(t) dt.$$

Hence, $\dfrac{n}{4} = 1$, which implies that $n = 4$.

(2) We use integration by parts. Let $u = x$ and $f''(2x) dx = dv$. Then, we have

$$\begin{aligned}
\int_0^1 x f''(2x) dx &= \frac{1}{2} x f'(2x) \Big|_0^1 - \frac{1}{2} \int_0^1 f'(2x) dx \\
&= \frac{1}{2} f'(2) - \frac{1}{4} f(2x) |_0^1 \\
&= \frac{1}{2} f'(2) - \frac{1}{4} f(2) + \frac{1}{4} f(0) \\
&= \frac{5}{2} - \frac{3}{4} + \frac{1}{4} = 2.
\end{aligned}$$

339. *Suppose that $f(\pi) = 2$ and*

$$\int_0^\pi \Big(f(x) + f''(x) \Big) \sin x \, dx = 5.$$

Compute $f(0)$.

Solution. We apply integration by parts. If $u = f(x)$ and $dv = \sin x \, dx$, then $du = f'(x) dx$ and $v = -\cos x$. So, we have

$$\int_0^\pi f(x) \sin x \, dx = -\cos x f(x) \Big|_0^\pi + \int_0^\pi f'(x) \cos x \, dx. \qquad (5.38)$$

Again, for the right integral we use integration by parts. If $u = f'(x)$ and $dv = \cos x \, dx$, then $du = f''(x) dx$ and $v = \sin x$. Hence, we obtain

$$\int_0^\pi f'(x) \cos x \, dx = \sin x f'(x) \Big|_0^\pi - \int_0^\pi f''(x) \sin x \, dx. \qquad (5.39)$$

We put (5.39) into (5.38). Then, we have

$$\int_0^\pi f(x) \sin x dx = -\cos x f(x)\Big|_0^\pi + \sin x f'(x)\Big|_0^\pi - \int_0^\pi f''(x) \sin x dx,$$

which implies that

$$\int_0^\pi \left(f(x) + f''(x) \right) \sin x dx = f(\pi) + f(0).$$

Therefore, we conclude that $f(0) = 3$.

340. *Let A denote the value of the integral*

$$\int_0^\pi \frac{\cos x}{(x+2)^2} dx.$$

Compute the following integral in terms of A:

$$\int_0^{\frac{\pi}{2}} \frac{\sin x \cos x}{x+1} dx.$$

Solution. We have

$$A = \int_0^{\frac{\pi}{2}} \frac{\cos 2y}{4(y+1)^2}(2dy)$$

$$= \frac{1}{2} \int_0^{\frac{\pi}{2}} \frac{\cos 2y}{(y+1)^2} dy$$

$$= \frac{1}{2} \int_0^{\frac{\pi}{2}} \frac{2\cos^2 y - 1}{(y+1)^2} dy$$

$$= \int_0^{\frac{\pi}{2}} \frac{\cos^2 y}{(y+1)^2} dy - \frac{1}{2} \int_0^{\frac{\pi}{2}} \frac{1}{(y+1)^2} dy$$

$$= \int_0^{\frac{\pi}{2}} \frac{\cos^2 y}{(y+1)^2} dy - \frac{1}{2}\left(\frac{-2}{\pi+2} + 1 \right)$$

So, we obtain

$$\int_0^{\frac{\pi}{2}} \frac{\cos^2 x}{(x+1)^2} dx = A + \frac{1}{2}(\frac{-2}{\pi+2} + 1). \qquad (5.40)$$

On the other hand, if $u = \dfrac{1}{x+1}$ and $dv = \sin x \cos x dx$, then $du = \dfrac{-1}{(x+1)^2} dx$ and $v = -\dfrac{1}{2}\cos^2 x$. So, by integration by parts, we obtain

$$\int_0^{\frac{\pi}{2}} \frac{\sin x \cos x}{x+1} dx = -\frac{1}{2} \frac{\cos^2 x}{x+1} \Big|_0^{\frac{\pi}{2}} - \frac{1}{2} \int_0^{\frac{\pi}{2}} \frac{\cos^2 x}{(x+1)^2} dx. \qquad (5.41)$$

If we put (5.40) into (5.41), we observe that

$$\int_0^{\frac{\pi}{2}} \frac{\sin x \cos x}{x+1} dx = -\frac{1}{2} \frac{\cos^2 x}{x+1} \Big|_0^{\frac{\pi}{2}} - \frac{1}{2}\left(A + \frac{1}{2}\left(\frac{-2}{\pi+2}+1\right)\right) = \frac{1}{2}\left(\frac{1}{2} + \frac{1}{\pi+2} - A\right).$$

341. *If* $0 < a < b$, *prove that*

$$\left| \int_a^b \frac{\sin x}{x} dx \right| \le \frac{2}{a}.$$

Solution. By using integration by parts, we obtain

$$\int \frac{\sin x}{x} dx = -\frac{\cos x}{x} - \int \frac{\cos x}{x^2} dx.$$

Therefore, we have

$$\left| \int_a^b \frac{\sin x}{x} dx \right| \le \left| \frac{\cos b}{b} \right| + \left| \frac{\cos a}{a} \right| + \left| \int \frac{\cos x}{x^2} dx \right|$$
$$\le \frac{1}{b} + \frac{1}{a} + \int_a^b \frac{1}{x^2} dx$$
$$\le \frac{1}{b} + \frac{1}{a} - \frac{1}{b} + \frac{1}{a} = \frac{2}{a}.$$

342. *Compute the following integral:*

$$\int \frac{x \ln x}{(1+x^2)^2} dx.$$

Solution. We apply integration by parts. Let $u = \ln x$ and $dv = \frac{x}{(1+x^2)^2} dx$. Then, $du = \frac{1}{x} dx$ and $v = \frac{-1}{2(1+x^2)}$. So, we obtain

$$\int \frac{x \ln x}{(1+x^2)^2} dx = -\frac{\ln x}{2(1+x^2)} + \frac{1}{2} \int \frac{1}{x(1+x^2)} dx.$$

Now, we use partial fractions decomposition to compute the integral on the right. If

$$\frac{1}{x(1+x^2)} = \frac{A}{x} + \frac{Bx+C}{1+x^2},$$

then we get $A = 1$, $B = -1$ and $C = 0$. Hence,

$$\int \frac{1}{x(1+x^2)} dx = \int \left(\frac{1}{x} - \frac{x}{1+x^2}\right) dx = \ln|x| - \frac{1}{2}\ln(1+x^2).$$

Consequently, we obtain

$$\int \frac{x\ln x}{(1+x^2)^2} dx = -\frac{\ln x}{2(1+x^2)} + \frac{1}{2}\ln|x| - \frac{1}{4}\ln(1+x^2) + C.$$

343. *Compute the following integral:*

$$\int \frac{\ln(x^2+x+1)}{x^2} dx.$$

Solution. We apply integration by parts. If $u = \ln(x^2+x+1)$ and $dv = \frac{1}{x^2}dx$, then $du = \frac{2x+1}{x^2+x+1}dx$ and $v = -\frac{1}{x}$. So, we get

$$\int \frac{\ln(x^2+x+1)}{x^2} dx = -\frac{1}{x}\ln(x^2+x+1) + \int \frac{2x+1}{x(x^2+x+1)} dx.$$

Now, we use partial fractions decomposition for the last integral. Let

$$\frac{2x+1}{x(x^2+x+1)} = \frac{A}{x} + \frac{Bx+C}{x^2+x+1} = \frac{Ax^2+Ax+A+Bx^2+Cx}{x(x^2+x+1)}$$
$$= \frac{(A+B)x^2+(A+C)x+A}{x(x^2+x+1)}.$$

So, we obtain $A = 1$, $B = -1$ and $C = 1$. Consequently, we have

$$\int \frac{2x+1}{x(x^2+x+1)} dx = \ln|x| + \int \frac{-x+1}{x^2+x+1} dx$$
$$= \ln|x| - \frac{1}{2}\int \frac{2x+1}{x^2+x+1} dx + \frac{3}{2}\int \frac{1}{x^2+x+1} dx$$
$$= \ln|x| - \frac{1}{2}\ln\left(x^2+x+1\right) + \frac{3}{2}\int \frac{1}{(x+\frac{1}{2})^2+\frac{3}{4}} dx$$
$$= \ln|x| - \frac{1}{2}\ln\left(x^2+x+1\right) + \sqrt{3}\tan^{-1}\left(\frac{2}{\sqrt{3}}(x+\frac{1}{2})\right) + C_1$$

Now, we can combine all the above parts.

344. *Compute the following integral:*

$$\int x (\ln x)^2 dx.$$

Solution. We apply integration by parts. If $u = (\ln x)^2$ and $dv = xdx$, then $du = \dfrac{2\ln x}{x}dx$ and $v = \dfrac{x^2}{2}$. Hence, we have

$$\int x(\ln x)^2 dx = \frac{1}{2}x^2(\ln x)^2 - \int x \ln x dx.$$

Again we use integration by parts. Suppose that $r = \ln x$ and $ds = xdx$. Then, $dr = \dfrac{1}{x}dx$ and $s = \dfrac{x^2}{2}$. Thus, we have

$$\int x \ln x dx = \frac{1}{2}x^2 \ln x - \int \frac{x}{2}dx = \frac{1}{2}x^2 \ln x - \frac{1}{4}x^2 + C_1.$$

Finally, we conclude that

$$\int x(\ln x)^2 dx = \frac{1}{2}x^2(\ln x)^2 - \frac{1}{2}x^2 \ln x + \frac{1}{4}x^2 + C.$$

345. *Compute the following integral:*

$$\int x \sin x \cos x dx.$$

Solution. We can write

$$\int x \sin x \cos x dx = \frac{1}{2}\int x \sin 2x dx.$$

By using substitution $t = 2x$ we have $dt = 2dx$ and so

$$\frac{1}{2}\int x \sin 2x dx = \frac{1}{8}\int t \sin t dt.$$

Now, we use integration by parts. If $u = t$ and $dv = \sin t dt$, then $du = dt$ and $v = -\cos t$. So, we have

$$\int t \sin t \, dt = -t \cos t + \int \cos t \, dt = -t \cos t + \sin t + C_1.$$

Consequently, we get

$$\int x \sin x \cos x \, dx = -\frac{1}{8} t \cos t + \frac{1}{8} \sin t + C$$
$$= -\frac{1}{4} x \cos 2x + \frac{1}{8} \sin 2x + C.$$

346. *Compute the following integral:*

$$\int e^{ax} \cos bx \, dx, \quad (ab \neq 0).$$

Solution. We put $u = e^{ax}$, $dv = \cos bx \, dx$, $du = ae^{ax} dx$ and $v = \dfrac{\sin bx}{b}$. Then, by integration by parts, we obtain

$$\int e^{ax} \cos bx \, dx = \frac{e^{ax} \sin bx}{b} - \frac{a}{b} \int e^{ax} \sin bx \, dx. \qquad (5.42)$$

Although the integral on the right is as hard as the one we are trying to evaluate, we integrate it by parts anyway, this time we choose $u = e^{ax}$, $dv = \sin bx \, dx$, $du = ae^{ax} dx$ and $v = -\dfrac{\cos bx}{b}$. Then, we have

$$\int e^{ax} \sin bx \, dx = -\frac{e^{ax} \cos bx}{b} + \frac{a}{b} \int e^{ax} \cos bx \, dx. \qquad (5.43)$$

By substitution of (5.43) into (5.42), we obtain

$$\int e^{ax} \cos bx \, dx = \frac{e^{ax} \sin bx}{b} + \frac{ae^{ax} \cos bx}{b^2} - \frac{a^2}{b^2} \int e^{ax} \cos bx \, dx.$$

Hence, we have

$$\left(1 + \frac{a^2}{b^2}\right) \int e^{ax} \cos bx \, dx = e^{ax} \frac{a \cos bx + b \sin bx}{b^2}.$$

Therefore, we conclude that

$$\int e^{ax} \cos bx \, dx = e^{ax} \frac{a \cos bx + b \sin bx}{a^2 + b^2} + C.$$

347. *Let*

$$I_n = \int_0^{\frac{\pi}{2}} x^n \sin x dx.$$

For every integer $n > 1$, prove that $I_n = -n(n-1)I_{n-2} + n\left(\frac{\pi}{2}\right)^{n-1}$.

Solution. We apply integration by parts. If $u = x^n$ and $dv = \sin x dx$, then $du = nx^{n-1}dx$ and $v = -\cos x$. Hence, we get

$$I_n = -x^n \cos x \Big|_0^{\frac{\pi}{2}} + \int_0^{\frac{\pi}{2}} nx^{n-1} \cos x dx$$

$$= n \int_0^{\frac{\pi}{2}} x^{n-1} \cos x dx.$$

In order to compute the right integral, again we integrate it by parts. We choose $u = x^{n-1}$ and $dv = \cos x dx$. Then, $du = (n-1)x^{n-2}dx$ and $v = \sin x$. Consequently, we have

$$\int_0^{\frac{\pi}{2}} x^{n-1} \cos x dx = x^{n-1} \sin x \Big|_0^{\frac{\pi}{2}} - \int_0^{\frac{\pi}{2}} (n-1)x^{n-2} \sin x dx$$

$$= \left(\frac{\pi}{2}\right)^{n-1} - (n-1)I_{n-2}.$$

Therefore, we conclude that

$$I_n = n\left(\left(\frac{\pi}{2}\right)^{n-1} - (n-1)I_{n-2}\right)$$

$$= -n(n-1)I_{n-2} + n\left(\frac{\pi}{2}\right)^{n-1}.$$

348. *Suppose that $Q(x) = (x - c_1)(x - c_2)\ldots(x - c_n)$, where $c_1, c_2, \ldots, c_n$ are distinct constant (i.e., $c_i \neq c_j$ if $i \neq j$), and let $P(x)$ be any polynomial of degree less than n. Show that*

$$\int \frac{P(x)}{Q(x)} dx = \sum_{i=1}^n \frac{P(c_i)}{Q'(c_i)} \ln|x - c_i| + C.$$

Solution. Let $Q(x) = Q_1(x)(x - c_1)$. We prove there is a polynomial $P_1(x)$ and a number A_1 such that

$$\frac{P(x)}{Q(x)} = \frac{A_1}{x - c_1} + \frac{P_1(x)}{Q_1(x)}.$$

Suppose that $t = x - c_1$. Then, $\widetilde{P}(t) = P(t + c_1)$ and $\widetilde{Q}(t) = Q(t + c_1)$ are polynomials too. Moreover, $\widetilde{Q}_1(0) = Q_1(c_1) \neq 0$, and we have to find

$$\frac{\widetilde{P}(t)}{\widetilde{Q}_1(t)t} = \frac{A_1}{t} + \frac{\widetilde{P}_1(t)}{\widetilde{Q}_1(t)} = \frac{A_1\widetilde{Q}_1(t) + t\widetilde{P}_1(t)}{\widetilde{Q}_1(t)t}.$$

This is equivalent to

$$\widetilde{P}(t) = A_1\widetilde{Q}_1(t) + t\widetilde{P}_1(t). \tag{5.44}$$

Every term in $t\widetilde{P}_1(t)$ has at least one power of t. So, the constant term of $\widetilde{P}(t)$ is exactly the constant term of $A_1\widetilde{Q}_1(t)$, which is $A_1\widetilde{Q}_1(0)$. Therefore, the constant terms of the right and left of (5.44) are the same if we choose

$$A_1 = \frac{\widetilde{P}(0)}{\widetilde{Q}_1(0)} = \frac{P(c_1)}{Q_1(c_1)}.$$

Since $Q(x) = (x - c_1)Q_1(x)$, it follows that $Q'(x) = Q_1(x) + (x - c_1)Q_1'(x)$, which implies that $Q'(c_1) = Q_1(c_1)$. So, we get

$$A_1 = \frac{P(c_1)}{Q'(c_1)}.$$

On the other hand, from (5.44), we have

$$t\widetilde{P}_1(t) = \widetilde{P}(t) - A_1\widetilde{Q}_1(t),$$

where the right side contains no constant term, do the right side is of the form $t\widetilde{h}(t)$. Since $\deg\widetilde{P}(t) \leq n$ and $\deg A_1\widetilde{Q}_1(t) = n$, we conclude that $\deg\widetilde{h}(t) = n - 1$. Hence, it is enough to choose $\widetilde{P}_1(t) = \widetilde{h}(t)$.

Now, we repeat the above process. Indeed, we have

$$\frac{P_1(x)}{Q_1(x)} = \frac{P_1(x)}{Q_2(x)(x - c_2)} = \frac{A_2}{x - c_2} + \frac{P_2(x)}{Q_2(x)}.$$

Then,

$$A_2 = \frac{P_1(c_2)}{Q_1'(c_2)}.$$

But $Q(x) = (x - c_1)Q_1(x)$, and so $Q'(x) = Q_1(x) + (x - c_1)Q_1'(x)$. Consequently, $Q'(c_2) = (c_2 - c_1)Q_1'(c_2)$. On the other hand, $P(x) = A_1Q_1x) + (x - c_1)P_1(x)$, and so $P(c_2) = (c_2 - c_1)P_1(c_2)$. Therefore, we obtain

$$A_2 = \frac{(c_2 - c_1)P_1(c_2)}{(c_2 - c_1)Q_1'(c_2)} = \frac{P(c_2)}{Q'(c_2)}.$$

If we repeat the above process again, we finally end up with

$$\frac{P(x)}{Q(x)} = \frac{P(c_1)}{Q'(c_1)} \cdot \frac{1}{x - c_1} + \frac{P(c_2)}{Q'(c_2)} \cdot \frac{1}{x - c_2} + \cdots + \frac{P(c_n)}{Q'(c_n)} \cdot \frac{1}{x - c_n}.$$

Therefore,

$$\int \frac{P(x)}{Q(x)} dx = \int \frac{P(c_1)}{Q'(c_1)} \cdot \frac{1}{x - c_1} dx + \cdots + \int \frac{P(c_n)}{Q'(c_n)} \cdot \frac{1}{x - c_n} dx$$
$$= \sum_{i=1}^{n} \frac{P(c_i)}{Q'(c_i)} \ln |x - c_i| + C.$$

349. Let $f(x) = \dfrac{x + 1}{x^2(x - 2)^3}$. *Find antiderivative F of f such that $F(1) = 0$.*

Solution. We can write

$$\frac{x + 1}{x^2(x - 2)^3} = \frac{A}{x^2} + \frac{B}{x} + \frac{C}{(x - 2)^3} + \frac{D}{(x - 2)^2} + \frac{E}{x - 2}. \qquad (5.45)$$

The above is an identity for all x (expect $x = 0, 2$). Multiplying on both sides of (5.45) by the lowest common denominator, we get

$$x + 1 = A(x - 2)^3 + Bx(x - 2)^3 + Cx^2 + Dx^2(x - 2) + Ex^2(x - 2)^2.$$

Hence, we must have

$$-8A = 1,$$
$$12A - 8B = 1,$$
$$-6A + 12B + C - 2D + 4E = 0,$$
$$A - 6B + D - 4E = 0,$$
$$B + E = 0.$$

Solving the above system of linear equations we obtain

$$A = -\frac{1}{8}, \; B = -\frac{5}{16}, \; C = \frac{3}{4}, \; D = -\frac{1}{2}, \; E = \frac{5}{16}.$$

Therefore, from (5.45) we have

$$\frac{x + 1}{x^2(x - 2)^3} = \frac{-\frac{1}{8}}{x^2} + \frac{-\frac{5}{16}}{x} + \frac{\frac{3}{4}}{(x - 2)^3} + \frac{-\frac{1}{2}}{(x - 2)^2} + \frac{\frac{5}{16}}{x - 2}.$$

Therefore, we find that

$$F(x) = \frac{1}{8x} - \frac{5}{16}\ln|x| - \frac{3}{8(x-2)^2} + \frac{1}{2(x-2)} + \frac{5}{16}\ln|x-2| + K.$$

If $F(1) = 0$, then $K = \frac{1}{4}$.

350. *Find the following integral:*

$$\int \frac{1}{x^3 + 1}\,dx.$$

Solution. Factoring the denominator of the integrand, we have

$$x^3 + 1 = (x + 1)(x^2 - x + 1).$$

Hence,

$$\frac{1}{x^3 + 1} = \frac{A}{x + 1} + \frac{Bx + C}{x^2 - x + 1}$$
$$= \frac{(A + B)x^2 + (-A + B + C)x + (A + C)}{(x + 1)(x^2 - x + 1)}.$$

So, we obtain a system of three linear equations as follows:

$$A + B = 0,$$
$$-A + B + C = 0,$$
$$A + C = 1.$$

Solving this system, we find that

$$A = \frac{1}{3}, \ B = -\frac{1}{3} \text{ and } C = \frac{2}{3}.$$

Thus, the partial fraction expansion of the integrand is

$$\frac{1}{x^3 + 1} = \frac{1}{3} \cdot \frac{1}{x + 1} - \frac{1}{3} \cdot \frac{x - 2}{x^2 - x + 1}$$
$$= \frac{1}{3} \cdot \frac{1}{x + 1} - \frac{1}{6} \cdot \frac{2x - 4}{x^2 - x + 1}$$
$$= \frac{1}{3} \cdot \frac{1}{x + 1} - \frac{1}{6} \cdot \frac{2x - 1 - 3}{x^2 - x + 1}$$
$$= \frac{1}{3} \cdot \frac{1}{x + 1} - \frac{1}{6} \cdot \frac{2x - 1}{x^2 - x + 1} + \frac{1}{2} \cdot \frac{1}{(x - \frac{1}{2})^2 + (\frac{\sqrt{3}}{2})^2}.$$

It follows that

$$\int \frac{1}{x^3+1}dx = \frac{1}{3}\int \frac{1}{x+1}dx - \frac{1}{6}\int \frac{2x-1}{x^2-x+1}dx + \frac{1}{2}\int \frac{1}{(x-\frac{1}{2})^2+(\frac{\sqrt{3}}{2})^2}dx$$

$$= \frac{1}{3}\ln|x+1| - \frac{1}{6}\ln|x^2-x+1| + \frac{1}{\sqrt{3}}\tan^{-1}\left(\frac{2}{\sqrt{3}}(x-\frac{1}{2})\right) + C.$$

351. *Compute the following integral:*

$$\int (2x^3 + x^2 - x + 4)e^{3x}dx.$$

Solution. We can write

$$\int (2x^3 + x^2 - x + 4)e^{3x}dx = (Ax^3 + Bx^2 + Cx + D)e^{3x} + C', \qquad (5.46)$$

where A, B, C and D are constants to be determined. We take derivative from both sides of (5.46). Then, we get

$$(2x^3 + x^2 - x + 4)e^{3x} = (3Ax^2 + 2Bx + C)e^{3x} + 3(Ax^3 + Bx^2 + Cx + D)e^{3x},$$

which implies that

$$2x^3 + x^2 - x + 4 = 3Ax^3 + (3A + 3B)x^2 + (2B + 3C)x + (C + 3D). \quad (5.47)$$

For (5.47) to be an identity, the coefficients on the left must be equal to the corresponding coefficients on the right. Hence

$$3A = 2,$$
$$3A + 3B = 1,$$
$$2B + 3C = -1,$$
$$C + 3D = 4.$$

Solving these equations simultaneously we obtain $A = \frac{2}{3}$, $B = -\frac{1}{3}$, $C = -\frac{1}{9}$ and $D = \frac{37}{27}$. Substituting these values in (5.46) we get

$$\int (2x^3 + x^2 - x + 4)e^{3x}dx = (\frac{2}{3}x^3 - \frac{1}{3}x^2 - \frac{1}{9}x + \frac{37}{27})e^{3x} + C'.$$

352. *If $P(x)$ is a polynomial of degree n, show that*

$$\int e^x P(x)dx = e^x \left(P(x) - P'(x) + P''(x) - \cdots + (-1)^n P^{(n)}(x)\right) + C.$$

Solution. Suppose that

$$f(x) = e^x P(x) \text{ and } F(x) = e^x \left(P(x) - P'(x) + P''(x) - \cdots + (-1)^n P^{(n)}(x) \right).$$

We show that $F(x)$ is an antiderivative of $f(x)$, and then by the definition

$$\int f(x)dx = F(x) + C.$$

We have

$$F'(x) = e^x \left(P(x) - P'(x) + P''(x) - \cdots + (-1)^n P^{(n)}(x) \right)$$
$$+ e^x \left(P'(x) - P''(x) + P'''(x) - \cdots + (-1)^n P^{(n+1)}(x) \right).$$

So, we get $F'(x) = e^x P(x) + (-1)^n P^{(n+1)}(x)$. Since $P(x)$ is a polynomial of degree n, it follows that $(-1)^n P^{(n+1)}(x) = 0$. Therefore, $F'(x) = e^x P(x)$.

353. *Compute the following integral:*

$$\int (x^2 + 5x - 1) \sin 3x dx.$$

Solution. We can write

$$\int (x^2 + 5x - 1) \sin 3x dx = (Ax^2 + Bx + C) \sin 3x + (A'x^2 + B'x + C') \cos 3x + C'',$$
$$(5.48)$$

where A, B, C, A', B' and C' are constants to be determined. If we take derivative from both sides of (5.48), then

$$(x^2 + 5x - 1) \sin 3x = (2Ax + B) \sin 3x + 3(Ax^2 + Bx + C) \cos 3x$$
$$+ (2A'x + B') \cos 3x - 3(A'x^2 + B'x + C') \sin 3x,$$

which implies that

$$(x^2 + 5x - 1) \sin 3x = \left(-3A'x^2 + (2A - 3B')x + (B - 3C') \right) \sin 3x$$
$$+ \left(3Ax^2 + (3B + 2A')x + (B' + 3C) \right) \cos 3x.$$
$$(5.49)$$

For (5.49) to be an identity, the coefficients on the left must equal the corresponding coefficients on the right. So,

$$-3A' = 1, \quad 2A - 3B' = 5, \quad B - 3C' = -1,$$
$$3A = 0, \quad 3B + 2A' = 0, \quad B' + 3C = 0.$$

Solving these equations simultaneously we get

$$A = 0, \quad B = \frac{2}{9}, \quad C = \frac{5}{9},$$
$$A' = -\frac{1}{3}, \quad B' = -\frac{5}{3}, \quad C' = \frac{11}{27}.$$

Consequently, we have

$$I = (\frac{2}{9}x + \frac{5}{9}) \sin 3x + (-\frac{1}{3}x^2 + -\frac{5}{3}x + \frac{11}{27}) \cos 3x + C''.$$

354. *Evaluate the integral*

$$\int \frac{\tan(\frac{x}{2})}{\sin x} dx$$

by two methods:

(1) let $z = \tan(\frac{x}{2})$;

(2) let $u = \frac{x}{2}$ and obtain an integral involving trigonometric functions of u.

Solution. (1) Suppose that $z = \tan(\frac{x}{2})$. Then, we obtain $dz = \frac{1}{2}(1 + \tan^2(\frac{x}{2}))dx$ or $dx = \frac{2}{1 + z^2}dz$. Also,

$$\sin x = 2 \sin(\frac{x}{2}) \cdot \cos(\frac{x}{2}) = 2\frac{\sin(\frac{x}{2}) \cdot \cos^2(\frac{x}{2})}{\cos(\frac{x}{2})}$$
$$= 2 \tan(\frac{x}{2}) \cdot \frac{1}{\sec^2(\frac{x}{2})} = \frac{2 \tan(\frac{x}{2})}{1 + \tan^2(\frac{x}{2})} = \frac{2z}{1 + z^2}.$$

Now, we find that

$$\int \frac{\tan(\frac{x}{2})}{\sin x}dx = \int \frac{\frac{2z}{1 + z^2}}{\frac{2z}{1 + z^2}}dz = \int dz = z + C = \tan(\frac{x}{2}) + C.$$

(2) If $u = \dfrac{1}{2}x$, then $dx = 2du$. Thus, we get

$$\int \frac{\tan(\frac{x}{2})}{\sin x}dx = \int \frac{\tan u}{\sin 2u}(2du) = \int \frac{\sin u}{\cos u} \cdot \frac{1}{2\sin u \cos u}(2du)$$
$$= \int \frac{1}{\cos^2 u}du = \tan u + C = \tan(\frac{x}{2}) + C.$$

355. *Prove that*

$$\int_0^{\frac{\pi}{4}} \tan^{2n+1} x\,dx = \frac{(-1)^n}{2}\left(\ln 2 + \sum_{k=1}^n \frac{(-1)^k}{k}\right).$$

Solution. Suppose that

$$I_{2n+1} = \int_0^{\frac{\pi}{4}} \tan^{2n+1} x\,dx.$$

Then, we have

$$I_{2n+1} = \int_0^{\frac{\pi}{4}} \tan^{2n-1} x \tan^2 x\,dx = \int_0^{\frac{\pi}{4}} \tan^{2n-1} x(-1+\sec^2 x)dx$$
$$= -\int_0^{\frac{\pi}{4}} \tan^{2n-1} x\,dx + \int_0^{\frac{\pi}{4}} \tan^{2n-1} x \sec^2 x\,dx$$
$$= -I_{2n-1} + \int_0^{\frac{\pi}{4}} \tan^{2n-1} x \sec^2 x\,dx.$$

In order to compute the last integral we use substitution $\tan x = u$, then $\sec^2 x\,dx = du$. So, we deduce that

$$I_{2n+1} = -I_{2n-1} + \int_0^1 u^{2n-1}du.$$

Hence we obtain the following recursive equation

$$I_{2n+1} = -I_{2n-1} + \frac{1}{2n}. \tag{5.50}$$

Consequently, by repeating Eq. (5.50), we get

$$I_{2n+1} = -I_{2n-1} + \frac{1}{2n}$$

$$= I_{2n-3} - \frac{1}{2(n-1)} + \frac{1}{2n}$$

$$\vdots$$

$$= (-1)^n I_1 + \frac{1}{2n} - \frac{1}{2(n-1)} + \cdots + (-1)^{n-1}\frac{1}{2}$$

$$= (-1)^n \int_0^{\frac{\pi}{4}} \tan x \, dx + \frac{1}{2n} - \frac{1}{2(n-1)} + \cdots + (-1)^{n-1}\frac{1}{2}$$

$$= \frac{(-1)^n}{2} \ln 2 + \frac{1}{2n} - \frac{1}{2(n-1)} + \cdots + (-1)^{n-1}\frac{1}{2}$$

$$= \frac{(-1)^n}{2} \left(\ln 2 + \sum_{k=1}^{n} \frac{(-1)^k}{k} \right).$$

356. *Prove that*

$$\int_0^\pi \frac{\sin(n + \frac{1}{2})x}{\sin(\frac{x}{2})} \, dx = \pi.$$

Solution. By Problem (37), we have

$$\frac{\sin(n + \frac{1}{2})x}{2\sin(\frac{x}{2})} = \frac{1}{2} + \cos x + \cos 2x + \cdots + \cos nx.$$

Therefore, we get

$$\frac{1}{2}\int_0^\pi \frac{\sin(n + \frac{1}{2})x}{\sin(\frac{x}{2})} \, dx = \int_0^\pi \left(\frac{1}{2} + \cos x + \cos 2x + \cdots + \cos nx \right) dx$$

$$= \left(\frac{1}{2}x + \sin x + \frac{1}{2}\sin 2x + \cdots + \frac{1}{n}\sin nx \right)\Big|_0^\pi = \frac{1}{2}\pi.$$

This completes the proof.

357. *Let*

$$I_n = \int_0^\pi \frac{1 - \cos nx}{1 - \cos x} \, dx.$$

If n is a non-negative integer, show that $I_{n+2} = 2I_{n+1} - I_n$.

Solution. We have

$$
\begin{aligned}
I_n + I_{n+2} &= \int_0^\pi \frac{1 - \cos nx}{1 - \cos x} dx + \int_0^\pi \frac{1 - \cos(n+2)x}{1 - \cos x} dx \\
&= \int_0^\pi \frac{1 - \cos nx + 1 - \cos(n+2)x}{1 - \cos x} dx \\
&= \int_0^\pi \frac{2 - \left(\cos nx + \cos(n+2)x\right)}{1 - \cos x} dx \\
&= \int_0^\pi \frac{2 - 2\cos(n+1)x \cos x}{1 - \cos x} dx \\
&= 2 \int_0^\pi \frac{1 - \cos(n+1)x + \cos(n+1)x - \cos(n+1)\cos x}{1 - \cos x} dx \\
&= 2 \int_0^\pi \frac{1 - \cos(n+1)x}{1 - \cos x} dx + 2 \int_0^\pi \cos(n+1)x\, dx \\
&= 2 I_{n+1}.
\end{aligned}
$$

358. *If a, b, a' and b' are given, with $ab \neq 0$, show that there exist constants A, B and C such that*

$$
\int \frac{a' \sin x + b' \cos x}{a \sin x + b \cos x} dx = Ax + B \ln |a \sin x + b \cos x| + C.
$$

Solution. First, we show that A and B exist such that

$$
a' \sin x + b' \cos x = A(a \sin x + b \cos x) + B(a \cos x - b \sin x). \qquad (5.51)
$$

Indeed, the following equations must be hold:

$$
a' = aA - bB \quad \text{and} \quad b' = bA + aB.
$$

The solutions of the above linear system with two variables are

$$
A = \frac{aa' + bb'}{a^2 + b^2} \quad \text{and} \quad B = \frac{ab' - ba'}{a^2 + b^2}.
$$

So, by applying (5.51), we obtain

$$
\begin{aligned}
\int \frac{a' \sin x + b' \cos x}{a \sin x + b \cos x} dx &= \int \frac{A(a \sin x + b \cos x) + B(a \cos x - b \sin x)}{a \sin x + b \cos x} dx \\
&= \int \left(A + B \frac{a \cos x - b \sin x}{a \sin x + b \cos x} \right) dx \\
&= Ax + B \ln |a \sin x + b \cos x| + C.
\end{aligned}
$$

359. *Compute the following integral:*

$$
I = \int_0^\pi \frac{x \sin x}{1 + \cos^2 x} dx.
$$

Solution. Applying the substitution $x = \pi - u$, then $dx = -du$. So,

$$I = -\int_{\pi}^{0} \frac{(\pi - u)\sin(\pi - u)}{1 + \cos^2(\pi - u)} du = \int_{0}^{\pi} \frac{(\pi - u)\sin u}{1 + \cos^2 u} du$$

$$= \pi \int_{0}^{\pi} \frac{\sin u}{1 + \cos^2 u} du - \int_{0}^{\pi} \frac{u\sin u}{1 + \cos^u} du$$

$$= \pi \int_{0}^{\pi} \frac{\sin u}{1 + \cos^2 u} du - I.$$

Consequently, we obtain

$$2I = \pi \int_{0}^{\pi} \frac{\sin u}{1 + \cos^2 u} du.$$

Now, we use the substitution $t = \cos u$, then $dt = -\sin u \, du$. Therefore, we obtain

$$I = -\frac{\pi}{2} \int_{1}^{-1} \frac{1}{1 + t^2} dt = \pi \int_{0}^{1} \frac{1}{1 + t^2} dt = \pi \tan^{-1} t \Big|_{0}^{1} = \frac{\pi^2}{4}.$$

360. *Show that for each continuous function f on $[-1, 1]$,*

$$\int_{0}^{\pi} xf(\sin x)dx = \frac{\pi}{2} \int_{0}^{\pi} f(\sin x)dx. \qquad (5.52)$$

Solution. We put $x = \pi - u$ and $dx = -du$. Then, we obtain

$$\int_{0}^{\pi} xf(\sin x)dx = \int_{\pi}^{0} (\pi - u)f(\sin(\pi - u))(-du)$$

$$= \int_{0}^{\pi} (\pi - u)f(\sin(\pi - u))du$$

$$= \pi \int_{0}^{\pi} f(\sin u)du - \int_{0}^{\pi} uf(\sin u)du$$

$$= \pi \int_{0}^{\pi} f(\sin x)dx - \int_{0}^{\pi} xf(\sin x)dx.$$

So, $2 \int_{0}^{\pi} xf(\sin x)dx = \pi \int_{0}^{\pi} f(\sin x)dx$. This completes the proof.

361. *Evaluate the following integral*

$$\int \sqrt{\tan x} dx.$$

Solution. Suppose that

$$A = \int \sqrt{\tan x} dx \text{ and } B = \int \sqrt{\cot x} dx.$$

Then, we have

$$A + B = \int \left(\sqrt{\tan x} + \sqrt{\cot x}\right) dx = \sqrt{2} \int \frac{\sin x + \cos x}{\sqrt{\sin 2x}} dx$$

$$= \sqrt{2} \int \frac{\sin x + \cos x}{\sqrt{1 - (\sin x - \cos x)^2}} dx = \sqrt{2} \sin^{-1} \left(\sin x - \cos x\right) + C_1$$

and

$$A - B = \int \left(\sqrt{\tan x} - \sqrt{\cot x}\right) dx = \sqrt{2} \int \frac{\sin x - \cos x}{\sqrt{\sin 2x}} dx$$

$$= -\sqrt{2} \int \frac{\cos x - \sin x}{\sqrt{(\sin x + \cos x)^2 - 1}} dx = -\sqrt{2} \cosh^{-1} x \left(\sin x + \cos x\right) + C_2.$$

Therefore, we conclude that

$$A = \frac{(A + B) + (A - B)}{2}$$

$$= \frac{\sqrt{2}}{2} \left(\sin^{-1} \left(\sin x - \cos x\right) - \cosh^{-1} x \left(\sin x + \cos x\right)\right) + C.$$

362. *For every positive real numbers m and n, compute the following integral*

$$\int_0^{\frac{\pi}{2}} \frac{\cos x}{m \sin x + n \cos x} dx.$$

Solution. Suppose that

$$A = \int_0^{\frac{\pi}{2}} \frac{\cos x}{m \sin x + n \cos x} dx \text{ and } B = \int_0^{\frac{\pi}{2}} \frac{\sin x}{m \sin x + n \cos x} dx.$$

Then, we deduce that

$$nA + mB = \int_0^{\frac{\pi}{2}} \frac{m \sin x + n \cos x}{m \sin x + n \cos x} dx = \frac{\pi}{2}$$

and

$$mA - nB = \int_0^{\frac{\pi}{2}} \frac{m \cos x - n \sin x}{m \sin x + n \cos x} dx$$

$$= \left(\ln |m \sin x + n \cos x| \right) \Big|_0^{\frac{\pi}{2}}$$

$$= \ln \left| m \sin \frac{\pi}{2} + n \cos \frac{\pi}{2} \right| - \ln |m \sin 0 + n \cos 0|$$

$$= \ln m - \ln n$$

Hence, we have the following equations

$$\begin{cases} nA + mB = \dfrac{\pi}{2} \\ mA - nB = \ln m - \ln n. \end{cases} \tag{5.53}$$

By solving (5.53), we obtain

$$A = \frac{\dfrac{n\pi}{2} + m \ln \dfrac{m}{n}}{m^2 + n^2}.$$

363. *For every real number m, compute the following integral:*

$$I = \int_0^{\frac{\pi}{2}} \frac{\sin^m x}{\sin^m x + \cos^m x} dx.$$

Solution. We put $y = \dfrac{\pi}{2} - x$. Then, we have

$$I = -\int_{\frac{\pi}{2}}^0 \frac{\cos^m y}{\cos^m y + \sin^m y} dy = \int_0^{\frac{\pi}{2}} \frac{\cos^m x}{\cos^m + \sin^m x} dx.$$

Thus, we obtain

$$I + I = \int_0^{\frac{\pi}{2}} \frac{\sin^m x + \cos^m x}{\sin^m x + \cos^m x} dx = \frac{\pi}{2},$$

and so $I = \dfrac{\pi}{4}$.

364. *Let $f : [0, 1] \to \mathbb{R}$ be a continuous function such that $xf(y) + yf(x) \leq 1$, for all $x, y \in [0, 1]$. Prove that*

$$\int_0^1 f(x) dx \leq \frac{\pi}{4}.$$

Solution. Suppose that $x = \sin t$ and $y = \cos t$. Then, by assumption we have $\sin t f(\cos t) + \cos t f(\sin t) \leq 1$, which implies that

$$\int_0^{\frac{\pi}{2}} \sin t f(\cos t)dt + \int_0^{\frac{\pi}{2}} \cos t f(\sin t)dt \leq \int_0^{\frac{\pi}{2}} dt = \frac{\pi}{2}. \tag{5.54}$$

On the other hand, it is easy to see that

$$\int_0^{\frac{\pi}{2}} \sin t f(\cos t)dt = \int_0^{\frac{\pi}{2}} \cos t f(\sin t)dt = \int_0^1 f(t)dt. \tag{5.55}$$

From (5.54) and (5.55), we reach to the desired inequality.

365. *If f is a continuous function, show that*

$$\int_0^{\frac{\pi}{2}} f(\sin x)dx = \int_0^{\frac{\pi}{2}} f(\cos x)dx$$

Solution. We make the substitution $x = \dfrac{\pi}{2} - t$. Then, we have

$$\int_0^{\frac{\pi}{2}} f(\sin x)dx = -\int_{\frac{\pi}{2}}^0 f\left(\sin(\frac{\pi}{2} - t)\right)dt$$

$$= \int_0^{\frac{\pi}{2}} f\left(\sin(\frac{\pi}{2} - t)\right)dt$$

$$= \int_0^{\frac{\pi}{2}} f(\cos t)dt$$

$$= \int_0^{\frac{\pi}{2}} f(\cos x)dx.$$

366. *Compute the following integral:*

$$I = \int_0^{\frac{\pi}{2}} \frac{1}{\left(\sqrt{\sin x} + \sqrt{\cos x}\right)^4}dx.$$

Solution. We can write

$$I = \int_0^{\frac{\pi}{2}} \frac{1}{\cos^2 x \left(\dfrac{\sqrt{\sin x}}{\sqrt{\cos x}} + 1\right)^4} dx = \int_0^{\frac{\pi}{2}} (1 + \tan^2 x) \frac{1}{\left(\sqrt{\tan x} + 1\right)^4} dx.$$

Making the substitution $\tan x = u^2$, then $(1 + \tan^2 x)dx = 2u\,du$. So,

$$I = \int_0^\infty \frac{2u}{(u + 1)^4} du.$$

Now, let $u + 1 = t$ and $du = dt$. Then, we have

$$I = \int_1^\infty \frac{2(t - 1)}{t^4} dt = \int_1^\infty 2(t^{-3} - t^{-4})dt = \frac{1}{3}.$$

367. *Use the substitution $x = a \cos^2 u + b \sin^2 u$, where $0 < u < \dfrac{\pi}{2}$, to show that*

$$\int \frac{1}{\sqrt{(x - a)(b - x)}} dx = a \tan^{-1} \sqrt{\frac{x - a}{b - x}} + C \quad (a < x < b).$$

Solution. By using the suggested substitution, we have

$$dx = (-2a \cos u \sin u + 2b \sin u \cos u) = 2(b - a) \sin u \cos u.$$

Hence, we get

$$
\begin{aligned}
\int \frac{1}{\sqrt{(x - a)(b - x)}} dx &= \int \frac{2(b - a) \sin u \cos u}{\sqrt{(a \cos^2 u + b \sin^2 u - a)(b - a \cos^2 u - b \sin^2 u)}} du \\
&= \int \frac{2(b - a) \sin u \cos u}{\sqrt{(b \sin^2 u - a \sin^2 u)(b \cos^2 u - a \cos^2 u)}} du \\
&= \int \frac{2(b - a) \sin u \cos u}{\sqrt{(b - a)^2 \sin^2 u \cos^2 u}} du \\
&= \int 2du = 2u + C.
\end{aligned}
$$

Since $x = a \cos^2 u + b \sin^2 u$, it follows that

$$
\begin{aligned}
\frac{x}{\cos^2 u} = a + b \frac{\sin^2 u}{\cos^2 u} &\Rightarrow x(1 + \tan^2 u) = a + b \tan^2 u \\
&\Rightarrow x - a = (b - x) \tan^2 u \\
&\Rightarrow \tan u = \sqrt{\frac{x - a}{b - x}} \\
&\Rightarrow u = \tan^{-1} \sqrt{\frac{x - a}{b - x}}.
\end{aligned}
$$

This completes the proof of our equality.

368. *Compute the following integral:*

$$I = \int_0^{\frac{\pi}{2}} \ln \sin x \, dx.$$

Solution. We have

$$I = \int_0^{\frac{\pi}{2}} \ln \sin(\frac{\pi}{2} - x) dx = \int_0^{\frac{\pi}{2}} \ln \cos x \, dx.$$

Hence, we obtain

$$I + I = \int_0^{\frac{\pi}{2}} (\ln \sin x + \ln \cos x) dx = \int_0^{\frac{\pi}{2}} \ln(\sin x \cos x) dx$$

$$= \int_0^{\frac{\pi}{2}} \ln \left(\frac{\sin 2x}{2} \right) dx = \int_0^{\frac{\pi}{2}} \ln \sin 2x \, dx - \int_0^{\frac{\pi}{2}} \ln 2 \, dx = I_1 - \frac{\pi}{2} \ln 2,$$

where $I_1 = \int_0^{\frac{\pi}{2}} \ln \sin 2x \, dx$. We set $2x = t$. Then, $2dx = dt$. So, we obtain

$$I_1 = \frac{1}{2} \int_0^{\pi} \ln \sin t \, dt = \frac{1}{2} \int_0^{\pi} \ln \sin x \, dx = \int_0^{\frac{\pi}{2}} \ln \sin x \, dx = I.$$

Therefore, $I = -\frac{1}{2} \pi \ln 2$.

369. *Compute the following integral:*

$$I = \int_0^{\frac{\pi}{2}} \left(\frac{x}{\sin x} \right)^2 dx.$$

Solution. We use integration by parts. We set $u = x^2$ and $dv = \frac{1}{\sin^2 x} dx$. Then, we obtain

$$I = -x^2 \cot x \Big|_0^{\frac{\pi}{2}} + \int_0^{\frac{\pi}{2}} 2x \cot x \, dx = 2 \int_0^{\frac{\pi}{2}} x \cot x \, dx.$$

We can use integration by parts again. We set $u = x$ and $dv = \cot x dx$. Then, we have

$$I = 2x \ln \sin x \Big|_0^{\frac{\pi}{2}} - 2 \int_0^{\frac{\pi}{2}} \ln \sin x dx = -2 \int_0^{\frac{\pi}{2}} \ln \sin x dx.$$

Now, by using Problem (368), we conclude that $I = -2(-\frac{1}{2}\pi \ln 2) = \pi \ln 2$.

370. *Let*

$$f(x) = -\int_0^x \ln \cos t dt.$$

Investigate the following equality:

$$f(x) = -2 \ln 2 + 2f\left(\frac{\pi}{4} + \frac{x}{2}\right) - 2f\left(\frac{\pi}{4} - \frac{x}{2}\right).$$

Solution. By substituting $t = \frac{\pi}{2} - u$, we get

$$f(x) = \int_{\frac{\pi}{2}}^{\frac{\pi}{2}-x} \ln \cos\left(\frac{\pi}{2} - u\right) du = \int_{\frac{\pi}{2}}^{\frac{\pi}{2}-x} \ln \sin u du.$$

Now, we apply the equality $\sin u = 2 \sin \frac{u}{2} \cdot \cos \frac{u}{2}$, then we obtain

$$f(x) = \int_{\frac{\pi}{2}}^{\frac{\pi}{2}-x} \ln\left(2 \sin \frac{u}{2} \cdot \cos \frac{u}{2}\right) du$$

$$= \int_{\frac{\pi}{2}}^{\frac{\pi}{2}-x} \ln 2 du + \int_{\frac{\pi}{2}}^{\frac{\pi}{2}-x} \ln \sin \frac{u}{2} du + \int_{\frac{\pi}{2}}^{\frac{\pi}{2}-x} \ln \cos \frac{u}{2} du$$

$$= -x \ln 2 + \int_{\frac{\pi}{2}}^{\frac{\pi}{2}-x} \ln \cos\left(\frac{\pi}{2} - \frac{u}{2}\right) du + \int_{\frac{\pi}{2}}^{\frac{\pi}{2}-x} \ln \cos \frac{u}{2} du$$

$$= -x \ln 2 - 2 \int_{\frac{\pi}{4}}^{\frac{\pi}{4}+\frac{x}{2}} \ln \cos t dt + 2 \int_{\frac{\pi}{4}}^{\frac{\pi}{4}-\frac{x}{2}} \ln \cos t dt$$

$$= -x \ln 2 + 2 \int_{\frac{\pi}{4}+\frac{x}{2}}^{\frac{\pi}{4}-\frac{x}{2}} \ln \cos t dt$$

$$= -x \ln 2 + 2f\left(\frac{\pi}{4} + \frac{x}{2}\right) - 2f\left(\frac{\pi}{4} - \frac{x}{2}\right).$$

371. *Find a non-zero differentiable function f such that for all $x \in \mathbb{R}$,*

$$f^2(x) = \int_0^x f(t) \frac{\sin t}{2 + \cos t} dt. \tag{5.56}$$

Solution. From (5.56), we obtain $2f(x)f'(x) = f(x)\dfrac{\sin x}{2 + \cos x}$. Hence,

$$2f(x) = \int_0^x \frac{\sin t}{2 + \cos t}dt.$$

If we put $u = \tan \dfrac{t}{2}$, then

$$\sin t = \frac{2u}{1 + u^2}, \quad \cos t = \frac{1 - u^2}{1 + u^2} \text{ and } dt = \frac{2}{1 + u^2}du.$$

We find that

$$\begin{aligned}
\int \frac{\sin t}{2 + \cos t}dt &= \int \frac{\dfrac{2u}{1 + u^2}}{2 + \dfrac{1 - u^2}{1 + u^2}} \cdot \frac{2}{1 + u^2}du \\
&= \int \frac{4u}{(3 + u^2)(1 + u^2)}du \\
&= \int \left(\frac{-2u}{3 + u^2} + \frac{2u}{1 + u^2}\right)du \\
&= -\ln(3 + u^2) + \ln(1 + u^2) + C \\
&= -\ln(2 + \frac{1}{\cos^2 \dfrac{t}{2}}) + \ln(\frac{1}{\cos^2 \dfrac{t}{2}}) + C \\
&= -\ln(2\cos^2 \frac{t}{2} + 1) + C \\
&= -\ln(2 + \cos t) + C.
\end{aligned}$$

Therefore, we obtain

$$f(x) = \frac{1}{2}\int_0^x \frac{\sin t}{2 + \cos t}dt = \frac{1}{2}\Big(-\ln(2 + \cos x) + \ln 3\Big) = \ln\sqrt{\frac{3}{2 + \cos x}}.$$

372. *The hyperbolic argument or Gudermannian function is the function $y = \mathrm{gd}x$ defined by the formula*

$$y = \mathrm{gd}x = \int_0^x \frac{1}{\cosh t}dt.$$

Show that

$$\mathrm{gd}x = \tan^{-1}(\sinh x) = 2\tan^{-1}(e^x) - \frac{\pi}{2},$$
$$\sinh x = \tan(\mathrm{gd}x), \quad \cosh x = \sec(\mathrm{gd}x), \quad \tanh x = \sin(\mathrm{gd}x), \quad \coth x = \csc(\mathrm{gd}x).$$

Solution. Using the substitution $u = \sinh t$, for which $du = \cosh t\, dt$ and $1 + u^2 = 1 + \sinh^2 t = \cosh^2 t$, we obtain

$$
\begin{aligned}
\text{gd}x &= \int_0^x \frac{1}{\cosh t}\, dt = \int_0^x \frac{\cosh t}{\cosh^2 t}\, dt \\
&= \int_0^{\sinh x} \frac{1}{1 + u^2}\, du = \tan^{-1} u \Big|_0^{\sinh x} = \tan^{-1}(\sinh x).
\end{aligned}
$$

Now, we show that

$$
\text{gd}x = 2\tan^{-1}(e^x) - \frac{\pi}{2}. \tag{5.57}
$$

Let $f(x) = 2\tan^{-1}(e^x) - \frac{\pi}{2}$. We take derivative of f and gd. We observe that the derivatives of both f and gd are the same. Indeed,

$$
\text{gd}'x = \frac{1}{\cosh x} = \frac{2}{e^x + e^{-x}} \quad \text{and} \quad f'(x) = 2 \cdot \frac{e^x}{1 + (e^x)^2} = \frac{2}{e^x + e^{-x}}.
$$

So, we must have $\text{gd}x = f(x) + C$, where C is a constant. Since $\text{gd}0 = \tan^{-1}(\sinh 0) = 0$ and $f(0) = 2\tan^{-1} 1 - \frac{\pi}{2} = 0$, we conclude that $C = 0$. This completes the proof of (5.57).

Since $\text{gd}x$ is an angle whose tangent is $\sinh x = \dfrac{\sinh x}{1}$, so we have a triangle ABC as Fig. 5.3, where the hypotenuse of the triangle is $\sqrt{1 + \sinh^2 x} = \cosh x$. From this triangle we get the following equalities.

$$
\begin{aligned}
\sinh x &= \tan(\text{gd}x), \\
\cosh x &= \sec(\text{gd}x), \\
\tanh x &= \sin(\text{gd}x), \\
\coth x &= \csc(\text{gd}x).
\end{aligned}
$$

Fig. 5.3 Angle whose tangent is $\sinh x$

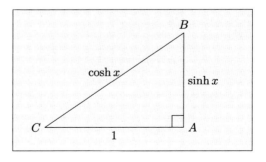

373. *Consider Problem (372), and prove that $y = \text{gd}x$ is a one to one function with inverse*

$$\text{gd}^{-1}x = \int_0^x \sec t\,dt = \ln|\sec x + \tan x|.$$

Solution. By the fundamental theorem of calculus, $\text{gd}'x = \text{sech}x > 0$, which implies that gd is strictly increasing and so is one to one.

Suppose that $y = \text{gd}^{-1}x$. Then, $x = \text{gd}y$ and we have

$$\frac{d}{dx}\left(\text{gd}^{-1}x\right) = \frac{dy}{dx} = \frac{1}{\dfrac{dx}{dy}} = \frac{1}{\dfrac{d}{dy}\text{gd}y}$$

$$= \frac{1}{\text{sech}y} = \cosh y = \sec(\text{gd}y) = \sec x.$$

So, by integration, we obtain $y = \text{gd}^{-1}x = \displaystyle\int_0^x \sec t\,dt$.

374. *Show that*

$$\int \sin^n x\,dx = -\frac{1}{n}\sin^{n-1}x\cos x + \frac{n-1}{n}\int \sin^{n-2}x\,dx, \quad (n = 2, 3, \ldots).$$

Solution. We use integration by parts. Let $u = \sin^{n-1}x$ and $dv = \sin x\,dx$. Then, $du = (n-1)\sin^{n-2}x\cos x\,dx$ and $v = -\cos x$. So, we obtain

$$\int \sin^n x\,dx = -\sin^{n-1}x\cos x + (n-1)\int \sin^{n-2}x\cos^2 x\,dx.$$

Since $\cos^2 x = 1 - \sin^2 x$, it follows that

$$\int \sin^n x\,dx = -\sin^{n-1}x\cos x + (n-1)\int \sin^{n-2}x\,dx - (n-1)\int \sin^n x\,dx,$$

where the last term on the right contains the integral we are trying to evaluate. Therefore, we have

$$n\int \sin^n x\,dx = -\sin^{n-1}x\cos x + (n-1)\int \sin^{n-2}x\,dx.$$

This completes the proof.

375. *Let*

$$I_n = \int_0^{\frac{\pi}{2}} (a \cos x + b \sin x)^n dx.$$

For every integer $n > 2$ prove that $n I_n = ab(a^{n-2} + b^{n-2}) + (n - 1)(a^2 + b^2)I_{n-2}$.

Solution. We can write

$$I_n = \int_0^{\frac{\pi}{2}} (a \cos x + b \sin x)^{n-1} (a \cos x + b \sin x) dx.$$

Now, we use integration by parts. If $u = (a \cos x + b \sin x)^{n-1}$ and $dv = (a \cos x + b \sin x) dx$, then $du = (n - 1)(a \cos x + b \sin x)^{n-2}(-a \sin x + b \cos x) dx$ and $v = a \sin x - b \cos x$. So, we have

$$I_n = \left((a \cos x + b \sin x)^{n-1}(a \sin x - b \cos x) \right) \Big|_0^{\frac{\pi}{2}}$$

$$- \int_0^{\frac{\pi}{2}} (n - 1)(a \cos x + b \sin x)^{n-2}(-a \sin x + b \cos x)(a \sin x - b \cos x) dx$$

$$= (a^{n-1}b + ab^{n-1}) + (n - 1) \int_0^{\frac{\pi}{2}} (a \cos x + b \sin x)^{n-2}(a \sin x - b \cos x)^2 dx$$

$$= (a^{n-1}b + ab^{n-1})$$

$$+ (n - 1) \int_0^{\frac{\pi}{2}} (a \cos x + b \sin x)^{n-2} \left(a^2 \sin^2 x + b^2 \cos^2 x - 2ab \sin x \cos x \right) dx$$

$$= (a^{n-1}b + ab^{n-1})$$

$$+ (n - 1) \int_0^{\frac{\pi}{2}} (a \cos x + b \sin x)^{n-2} \left((a^2 + b^2) - (a \cos x + b \sin x)^2 \right) dx$$

$$= (a^{n-1}b + ab^{n-1}) + (n - 1)(a^2 + b^2) \int_0^{\frac{\pi}{2}} (a \cos x + b \sin x)^{n-2} dx$$

$$- (n - 1) \int_0^{\frac{\pi}{2}} (a \cos x + b \sin x)^n dx$$

$$= ab(a^{n-2} + b^{n-2}) + (n - 1)(a^2 + b^2)I_{n-2} - (n - 1)I_n.$$

Therefore, the desired equality is proved.

376. *For every positive integers m, n, let*

$$I_{m,n} = \int_0^{\frac{\pi}{2}} \cos^m x \sin nx dx.$$

Prove that $I_{m,n} = \dfrac{1}{m+n} + \dfrac{m}{m+n} I_{m-1,n-1}.$

Solution. We apply integration by parts. If $u = \cos^m x$ and $dv = \sin nx dx$, then $du = -m\cos^{m-1} x \sin x dx$ and $v = -\dfrac{1}{n}\cos nx$. Hence, we have

$$I_{m,n} = \cos^m x \left(-\frac{\cos nx}{n} \right) \Big|_0^{\frac{\pi}{2}} - \frac{1}{n} \int_0^{\frac{\pi}{2}} m \cos^{m-1} x \sin x \cos nx dx$$

$$= \frac{1}{n} - \frac{m}{n} \int_0^{\frac{\pi}{2}} \cos^{m-1} x (\cos nx \sin x) dx$$

$$= \frac{1}{n} - \frac{m}{n} \int_0^{\frac{\pi}{2}} \cos^{m-1} x \left(\cos x \sin nx - \sin(n-1)x \right) dx$$

$$= \frac{1}{n} - \frac{m}{n} \left(\int_0^{\frac{\pi}{2}} \cos^m x \sin nx dx - \int_0^{\frac{\pi}{2}} \cos^{m-1} \sin(n-1)x dx \right)$$

$$= \frac{1}{n} - \frac{m}{n} I_{m,n} + \frac{m}{n} I_{m-1,n-1}.$$

So, we get $\left(1 + \dfrac{m}{n}\right) I_{m,n} = \dfrac{1}{n} + \dfrac{m}{n} I_{m-1,n-1}$. This completes the proof of our equality.

377. *Let a be a positive real number. Show that*

$$\int_0^\infty e^{-a\sqrt{x}} dx = \frac{2}{a^2}.$$

Solution. Substituting $\sqrt{x} = t$ we get $\dfrac{1}{2\sqrt{x}} dx = dt$ and

$$\int_0^k e^{-a\sqrt{x}} dx = 2 \int_0^{\sqrt{k}} t e^{-at} dt.$$

Now, we apply integration by parts. If $u = t$ and $dv = e^{-at} dt$, then $du = dt$ and $v = -\dfrac{1}{a}e^{-at}$. Hence, we obtain

$$\int_0^{\sqrt{k}} t e^{-at} dt = -\frac{1}{a} t e^{-at} \Big|_0^{\sqrt{k}} + \frac{1}{a} \int_0^{\sqrt{k}} e^{-at} dt$$

$$= -\frac{\sqrt{k}}{a} e^{-a\sqrt{k}} - \frac{1}{a^2} e^{-at} \Big|_0^{\sqrt{k}}$$

$$= -\frac{\sqrt{k}}{a} e^{-a\sqrt{k}} - \frac{1}{a^2} e^{-a\sqrt{k}} + \frac{1}{a^2}.$$

Therefore, we find that

$$\int_0^\infty e^{-a\sqrt{x}}dx = \lim_{k\to\infty}\left(-\frac{2}{a}e^{-a\sqrt{k}}\cdot\sqrt{k} - \frac{2}{a^2}e^{-a\sqrt{k}} + \frac{2}{a^2}\right) = \frac{2}{a^2}.$$

378. *Give an example of a function f for which* $\lim_{u\to\infty}\int_{-u}^{u} f(x)dx = 0$, *even though*

$\lim_{u\to\infty}\int_{-\infty}^{\infty} f(x)dx$ *is divergent.*

Solution. Consider the function $f(x) = x$. Then,

$$\lim_{u\to\infty}\int_{-u}^{u} xdx = \lim_{u\to\infty}\frac{1}{2}x^2\Big|_{-u}^{u} = 0,$$

while both

$$\int_{-\infty}^{0} xdx = \lim_{u\to\infty}\frac{1}{2}x^2\Big|_{-u}^{0} \quad\text{and}\quad \int_0^\infty xdx = \lim_{u\to\infty}\frac{1}{2}x^2\Big|_0^u$$

are divergent.

379. *Let p be a positive number. Show that the improper integral*

$$\int_1^\infty \frac{\ln x}{x^p}dx$$

is convergent if $p > 1$ *and divergent if* $p \le 1$.

Solution. Suppose that $p > 1$. We apply integration by parts. If $u = \ln x$ and $dv = \frac{1}{x^p}dx$, then $du = \frac{1}{x}dx$ and $v = \frac{1}{(1-p)x^{p-1}}$. Hence, we have

$$\int \frac{\ln x}{x^p}dx = \frac{\ln x}{(1-p)x^{p-1}} - \int \frac{1}{(1-p)x^p}dx$$
$$= \frac{\ln x}{(1-p)x^{p-1}} - \frac{1}{(1-p)^2x^{p-1}}.$$

Consequently, we get

$$\int_1^\infty \frac{\ln x}{x^p} dx = \lim_{u \to \infty} \int_1^u \frac{\ln x}{x^p} dx$$

$$= \lim_{u \to \infty} \left(\frac{\ln x}{(1-p)x^{p-1}} - \frac{1}{(1-p)^2 x^{p-1}} \right) \Big|_1^u$$

$$= \lim_{u \to \infty} \left(\frac{\ln u}{(1-p)u^{p-1}} - \frac{1}{(1-p)^2 u^{p-1}} \right) - \left(- \frac{1}{(1-p)^2} \right)$$

$$= \frac{1}{(1-p)^2}.$$

Thus, the integral is convergent for all $p > 1$.

Now, let $p \le 1$. For $e \le x$, we have

$$\frac{1}{x^p} \le \frac{\ln x}{x^p}.$$

It is easy to see that

$$\int_e^\infty \frac{1}{x^p} dx = \begin{cases} \lim_{u \to \infty} \left(\dfrac{u^{1-p}}{1-p} - \dfrac{e^{1-p}}{1-p} \right) = \infty & \text{if } 0 < p < 1 \\ \lim_{u \to \infty} \int_e^u \dfrac{1}{x} dx = \lim_{u \to \infty} \ln x \Big|_e^u = \lim_{u \to \infty} (\ln u - 1) = \infty & \text{if } p = 1. \end{cases}$$

So, $\int_e^\infty \frac{\ln x}{x^p} dx$ is divergent and consequently the integral $\int_1^\infty \frac{\ln x}{x^p} dx$ is divergent for all $0 < p \le 1$.

380. *Show that the improper integral*

$$\int_0^\infty \frac{e^{-x^2}}{\sqrt{x^2 + x}} dx$$

is convergent.

Solution. We can write

$$\int_0^\infty \frac{e^{-x^2}}{\sqrt{x^2 + x}} dx = \int_0^1 \frac{e^{-x^2}}{\sqrt{x^2 + x}} + \int_1^\infty \frac{e^{-x^2}}{\sqrt{x^2 + x}} dx.$$

Since

$$\lim_{x \to 0} \frac{e^{-x^2}}{\sqrt{x^2 + x}} = \infty,$$

it follows that the first integral in the right side is improper too. The functions $\dfrac{1}{\sqrt{x}}$ and $\dfrac{e^{-x^2}}{\sqrt{x^2 + x}}$ are positive on $(0, 1]$, and

$$\lim_{x \to 0} \frac{\dfrac{e^{-x^2}}{\sqrt{x^2 + x}}}{\dfrac{1}{\sqrt{x}}} = \lim_{x \to 0} \frac{\sqrt{x}\, e^{-x^2}}{\sqrt{x^2 + x}} = \lim_{x \to 0} \frac{e^{-x^2}}{\sqrt{x + 1}} = 1.$$

Since $\displaystyle\int_0^1 \frac{1}{\sqrt{x}} dx$ is convergent, by the limit comparison test, $\displaystyle\int_0^1 \frac{e^{-x^2}}{\sqrt{x^2 + x}} dx$ is convergent.

Analogously, the functions $\dfrac{1}{x^2}$ and $\dfrac{e^{-x^2}}{\sqrt{x^2 + x}}$ are positive on $[1, \infty)$, and

$$\lim_{x \to \infty} \frac{\dfrac{e^{-x^2}}{\sqrt{x^2 + x}}}{\dfrac{1}{x^2}} = \lim_{x \to \infty} \frac{x^2}{e^{x^2} \sqrt{x^2 + x}} = 0.$$

Now, since $\displaystyle\int_1^\infty \frac{1}{x^2} dx$ is convergent, by the limit comparison test $\displaystyle\int_1^\infty \frac{e^{-x^2}}{\sqrt{x^2 + x}} dx$ is convergent.

381. Let f be a continuous function on $[0, \infty)$ such that $\lim\limits_{x \to \infty} f(x) = 1$. Prove that

$$\lim_{t \to 0^+} t \int_0^\infty e^{-tx} f(x)dx = 1.$$

Solution. Suppose that $\epsilon > 0$ is given. Then, there exists $M > 0$ such that $|f(x) - 1| < \dfrac{\epsilon}{2}$, for all $x > M$. On the other hand, we have

$$\begin{aligned}
t \int_0^\infty e^{-tx} dx &= \lim_{u \to \infty} \int_0^u e^{-tx} dx \\
&= \lim_{u \to \infty} \left(-e^{-tx} \Big|_0^u \right) \\
&= \lim_{u \to \infty} (1 - e^{-tu}) \\
&= 1.
\end{aligned}$$

Therefore,

$$\begin{aligned}
\left| t \int_0^\infty e^{-tx} f(x)dx - 1 \right| &= \left| t \int_0^\infty e^{-tx} f(x)dx - t \int_0^\infty e^{-tx} dx \right| \\
&= t \left| \int_0^\infty (f(x) - 1)e^{-tx} dx \right| \\
&\le t \left| \int_0^M (f(x) - 1)e^{-tx} dx \right| + t \left| \int_M^\infty (f(x) - 1)e^{-tx} dx \right|
\end{aligned}$$

$$\leq t \int_0^M |f(x) - 1| e^{-tx} dx + t \int_M^\infty |f(x) - 1| e^{-tx} dx$$

$$\leq t \int_0^M |f(x) - 1| dx + \frac{\epsilon}{2} t \int_M^\infty e^{-tx} dx$$

$$\leq t \int_0^M |f(x) - 1| dx + \frac{\epsilon}{2} t \int_0^\infty e^{-tx} dx$$

$$\leq t \int_0^M |f(x) - 1| dx + \frac{\epsilon}{2}.$$

Since $|f(x) - 1|$ is continuous on $[0, M]$, it follows that the integral $\int_0^M |f(x) - 1| dx$ exists and so

$$\lim_{t \to 0} t \int_0^M |f(x) - 1| dx = 0.$$

Hence, there is $\delta > 0$ such that

$$t \int_0^M |f(x) - 1| dx < \frac{\epsilon}{2},$$

whenever $0 < t < \delta$. Therefore, we get

$$\left| t \int_0^\infty e^{-tx} f(x) dx - 1 \right| \leq t \int_0^M |f(x) - 1| dx + \frac{\epsilon}{2}$$
$$\leq \frac{\epsilon}{2} + \frac{\epsilon}{2} = \epsilon,$$

whenever $0 < t < \delta$.

382. *If $x > 0$, show that the improper integral*

$$\int_0^\infty t^{x-1} e^{-t} dt$$

is convergent.

Solution. We can write

$$\int_0^\infty t^{x-1} e^{-t} dt = \int_0^1 t^{x-1} e^{-t} dt + \int_1^\infty t^{x-1} e^{-t} dt.$$

Now, if $0 \leq t \leq 1$, then $t^{x-1} e^{-t} \leq t^{x-1}$ and

$$\int_0^1 t^{x-1} dt = \frac{t^x}{x} \Big|_0^1 = \frac{1}{x}.$$

Hence, $\int_0^1 t^{x-1}e^{-t}dt$ is convergent for $x > 0$.

Also, we observe that $\dfrac{t^{x-1}e^{-t}}{t^{-2}} \to 0$ as $t \to \infty$ and $\int_1^\infty t^{-2}dt$ are convergent.

Hence, $\int_1^\infty t^{x-1}e^{-t}dt$ is convergent. Therefore, $\int_0^\infty t^{x-1}e^{-t}dt$ is convergent for $x > 0$.

383. *For $x > 0$ the integral in Problem (382) defines a function of x, known as the Gamma function and denoted by $\Gamma(x)$. Thus, $\Gamma(x) = \int_0^\infty t^{x-1}e^{-t}dt$. Show that $\Gamma(x+1) = x\Gamma(x)$ for $x > 0$ and in particular $\Gamma(n+1) = n!$ $(n = 1, 2, \ldots)$.*

Solution. For any $u > 0$, using integration by parts, we obtain

$$\int_0^u t^x e^{-t}dt = -t^x e^{-t}\Big|_0^u + x\int_0^u t^{x-1}e^{-t}dt$$
$$= -u^x e^{-u} + x\int_0^u t^{x-1}e^{-t}dt.$$

So, we have

$$(x+1) = \lim_{u\to\infty}\int_0^u t^x e^{-t}dt$$
$$= \lim_{u\to\infty}\left(-t^x e^{-t}\Big|_0^u + x\int_0^u t^{x-1}e^{-t}dt\right)$$
$$= x\int_0^\infty t^{x-1}e^{-t}dt$$
$$= x(x).$$

Finally, it is easy to see that $\Gamma(1) = 1$. Then, we can prove $\Gamma(n+1) = n!$ by induction.

384. *With the help of Problem (383) show that*

$$\int_0^1 x^m (\ln x)^n dx = \frac{(-1)^n n!}{(m+1)^{n+1}},$$

where m and n are any non-negative integers.

Solution. We use the substitution $x = -(m+1)\ln t$ in Problem (383). We get

$$n! = \int_0^\infty x^n e^{-x} dx$$

$$= \int_1^0 (-1)^n (m+1)^n (\ln t)^n e^{(m+1)\ln t} \cdot \frac{-(m+1)}{t} dt$$

$$= (-1)^n (m+1)^{n+1} \int_0^1 t^m (\ln t)^n dt$$

$$= (-1)^n (m+1)^{n+1} \int_0^1 x^m (\ln x)^n dx.$$

This completes the proof.

385. *Show that for $x > 0$, $y > 0$, the improper integral*

$$\beta(x, y) = \int_0^1 t^{x-1} (1-t)^{y-1} dt \tag{5.58}$$

is convergent. The function $\beta(x, y)$ for $x > 0$, $y > 0$ is called the Beta function.

Solution. Clearly, integral (5.58) is proper for $x \geq 1$ and $y \geq 1$. Hence, it is enough to consider the case $0 < x < 1$ and $0 < y < 1$. In this case both the points $t = 0$ and $t = 1$ are problematic. So, we consider the integrals:

$$\int_0^{\frac{1}{2}} t^{x-1} (1-t)^{y-1} dt \text{ and } \int_{\frac{1}{2}}^1 t^{x-1} (1-t)^{y-1} dt.$$

If $0 < t \leq \frac{1}{2}$, then $(\frac{1}{1-t})^{1-y} < 2^{1-y}$, which implies that

$$t^{x-1} (1-t)^{y-1} < 2^{1-y} t^{x-1}.$$

Since $\int_0^{\frac{1}{2}} t^{x-1} dt$ is convergent, it follows that $\int_0^{\frac{1}{2}} t^{x-1} (1-t)^{y-1} dt$ is convergent. For the second integral, we consider the change of variable $u = 1 - t$. Then, we have

$$\int_{\frac{1}{2}}^1 t^{x-1} (1-t)^{y-1} dt = \int_0^{\frac{1}{2}} u^{y-1} (1-u)^{x-1} du,$$

which is convergent by the above argument. Therefore, the improper integral (5.58) is convergent for all $x > 0$, $y > 0$.

386. *Determine whether the improper integral*

$$\int_0^\infty \sin(x^2) dx$$

is convergent or divergent.

Solution. If $x = \sqrt{t}$, then

$$\int_0^\infty \sin(x^2)dx = \frac{1}{2}\int_0^\infty \frac{\sin t}{\sqrt{t}}dt.$$

On the other hand, we can write

$$\int_0^\infty \frac{\sin t}{\sqrt{t}}dt = \int_0^{\frac{\pi}{2}} \frac{\sin t}{\sqrt{t}}dt + \int_{\frac{\pi}{2}}^\infty \frac{\sin t}{\sqrt{t}}dt.$$

Since $\lim\limits_{t\to 0^+} \dfrac{\sin t}{\sqrt{t}} = 0$, it follows that the first integral in the right side is proper. For

the second integral, we apply integration by parts. If $u = \dfrac{1}{\sqrt{t}}$ and $dv = \sin t\, dt$, then

$du = -\dfrac{1}{2}t^{-\frac{3}{2}}dt$ and $v = -\cos t\, dt$. Consequently, we obtain

$$\int_{\frac{\pi}{2}}^\infty \frac{\sin t}{\sqrt{t}}dt = \lim_{u\to\infty} -\frac{\cos t}{\sqrt{t}}\Big|_{\frac{\pi}{2}}^u - \frac{1}{2}\int_{\frac{\pi}{2}}^\infty \frac{\cos t}{t^{\frac{3}{2}}}dt.$$

Since

$$\frac{|\cos t|}{t^{\frac{3}{2}}} \le \frac{1}{t^{\frac{3}{2}}}$$

and $\displaystyle\int_{\frac{\pi}{2}}^\infty \frac{1}{t^{\frac{3}{2}}}dt$ is convergent, we conclude that $\displaystyle\int_{\frac{\pi}{2}}^\infty \frac{\cos t}{t^{\frac{3}{2}}}dt$ is convergent.

387. *Prove that*

$$\int_0^\infty \frac{\sin x}{x}dx$$

is convergent.

Solution. Since $\lim\limits_{x\to 0} \dfrac{\sin x}{x} = 1$, the function

$$f(x) = \begin{cases} \dfrac{\sin x}{x} & \text{if } x \ne 0 \\ 1 & \text{if } x = 0 \end{cases}$$

is continuous and $\displaystyle\int_0^1 \frac{\sin x}{x}dx$ is proper. So, it is enough to prove that $\displaystyle\int_1^\infty \frac{\sin x}{x}dx$
is convergent.
   Applying the integration by parts we get

$$\int_1^\infty \frac{\sin x}{x}\,dx = \lim_{a\to\infty}\left(\cos 1 - \frac{\cos \alpha}{\alpha}\right) - \int_1^\infty \frac{\cos x}{x^2}\,dx.$$

Clearly, we have

$$\left|\frac{\cos x}{x^2}\right| \le \frac{1}{x^2}.$$

Since $\int_1^\infty \frac{1}{x^2}\,dx$ is convergent, it follows that $\int_1^\infty \frac{\sin x}{x}\,dx$ is convergent.

388. *Let f be a continuous function on $[0, \infty)$ such that $\lim_{t\to\infty} f(t) = 0$ and let $b > a > 0$. Prove that*

$$\int_0^\infty \frac{f(at) - f(bt)}{t}\,dt = f(0)\ln\left(\frac{b}{a}\right).$$

Solution. Assume that $x, y > 0$ and $u = at$. We have

$$\int_x^y \frac{f(at)}{t}\,dt = \int_{ax}^{ay} \frac{f(u)}{u}\,du.$$

Hence, we get

$$\begin{aligned}
\int_x^y \frac{f(at) - f(bt)}{t}\,dt &= \int_{ax}^{ay} \frac{f(u)}{u}\,du - \int_{bx}^{by} \frac{f(u)}{u}\,du \\
&= \int_{ax}^{bx} \frac{f(u)}{u}\,du + \int_{bx}^{ay} \frac{f(u)}{u}\,du - \int_{bx}^{ay} \frac{f(u)}{u}\,du - \int_{ay}^{by} \frac{f(u)}{u}\,du \\
&= \int_{ax}^{bx} \frac{f(u)}{u}\,du - \int_{ay}^{by} \frac{f(u)}{u}\,du.
\end{aligned}$$

Therefore,

$$\int_0^\infty \frac{f(at) - f(bt)}{t}\,dt = \lim_{x\to 0^+}\int_{ax}^{bx} \frac{f(u)}{u}\,du - \lim_{y\to\infty}\int_{ay}^{by} \frac{f(u)}{u}\,du. \qquad (5.59)$$

For the first term in the right side of Eq. (5.59), let

$$M(x) = \max\{f(u) : u \in [ax, bx]\} \text{ and } m(x) = \min\{f(u) : u \in [ax, bx]\}.$$

Then, we have

$$m(x)\ln\left(\frac{b}{a}\right) \le \int_{ax}^{bx} \frac{f(u)}{u}\,du \le M(x)\ln\left(\frac{b}{a}\right).$$

Since f is continuous at $x = 0$, we get $\lim_{x\to 0^+} m(x) = \lim_{x\to 0^+} M(x) = f(0)$. This implies that

$$\lim_{x \to 0^+} \int_{ax}^{bx} \frac{f(u)}{u} du = f(0) \ln \left(\frac{b}{a} \right).$$

Now, we show that the second term of Eq. (5.59) is equal to 0. Let $\epsilon > 0$ be arbitrary. Since $f(u) \to 0$ as $u \to \infty$, it follows that there is $N > 0$ such that $|f(u)| < \epsilon$, whenever $u > N$. If $y > \dfrac{N}{a}$, then we get

$$\left| \int_{ay}^{by} \frac{f(u)}{u} du \right| < \epsilon \ln \left(\frac{b}{a} \right).$$

This implies that

$$\lim_{y \to \infty} \int_{ay}^{by} \frac{f(u)}{u} du = 0.$$

5.3 Exercises

Easier Exercises

Use integration by substitution to evaluate:

1. $\displaystyle \int x\sqrt{1+4x}\,dx,$

2. $\displaystyle \int \frac{x^5}{\sqrt{1-x^6}}\,dx,$

3. $\displaystyle \int \frac{x}{\sqrt{1+x^2+\sqrt{(1+x^2)^3}}}\,dx,$

4. $\displaystyle \int \frac{e^x}{e^{2x}+1}\,dx,$

5. $\displaystyle \int \frac{\sqrt{\tan^{-1} x}}{x^2+1}\,dx,$

6. $\displaystyle \int \frac{\sinh \sqrt{x} \cosh \sqrt{x}}{\sqrt{x}}\,dx.$

7. By using a substitution, prove that for all positive numbers x and y,

$$\int_x^{xy} \frac{1}{t}\,dt = \int_1^y \frac{1}{t}\,dt.$$

8. If n is any positive integer, show that

$$\int_0^1 (1-x^2)^{n-\frac{1}{2}}\,dx = \int_0^{\frac{\pi}{2}} \cos^{2n} x\,dx.$$

Use integration by parts to evaluate:

9. $\int x^2 \sinh x dx,$

12. $\int \sin(\ln x) dx,$

10. $\int \csc^3 x dx,$

13. $\int (\ln x)^3 dx,$

11. $\int x \sec^2 x dx,$

14. $\int \dfrac{\cot^{-1} \sqrt{x}}{\sqrt{x}} dx.$

15. (a) Drive the following formula, where a and b are any real numbers:

$$\int x^a (\ln x)^b dx = \begin{cases} \dfrac{x^{a+1} (\ln x)^b}{a+1} - \dfrac{b}{a+1} \displaystyle\int x^a (\ln x)^{b-1} dx & \text{if } a \neq -1 \\ \dfrac{(\ln x)^{b+1}}{b+1} + C & \text{if } a = -1 \text{ and } b \neq -1. \end{cases}$$

(b) Use the formula derived in (a) to find $\int x^4 (\ln x)^2 dx.$

Verify the given reduction formula, where n is any positive integer:

16. $\int x^n \cos x dx = x^n \sin x - n \int x^{n-1} \sin x dx,$

17. $\int \tan^n x dx = \dfrac{1}{n-1} \tan^{n-1} x - \int \tan^{n-2} x dx,$

18. $\int (\ln x)^n dx = x(\ln x)^n - n \int (\ln x)^{n-1} dx.$

Evaluate each trigonometric integral:

19. $\int \sin \dfrac{5x}{2} \sin \dfrac{x}{2} dx,$

22. $\int \cot^5 x \csc x dx,$

20. $\int \dfrac{\cos^3 x}{\sin^4 x} dx,$

23. $\int_{\frac{1}{6}}^{\frac{5}{6}} \cot^2 \pi x \csc^2 \pi x dx,$

21. $\int_0^{\frac{\pi}{4}} (\tan x)^{\frac{3}{2}} \sec^4 x dx,$

24. $\int \sec^5 x dx.$

Use trigonometric substitution to evaluate:

25. $\int (x^2 - 1)^{\frac{3}{2}} dx,$

28. $\int \sqrt{2x - x^2} dx,$

26. $\int_1^3 \dfrac{x}{\sqrt{x^2 + 2x + 5}} dx,$

29. $\int \dfrac{\sec^2 x}{(4 - \tan^2 x)^{\frac{3}{2}}} dx,$

27. $\int \dfrac{\ln^3 x}{x\sqrt{\ln^2 x - 4}} dx,$

30. $\int \dfrac{e^x}{(e^{2x} + 8e^x + 7)^{\frac{3}{2}}} dx.$

31. Verify the equality

$$\int \frac{1}{x\sqrt{x^2 - a^2}} dx = \frac{1}{a} \sec^{-1} \frac{|x|}{a} + C, \quad (|x| > a > 0).$$

Obtain the given result by a hyperbolic function substitution:

32. $\displaystyle\int \frac{1}{x^2\sqrt{a^2 + x^2}} dx = -\frac{1}{a^2} \coth\left(\sinh^{-1}\frac{x}{a}\right) + C, \quad a > 0,$

33. $\displaystyle\int \frac{1}{\left(a^2 - x^2\right)^{\frac{3}{2}}} dx = \frac{1}{a^2} \sinh\left(\tanh^{-1}\frac{x}{a}\right) + C, \quad a > 0.$

Use partial fractions to evaluate:

34. $\displaystyle\int \frac{x^4 + 3x^3 + 2x + 1}{x^3 - 2x} dx,$

35. $\displaystyle\int \left(\frac{x-1}{x+3}\right)^2 dx,$

36. $\displaystyle\int \frac{x^2 + 3}{(x^2 - x - 2)(x - 1)^2} dx,$

37. $\displaystyle\int \frac{x+4}{x^3 - x^2 - 2x} dx,$

38. $\displaystyle\int \frac{1}{x^2(x^2 + 3)^2} dx,$

39. $\displaystyle\int \frac{x^7}{(x^4 - 1)^2} dx.$

40. Show that

$$\int \frac{ax^2 + bx + c}{x^3(x - 1)^2} dx$$

is a fraction if and only if $a + 2b + 3c = 0$.

Evaluate each integral of fraction of trigonometric functions:

41. $\displaystyle\int \frac{1}{\sin x + \cos x + 2} dx,$

42. $\displaystyle\int \frac{1}{a + b\tan x} dx,$

43. $\displaystyle\int \frac{1}{a^2 \sin^2 x + b^2 \cos^2 x} dx,$

44. $\displaystyle\int \frac{\cos^2 x \sin x}{\sin x + \cos x} dx,$

45. $\displaystyle\int \frac{\cos^3 x}{2 \sin x - 1} dx,$

46. $\displaystyle\int \frac{1}{(3 + 4\cos 2x)\cot x} dx.$

47. Compute the following definite integral:

$$\int_{-\pi}^{\pi} |[\sin x + \cos x]| dx.$$

48. Verify that

$$\int_{1}^{\infty} \left(\frac{1}{x} + \frac{1}{x^2} - \frac{4}{4x - 1}\right) dx \neq \int_{1}^{\infty} \frac{1}{x} dx + \int_{1}^{\infty} \frac{1}{x^2} dx - \int_{1}^{\infty} \frac{4}{4x - 1} dx.$$

Determine the convergence/divergence of improper integral:

49. $\displaystyle\int_0^1 \frac{\sqrt{x}}{e^{\sin x - 1}}dx,$

50. $\displaystyle\int_1^\infty \frac{1 - 5\sin 2x}{x^2 + \sqrt{x}}dx,$

51. $\displaystyle\int_0^1 \cos(\frac{1}{x^2})dx,$

52. $\displaystyle\int_0^{\frac{\pi}{4}} \frac{1}{x - \sin x}dx,$

53. $\displaystyle\int_1^\infty \frac{1 + e^{-x}}{x}dx,$

54. $\displaystyle\int_1^\infty \frac{\sin^2 x}{\sqrt{x}}dx,$

55. $\displaystyle\int_0^1 \frac{\cos x}{x^2}dx,$

56. $\displaystyle\int_1^\infty \frac{1}{x \ln x}dx,$

57. $\displaystyle\int_0^1 \frac{1}{\sqrt{1 - x^2}}dx,$

58. $\displaystyle\int_{-\infty}^\infty \frac{1}{x^2(1 + e^x)}dx,$

59. $\displaystyle\int_0^\infty (x^{27} + \sin x)e^{-x}dx,$

60. $\displaystyle\int_1^\infty \frac{\ln x + \sin x}{\sqrt{x}}dx,$

61. $\displaystyle\int_{-\infty}^\infty \frac{(x^2 + 3)^{\frac{3}{2}}}{(x^4 + 1)^{\frac{3}{2}}} \sin^2 xdx,$

62. $\displaystyle\int_0^\infty \frac{1 + \cos^2 x}{\sqrt{1 + x^2}}dx.$

63. Determine all values of p for which $\displaystyle\int_0^\infty \frac{1 - e^x}{x^p}dx$ is convergent.

64. Determine all values of p for which $\displaystyle\int_0^\infty \frac{x^{2p}}{e^x}dx$ is convergent.

Find conditions on p and q such that the integral converges:

65. $\displaystyle\int_{-1}^1 (1 - x)^p (1 + x)^q dx,$

66. $\displaystyle\int_0^\infty \frac{x^p}{(1 + x^2)^q}dx,$

67. $\displaystyle\int_1^\infty \frac{\left(\ln(1 + x) - \ln x\right)^q}{x^p}dx,$

68. $\displaystyle\int_0^\infty \frac{(x - \sin x)^q}{x^p}dx.$

69. Suppose that f and g are two polynomials such that g has no real zeros. Find necessary and sufficient conditions for convergence of

$$\int_{-\infty}^\infty \frac{f(x)}{g(x)}dx.$$

Harder Exercises

70. Prove directly from the definition of definite integral that

$$\int_a^b x^2 dx = \frac{b^3 - a^3}{3}.$$

71. Show that

$$\lim_{n \to \infty} \sum_{k=1}^n \frac{k}{k^2 + n^2} = \ln \sqrt{2}.$$

Use definite integral to find:

72. $\lim\limits_{n\to\infty} \dfrac{1}{n}\left(\left(\dfrac{1}{n}\right)^2 + \left(\dfrac{2}{n}\right)^2 + \cdots \left(\dfrac{n}{n}\right)^2\right),$

73. $\lim\limits_{n\to\infty} \dfrac{1}{n}\left(e^{\frac{3}{n}} + e^{\frac{6}{n}} + \cdots + e^{\frac{3n}{n}}\right),$

74. $\lim\limits_{n\to\infty} \dfrac{\pi}{2n}\left(1 + \cos\dfrac{\pi}{2n} + \cos\dfrac{2\pi}{2n} + \cdots + \cos\dfrac{(n-1)\pi}{2n}\right).$

Find the value of the given limit:

75. $\lim\limits_{n\to\infty} \displaystyle\int_0^{\frac{\pi}{2}} \sin^n x\, dx,$

76. $\lim\limits_{n\to\infty} \displaystyle\int_0^1 \sqrt[n]{1 - x^n}\, dx.$

77. Prove that if f is a continuous function and u and v are differentiable functions, then

$$\frac{d}{dx}\int_{v(x)}^{u(x)} f(t)dt = f\big(u(x)\big)\frac{d}{dx}u(x) - f\big(v(x)\big)\frac{d}{dx}v(x).$$

Find the derivative of the given function:

78. $f(x) = \displaystyle\int_{-x}^{x} \dfrac{1}{t^2 + x^2}dt$

79. $f(x) = \displaystyle\int_{-\frac{x}{2}}^{x^2} \dfrac{\sin(xt)}{t}dt.$

80. Suppose that $f : \mathbb{R} \to \mathbb{R}$ is defined as follows:

$$fx) = \int_0^x e^{t^2}dt.$$

If $g(x) = f\big(f(x)\big)$, find $g'(0)$.

Find the value of the given limit:

81. $\lim\limits_{x\to\infty} \dfrac{\left(\displaystyle\int_0^x e^{-t^2}dt\right)^2}{\displaystyle\int_0^x e^{2t^2}dt},$

83. $\lim\limits_{x\to1} \dfrac{\displaystyle\int_0^{(x-1)^4} \sin t^3 dt}{(x-1)^{16}},$

82. $\lim\limits_{x\to0^+} \dfrac{\displaystyle\int_0^{x^2} \sin\sqrt{t}\, dt}{x^3},$

84. $\lim\limits_{x\to0} \dfrac{\displaystyle\int_0^{\tan x} (1 + \sin^2 t)dt}{\tan x}.$

85. A function $f : \mathbb{R} \to \mathbb{R}$ is defined by

$$f(x) = \begin{cases} \dfrac{1}{x}\displaystyle\int_0^x \sin t^2 dt & \text{if } x \neq 0 \\ 0 & \text{if } x = 0. \end{cases}$$

In which points, f is differentiable and determine its derivative.

86. Let f be a continuous function on $[0, 1]$. Show that

$$\int_0^1 x^2 f(x)dx = \frac{1}{3} f(c)$$

for some $c \in [0, 1]$.

87. Suppose that $f : [0, 1] \to \mathbb{R}$ is a continuous function and $\int_0^1 f(x)dx = 1$. Show that

   (a) there exists $c \in (0, 1)$ such that $f(c) = 1$;
   (b) there exist two distinct numbers $c_1, c_2 \in (0, 1)$ such that $f(c_1) + f(c_2) = 2$.

88. Let $f, g : [a, b] \to \mathbb{R}$ be two integrable functions. Suppose that f is increasing and g is non-negative on $[a, b]$. Show that there exists $c \in [a, b]$ such that

$$\int_a^b f(x)g(x)dx = f(b) \int_a^c g(x)dx + f(a) \int_c^b g(x)dx.$$

89. Suppose that $f : [1, 5] \to \mathbb{R}$ is a continuous function. Show that there exists $c \in [1, e]$ such that $\int_1^e f(x)dx = cf(c)$.

90. Let $f : [0, 1] \to \mathbb{R}$ be a continuous function such that $\int_0^1 f(x)dx = 1$. Show that there exists $c \in (0, 1)$ such that $f(c) = 3c^2$.

91. Let f be a continuous function on $\mathbb{R}$ and $a \neq 0$. If

$$g(x) = \frac{1}{a} \int_0^x f(t) \sin a(x - t)dt,$$

show that $f(x) = g''(x) + a^2 g(x)$.

92. Let n be a positive integer and a be a real number such that $\sqrt{n\pi} \leq a \leq \sqrt{(n + 1)\pi}$. Show that

$$\int_{\sqrt{n\pi}}^{\sqrt{(n+1)\pi}} \sin x^2 dx = \frac{(-1)^n}{a}.$$

Prove that

93. $\dfrac{\sqrt{3}}{8} \leq \displaystyle\int_{\frac{\pi}{4}}^{\frac{\pi}{3}} \dfrac{\sin x}{x} dx \leq \dfrac{\sqrt{2}}{6}$,

94. $\dfrac{2\pi^2}{9} \leq \displaystyle\int_{\frac{\pi}{6}}^{\frac{\pi}{2}} \dfrac{2x}{\sin x} dx \leq \dfrac{4\pi^2}{9}$,

95. $\dfrac{1}{3\sqrt{2}} \leq \displaystyle\int_0^1 \dfrac{x^2}{\sqrt{1+x}} dx \leq \dfrac{1}{3}$.

96. Show that

$$\int_n^{n+1} \frac{1}{x}\,dx < \frac{1}{n},$$

for all $n \in \mathbb{N}$.

97. Suppose that $f : [0, 1] \to \mathbb{R}$ is defined by

$$f(x) = \begin{cases} \dfrac{1}{n} & \text{if } \dfrac{1}{n+1} < x \le \dfrac{1}{n}, \text{ where } n \in \mathbb{N} \\ 0 & \text{if } x = 0. \end{cases}$$

Show that

$$\int_0^1 f(x)\,dx = \frac{\pi^2}{6} - 1.$$

98. Suppose that a real function f is defined by

$$f(x) = \begin{cases} \dfrac{1}{2^n} & \text{if } \dfrac{1}{2^{n+1}} < x \le \dfrac{1}{2^n}, \text{ where } n \ge 0 \\ 0 & \text{if } x = 0. \end{cases}$$

Show that

$$\int_0^1 f(x)\,dx = \frac{2}{3}.$$

99. Let $f^{(n+1)}$ be integrable on $[a, b]$. Show that

$$f(b) = \sum_{k=0}^n \frac{f^{(k)}(a)}{k!}(b-a)^k + \frac{1}{n!}\int_a^b f^{(n+1)}(x)(b-x)^n\,dx.$$

Hint: Integrate by parts and use induction.

100. Determine whether the improper integral

$$\int_0^\infty \cos(x^2)\,dx$$

is convergent or divergent.

101. Determine whether the improper integrals

$$\int_1^\infty \frac{\cos x \sin x}{x}\,dx \text{ and } \int_1^\infty \left| \frac{\cos x \sin x}{x} \right| dx$$

are convergent or divergent..

102. Suppose that a and b are real numbers such that $0 < a < b$. Find the exact value of the following integral:

$$\int_0^\infty \frac{x^{a-1}}{1+x^b} dx.$$

103. Discuss the convergence/divergence of the improper integral $\int_1^\infty \frac{x^3 \sin x}{e^x \ln x} dx.$

104. It is proved that

$$\int_0^\infty \frac{\sin x}{x} dx = \frac{\pi}{2}. \tag{5.60}$$

By using (5.60), prove that

(a) $\displaystyle\int_0^\infty \frac{\sin x \cos x}{x} dx = \frac{\pi}{4};$

(b) $\displaystyle\int_0^\infty \frac{\sin^2 x}{x^2} dx = \frac{\pi}{2};$

(c) $\displaystyle\int_0^\infty \frac{\sin^4 x}{x^2} dx = \frac{\pi}{4}.$

105. Let f be a continuous function and $\displaystyle\int_a^x f(t)dt$ be bounded on $[a, b]$. Suppose that g is a positive integrable function on $[a, b)$, g' is non-negative and
$\displaystyle\lim_{x \to b^-} g(x) = \infty$. Show that

$$\lim_{x \to b^-} \frac{1}{\left(g(x)\right)^p} \int_a^x f(t)g(t)dt = 0, \quad p > 1.$$

Hint: Integrate by parts.

Chapter 6
Applications of Integrations

6.1 Basic Concepts and Theorems

Area of a Region in a Plane

Area under a curve: If $y = f(x)$ is a non-negative continuous function on $[a, b]$, then the area under the curve $y = f(x)$ from $x = a$ to $x = b$ is equal to

$$A = \int_a^b f(x)dx.$$

To compute the area of the region bounded by the graph of a function $y = f(x)$ and the x-axis when the function takes on both positive and negative values, we must be careful to break up the interval $[a, b]$ into subintervals on which the function does not change sign. So, in order to find the area between the graph of $y = f(x)$ and the x-axis on $[a, b]$:

(1) Subdivide $[a, b]$ at the zeros of f;
(2) Integrate f over each subinterval;
(3) Add the absolute values of the integrals.

    Area between two curves: Let f and g be two continuous functions on $[a, b]$ such that $f(x) \geq g(x) \geq 0$, for all $x \in [a, b]$. Then, the area between $y = f(x)$ and $y = g(x)$ bounded by the vertical lines $x = a$ and $x = b$ is equal to

$$A = \int_a^b \big(f(x) - g(x)\big)dx.$$

© The Author(s), under exclusive license to Springer Nature Singapore Pte Ltd. 2020
B. Davvaz, *Examples and Problems in Advanced Calculus: Real-Valued Functions*,
https://doi.org/10.1007/978-981-15-9569-1_6

Length of Arc of a Plane Curve

Let C be a plane curve with *parametric equations* $x = x(t)$ and $y = y(t)$, where $a \le t \le b$. Suppose that the functions $x(t)$ and $y(t)$ are *continuously differentiable* on $[a, b]$, which means that the derivatives $x'(t)$ and $y'(t)$ are continuous on $[a, b]$. Then, the length of C is equal to

$$L = \int_a^b \sqrt{\left(x'(t)\right)^2 + \left(y'(t)\right)^2}\, dt.$$

Now, suppose that C is a graph of a continuously differentiable function $y = f(x)$, where $a \le x \le b$. Then, C has the parametric representation $x = t$ and $y = f(t)$, where $a \le x \le b$. So, the length of arc of the curve $y = f(x)$ from the point $(a, f(a))$ to the point $(b, f(b))$ takes the form

$$L = \int_a^b \sqrt{\left(1 + (f'(x))^2\right)}\, dx.$$

Analogously, if C is the graph of a continuously differentiable function $x = g(y)$, where $c \le y \le d$, then the length of arc of the curve $x = g(y)$ from the point $(g(c), c)$ to the point $(g(d), d)$ is given by

$$L = \int_c^d \sqrt{\left(g'(y)\right)^2 + 1}\, dy.$$

Volume Using Cross Sections

A *cross section* of a solid S is the plane region formed by intersecting S with a plane.

The volume of a solid of integrable cross-sectional area $A(x)$ from $x = a$ to $x = b$ is

$$V = \int_a^b A(x)\, dx. \tag{6.1}$$

This formula applies whenever $A(x)$ is integrable, and in particular when it is continuous. In order to apply (6.1) to calculate the volume of a solid, take the following steps:

(1) Sketch the solid and a typical cross section;
(2) Find a formula for $A(x)$, the area of a typical cross section;
(3) Find the limits of integration;
(4) Integrate $A(x)$ to find the volume.

Volume of a Solid of Revolution: Circular Disk and Circular Ring Method

Circular disk method: Let f be a non-negative continuous function on $[a, b]$. Suppose that S is the solid of revolution obtained by revolving about the x-axis the region R bounded by the curve $y = f(x)$, the x-axis, and the lines $x = a$ and $x = b$. If V is the number of cubic units in the volume of S, then

$$V = \pi \int_a^b \left(f'(x)\right)^2 dx.$$

Sometimes the axis of revolution is chosen to be y-axis, rather than the x-axis. Thus, assume that the region R bounded by the graph of a non-negative continuous function $x = g(y)$, where $c \leq y \leq d$, the y-axis and the horizontal lines $y = c$ and $y = d$, is revolved about the y-axis, generating a solid of revolution. Then, the volume of the solid is equal to

$$V = \pi \int_c^d \left(g'(y)\right)^2 dy.$$

   Circular ring method: Let f and g be two non-negative continuous functions on $[a, b]$ such that $f(x) \geq g(x)$, for all $x \in [a, b]$. If V cubic units are the volume of the solid of the revolution generated by revolving about the x-axis the region R bounded by the curve $y = f(x)$ and $y = g(x)$ and the lines $x = a$ and $x = b$, then

$$V = \pi \int_a^b \left(\left(f'(x)\right)^2 - \left(g'(x)\right)^2\right) dx.$$

Volume of a Solid of Revolution: Shell Method

Shell method: Let f be a non-negative continuous function on $[a, b]$, where $a \geq 0$, and let R be the region bounded by the curve $y = f(x)$, the x-axis, and the lines $x = a$ and $x = b$. If S is the solid of revolution obtained by revolving R about the y-axis and V cubic units are the volume of S, then

$$V = 2\pi \int_a^b x f(x) dx.$$

More generally, if R is the region bounded by two curves $y = f(x)$ and $y = g(x)$ with $f(x) \geq g(x) \geq 0$, and two vertical lines $x = a$ and $x = b$ ($a < b$, the volume of the solid of revolution S generated by revolving R about the y-axis is given by

$$V = 2\pi \int_a^b x\left(f(x) - g(x)\right) dx.$$

Area of a Surface of Revolution

Let $x = x(t)$ and $y = y(t)$ ($a \leq t \leq b$) be the parametric equations of a curve C in the xy plane, where the functions $x(t)$ and $y(t)$ are continuously differentiable and $y(t)$ is non-negative. Suppose that we revolve C about the x-axis. This generates a surface, called a surface of revolution, with the x-axis as its axis of the revolution. Then, the area of the surface is given by the formula

$$S = 2\pi \int_a^b y(t)\sqrt{\left(x'(t)\right)^2 + \left(y'(t)\right)^2}\,dt.$$

Similarly, if $x(t) \geq 0$ and the curve C revolved about the y-axis instead of the x-axis, the area of the surface is given by the formula

$$S = 2\pi \int_a^b x(t)\sqrt{\left(x'(t)\right)^2 + \left(y'(t)\right)^2}\,dt.$$

If $f(x) \geq 0$ is continuously differentiable on $[a, b]$, the area of the surface generated by revolving the graph of $y = f(x)$ about the x-axis is

$$S = 2\pi \int_a^b f(x)\sqrt{1 + \left(f'(x)\right)^2}\,dx$$

If $x = g(y) \geq 0$ is continuously differentiable on $[c, d]$, the area of the surface generated by revolving the graph of $x = g(y)$ about the y-axis is

$$S = 2\pi \int_c^d g(y)\sqrt{1 + \left(g'(y)\right)^2}\,dy$$

Work and Kinetic Energy

Work is defined as the amount of energy required to perform a physical task. When force is constant, work can simply be calculated using the equation $w = f \cdot d$, where w is work, f is a constant force and d is the distance through which the force acts. Frequently, the force is not constant and will change over time.

With the help of integration, we can use the physical idea of Sir Isaac Newton to make a detailed study of motion along a straight line (rectilinear motion). Let $x = x(t)$ be the position at time t of a particle of mass m moving along a line L. The *particle's velocity* $v = v(t)$ and *acceleration* $a = a(t)$ are given by the derivative

$$v = \frac{dx}{dt} \quad \text{and} \quad a = \frac{dv}{dt} = \frac{d^2x}{dt^2}.$$

Suppose that the particle is subject to a force f, acting along the x-axis from a to b. Then, *Newton's second law* of motion tells us that

$$m\frac{d^2x}{dt^2} = f. \tag{6.2}$$

Thus once f is known, the particle's position as a function of time can be determined by solving (6.2), subject to appropriate initial conditions. According to Newton's second law and the chain rule, we have

$$m\frac{dv}{dt} = m\frac{dv}{dx}\frac{dx}{dt} = mv\frac{dv}{dx} = f. \tag{6.3}$$

Let $v_a = v(a)$ and $v_b = v(b)$ be the particle's velocity at two different positions a and b. Then, by integrating from (6.3) we obtain

$$\int_a^b f(x)dx = \frac{1}{2}mv_b^2 - \frac{1}{2}mv_a^2.$$

In physics, the expression $\frac{1}{2}mv^2$ is called *kinetic energy* of the moving particle. Therefore, the work done by the force equals the change particle's kinetic energy, and we can find the work by calculating this change.

  Definition of work: Let f be a continuous function on $[a, b]$ and $f(x)$ be the number of units in the force acting on an object at the point x on the x-axis. Then, if w units are the *work* done by the force as the object moves from a to b, then W is given by

$$w = \int_a^b f(x)dx.$$

  If the reader is interested to see the proof of theorems that are presented in this chapter, we refer him/her to [1–16].

6.2 Problems

389. *Find the area of the region between the curves* $y = x^3$ *and* $y = x$.

Solution. We draw the curves (see Fig. 6.1). We determine the intersections of the two curves. If $x^3 = x$, then $x = -1,\ 0,\ 1$. Hence, the coordinate of the point P is $(1, 1)$.

  Thus, the desired area is equal to

$$2\int_0^1 (x - x^3)dx = 2\left(\frac{1}{2}x^2 - \frac{1}{4}x^4\right)\Big|_0^1 = \frac{1}{2}.$$

390. *Find the area of the region bounded by the curve* $y = (x + 3)(x - 1)(\frac{1}{2}x - 2)$ *and the* x-axis.

Fig. 6.1 Region between the curves $y = x^3$ and $y = x$

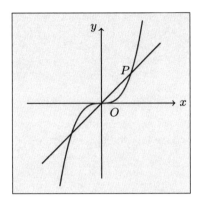

Fig. 6.2 Region bounded by the curve $y = (x + 3)(x - 1)(\frac{1}{2}x - 2)$ and the x-axis

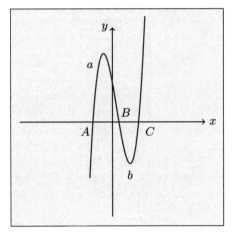

Solution. Solving the equation $(x + 3)(x - 1)(\frac{1}{2}x - 2) = 0$ we conclude that the coordinates of the points A, B and C are $(-3, 0)$, $(1, 0)$ and $(4, 0)$, respectively (see Fig. 6.2).

The area of the region $AaBA$ is equal to

$$\int_{-3}^{1} (x + 3)(x - 1)(\frac{1}{2}x - 2)dx = \int_{-3}^{1} \left(\frac{1}{2}x^3 - x^2 - \frac{11}{2}x + 6\right)dx$$
$$= \left(\frac{1}{8}x^4 - \frac{1}{3}x^3 - \frac{11}{4}x^2 + 6x\right)\Big|_{-3}^{1}$$
$$= \frac{640}{24}.$$

Similarly, the area of the region $BbCB$ is equal to

$$-\int_{1}^{4} (x + 3)(x - 1)(\frac{1}{2}x - 2)dx = \frac{297}{24}.$$

Fig. 6.3 Region bounded by $y = x^2 - 6x + 8$, the x-axis, and the vertical lines $x = 0$ and $x = 5$

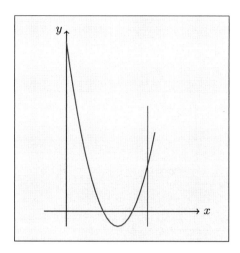

The desired area is equal $\dfrac{640}{24} + \dfrac{297}{24} = \dfrac{937}{24}$.

391. *Find the area of the region bounded by $y = x^2 - 6x + 8$, the x-axis, and the vertical lines $x = 0$ and $x = 5$.*

Solution. First we draw a sketch of the area to be determined (see Fig. 6.3). Next, we determine where the graph crosses the x-axis. If $x^2 - 6x + 8 = 0$, then $x = 2$ and $x = 4$. We must integrate on three intervals $[0, 2]$, $[2, 4]$ and $[4, 5]$. Thus, we get

$$A_1 = \int_0^2 (x^2 - 6x + 8)dx = \left(\frac{1}{3}x^3 - 3x^2 + 8x\right)\Big|_0^2 = \frac{20}{3},$$

$$A_2 = \int_2^4 (x^2 - 6x + 8)dx = -\frac{4}{3},$$

$$A_3 = \int_4^5 (x^2 - 6x + 8)dx = \frac{4}{3}.$$

In order to obtain the desired area, we add the absolute values of A_1, A_2 and A_3. Therefore, we have

$$S = |A_1| + |A_2| + |A_3| = \frac{20}{3} + \frac{4}{3} + \frac{4}{3} = \frac{28}{3}.$$

392. *Find the area of the region enclosed by $y = \cos x$, $y = \sin 2x$, $x = 0$ and $x = \dfrac{\pi}{2}$.*

Solution. First, we find the intersections of the curves $y = \cos x$ and $y = \sin 2x$. If $\cos x = \sin 2x$, then $\cos x = 0$ or $\sin x = \dfrac{1}{2}$. Hence, the coordinates of the points A

Fig. 6.4 Region enclosed by
$y = \cos x$, $y = \sin 2x$, $x = 0$
and $x = \dfrac{\pi}{2}$

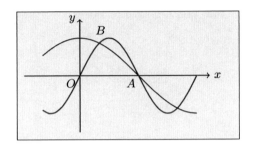

Fig. 6.5 Region between the
curves $y = xe^{1-x}$ and
$y = 4x^2 - 3x$

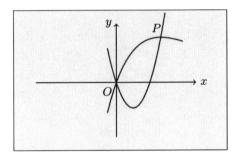

and B are $(\dfrac{\pi}{2}, 0)$ and $(\dfrac{\pi}{6}, \dfrac{\sqrt{3}}{2})$, respectively. See Fig. 6.4. Thus, the desired area is
equal to

$$\int_0^{\frac{\pi}{6}} (\cos x - \sin 2x)dx + \int_{\frac{\pi}{6}}^{\frac{\pi}{2}} (\sin 2x - \cos x)dx$$

$$= \left(\sin x + \frac{1}{2} \cos 2x \right)\Big|_0^{\frac{\pi}{6}} + \left(-\frac{1}{2} \cos 2x - \sin x \right)\Big|_{\frac{\pi}{6}}^{\frac{\pi}{2}} = \frac{1}{2}.$$

393. *Find the area of the region between the curves $y = xe^{1-x}$ and $y = 4x^2 - 3x$.*

Solution. First, we determine the intersections of the curves $y = xe^{1-x}$ and $y = 4x^2 - 3x$. If $xe^{1-x} = 4x^2 - 3x$, then $x = 0$ or $x = 1$. We sketch the curves (see Fig. 6.5). The coordinate of P is $(1, 1)$.

Thus, the required area is equal to

$$\int_0^1 (xe^{1-x} - 4x^2 + 3x)dx = \left(-xe^{1-x} - e^{1-x} - \frac{4}{3}x^3 + \frac{3}{2}x^2 \right)\Big|_0^1 = -\frac{11}{6} + e.$$

394. *Find the area of the region inside the loop of the curve $y^2 = x(x - 1)^2$.*

Solution. If $x(x - 1)^2 = 0$, then $x = 0$ or $x = 1$. We draw the curve (see Fig. 6.6). The given curve is symmetrical with respect to the x-axis. Moreover, the curve is divided to two functions: $f(x) = -\sqrt{x}(x - 1)$ and $g(x) = \sqrt{x}(x - 1)$. Note that if $x \in [0, 1]$, then $f(x) \geq 0$ and $g(x) \leq 0$.

Fig. 6.6 Region inside the
loop of the curve
$y^2 = x(x - 1)^2$

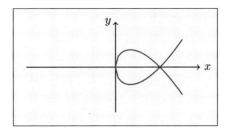

Fig. 6.7 The two parabolas
$y^2 = 4x$ and $x^2 = 4y$ divide
the square bounded by
$x = 0, x = 4, y = 0$ and
$y = 4$ into three equal areas

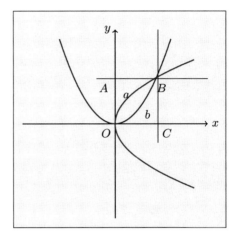

Now, the desired area is equal to

$$2 \int_0^1 -\sqrt{x}(x - 1)dx = 2\left(\frac{2}{3}x^{\frac{3}{2}} - \frac{2}{5}x^{\frac{5}{2}}\right)\Big|_0^1 = \frac{8}{15}.$$

395. *Prove that the two parabolas* $y^2 = 4x$ *and* $x^2 = 4y$ *divide the square bounded by* $x = 0, x = 4, y = 0$ *and* $y = 4$ *into three equal areas.*

Solution. Solving $y^2 = 4x$ and $x^2 = 4y$ we see that the coordinate of B is $(4, 4)$. Let S, S' and S'' be the areas of regions $OCBbO$, $OCBaO$ and $OABaO$, respectively (see Fig. 6.7).
    We have

$$S = \int_0^4 \frac{x^2}{4}dx = \frac{16}{3}$$

and

$$S' = \int_0^4 2\sqrt{x}dx = \frac{4}{3}\left(x^{\frac{3}{2}}\right)\Big|_0^4 = \frac{32}{3}.$$

Fig. 6.8 The circle $x^2 + y^2 = a^2$ is divided into three parts by hyperbola $x^2 - 2y^2 = \frac{a^2}{4}$

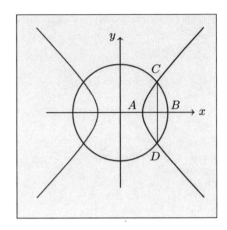

So, the area of the region $OaBbO$ is equal to $S' - S = \dfrac{16}{3}$. On the other hand, the area of the square $OABC$ is equal to 16. Consequently, $S'' = 16 - \dfrac{32}{3} = \dfrac{16}{3}$.

396. *The circle $x^2 + y^2 = a^2$ is divided into three parts by hyperbola $x^2 - 2y^2 = \dfrac{a^2}{4}$. Find the area of these three parts.*

Solution. With a simple calculation we see that the coordinate of B is $(a, 0)$, the coordinate of A is $(\dfrac{a}{2}, 0)$ and the coordinate of C is $(\dfrac{\sqrt{3}a}{2}, \dfrac{a}{2})$; see Fig. 6.8.

Now, let S be the area of region $BCADB$. Then, we have

$$S = 2\left(\int_{\frac{a}{2}}^{\frac{\sqrt{3}a}{2}} \frac{1}{2}\sqrt{x^2 - \frac{a^2}{4}}\,dx + \int_{\frac{\sqrt{3}a}{2}}^{a} \sqrt{a^2 - x^2}\,dx \right)$$

$$= \left(\frac{x}{2}\sqrt{x^2 - \frac{a^2}{4}} + \frac{a^2}{8}\ln\left|x + \sqrt{x^2 - \frac{a^2}{4}}\right| \right)\Big|_{\frac{a}{2}}^{\frac{\sqrt{3}a}{2}}$$

$$+ 2\left(\frac{x}{2}\sqrt{a^2 - x^2} + \frac{a^2}{2}\sin^{-1}\frac{x}{a} \right)\Big|_{\frac{\sqrt{3}a}{2}}^{a}$$

$$= \frac{a^2}{4\sqrt{2}}\ln(\sqrt{3} - \sqrt{2}) + \frac{\pi a^2}{6}.$$

Note that the area of the right part is equal to the area of the left part. Therefore, the area of the middle part is equal to

$$\pi a^2 - 2\left(\frac{a^2}{4\sqrt{2}}\ln(\sqrt{3} - \sqrt{2}) + \frac{\pi a^2}{6} \right) = \frac{2\pi a^2}{3} - \frac{a^2}{2\sqrt{2}}\ln(\sqrt{3} - \sqrt{2}).$$

Fig. 6.9 Region above
x-axis and included between
the circle $x^2 + y^2 = 8x$ and
inside of the parabola
$y^2 = 4x$

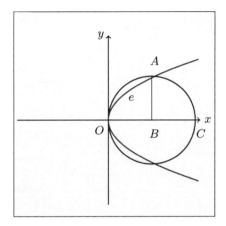

397. *Find the area of the region above x-axis and included between the circle*
$x^2 + y^2 = 8x$ *and inside of the parabola* $y^2 = 4x$.

Solution. The equation of the circle $x^2 + y^2 = 8x$ can be expressed as $(x - 4)^2 +$
$y^2 = 16$. Thus, the center of the circle is $(4, 0)$ and radius is 4. Its intersections with
parabola $y^2 = 4x$ are the points $O(0, 0)$ and $A(4, 4)$; see Fig. 6.9. Let S and S' be
the areas of the regions $OBAeO$ and $BACB$, respectively. We have

$$S = \int_0^4 2\sqrt{x}\,dx = \frac{32}{3}$$

and

$$S' = \int_4^8 \sqrt{16 - (x - 4)^2}\,dx$$
$$= \int_0^4 \sqrt{16 - u^2}\,du$$
$$= \left(\frac{u}{2}\sqrt{16 - u^2} + 8\sin^{-1}\frac{u}{4}\right)\Big|_0^4$$
$$= \frac{4}{3}(8 + 3\pi).$$

Therefore, the required area of the region $OeACBO$ is equal to $S + S' = \frac{4}{3}(16 + 3\pi)$.

398. *Show that the area common to the two ellipses* $\dfrac{x^2}{a^2} + \dfrac{y^2}{b^2} = 1$ *and* $\dfrac{x^2}{b^2} + \dfrac{y^2}{a^2} = 1$
$(a > b)$ *is* $2ab\tan^{-1}\left(\dfrac{2ab}{a^2 + b^2}\right)$.

Fig. 6.10 Area common to
the two ellipses
$$\frac{x^2}{a^2} + \frac{y^2}{b^2} = 1 \text{ and}$$
$$\frac{x^2}{b^2} + \frac{y^2}{a^2} = 1$$

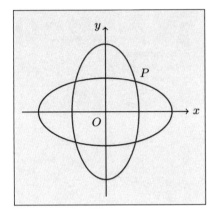

Solution. Solving the given two ellipses, the coordinate of the point P is
$\left(\dfrac{ab}{\sqrt{a^2 + b^2}}, \dfrac{ab}{\sqrt{a^2 + b^2}} \right)$ (see Fig. 6.10).
So, the required area is equal to

$$4\left(\int_0^{\frac{ab}{\sqrt{a^2+b^2}}} \frac{b}{a}\sqrt{a^2 - x^2}dx + \int_{\frac{ab}{\sqrt{a^2+b^2}}}^b \frac{a}{b}\sqrt{b^2 - x^2}dx \right)$$

$$= 4\left(\frac{b}{a}\left(\frac{x\sqrt{a^2-x^2}}{2} + \frac{a^2}{2}\sin^{-1}\frac{x}{a} \right)\Big|_0^{\frac{ab}{\sqrt{a^2+b^2}}} + \frac{a}{b}\left(\frac{x\sqrt{b^2-x^2}}{2} + \frac{b^2}{2}\sin^{-1}\frac{x}{b} \right)\Big|_{\frac{ab}{\sqrt{a^2+b^2}}}^b \right)$$

$$= 4\left(\frac{b}{a}\left(\frac{a^3 b}{2(a^2+b^2)} + \frac{a^2}{2}\sin^{-1}\frac{b}{\sqrt{a^2+b^2}} \right) \right.$$

$$\left. + \frac{a}{b}\left(\frac{b^2}{2}\sin^{-1} 1 - \frac{ab^3}{2(a^2+b^2)} - \frac{b^2}{2}\sin^{-1}\frac{a}{\sqrt{a^2+b^2}} \right) \right)$$

$$= 2\left(\frac{a^2 b^2}{a^2+b^2} + ab\sin^{-1}\frac{b}{\sqrt{a^2+b^2}} + ab\left(\frac{\pi}{2} - \sin^{-1}\frac{a}{\sqrt{a^2+b^2}} \right) - \frac{a^2 b^2}{a^2+b^2} \right)$$

$$= 2ab\left(\sin^{-1}\frac{b}{\sqrt{a^2+b^2}} + \cos^{-1}\frac{a}{\sqrt{a^2+b^2}} \right)$$

$$= 2ab\left(\tan^{-1}\frac{b}{a} + \tan^{-1}\frac{b}{a} \right)$$

$$= 2ab\tan^{-1}\left(\frac{2\frac{b}{a}}{1 - \frac{b^2}{a^2}} \right)$$

$$= 2ab\tan^{-1}\left(\frac{2ab}{a^2+b^2} \right).$$

Fig. 6.11 A circle with
radius 1 touches the curve
$y = 2|x|$ twice

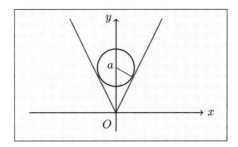

399. *A circle with radius* 1 *touches the curve* $y = 2|x|$ *twice. Determine the area of the region that lies between the two curves.*

Solution. The equation of the lower semicircle is $y = a - \sqrt{1 - x^2}$. At the point of tangency, the slopes of the line and semicircle must be equal. Suppose that $x > 0$. Since $\dfrac{x}{\sqrt{1 - x^2}} = 2$, it follows that $x = \dfrac{2}{5}\sqrt{5}$, and hence $y = \dfrac{4}{5}\sqrt{5}$. The slope of the perpendicular line segment is $-\dfrac{1}{2}$. Thus the equation of the line segment is $y - \dfrac{4}{5}\sqrt{5} = -\dfrac{1}{2}(x - \dfrac{2}{5}\sqrt{5})$, or equivalently $y = -\dfrac{1}{2}x + \sqrt{5}$. Now, we conclude $a = \sqrt{5}$. Hence, the equation of the lower semicircle is $y = \sqrt{5} - \sqrt{1 - x^2}$. See Fig. 6.11.

Therefore, the desired area is equal to

$$2 \int_0^{\frac{2}{5}\sqrt{5}} \left((\sqrt{5} - \sqrt{1 - x^2}) - 2x \right) dx$$
$$= 2\left(\sqrt{5}x - \frac{x}{2}\sqrt{1 - x^2} - \frac{1}{2}\sin^{-1} x - x^2 \right)\Big|_0^{\frac{2}{5}\sqrt{5}}$$
$$= 2 - \sin^{-1}\left(\frac{2}{\sqrt{5}} \right).$$

400. *Given that the area enclosed by the unit circle, described by the equation* $x^2 + y^2 = 1$, *is* π. *Use Problem (264) to show that the area enclosed by the ellipse described by the equation* $\dfrac{x^2}{a^2} + \dfrac{y^2}{b^2} = 1$ *is* $ab\pi$.

Solution. By assumption, we have

$$\int_{-1}^1 \sqrt{1 - x^2}\, dx = \frac{\pi}{2}.$$

Suppose that S is the area of ellipse. Then, we get

$$S = 2 \int_{-a}^{a} b \sqrt{1 - \frac{x^2}{a^2}} dx = 2b \int_{-a}^{a} \sqrt{1 - \frac{x^2}{a^2}} dx.$$

We use Problem (264). If $\dfrac{x}{a} = u$, then $dx = a\,du$. Hence, we deduce that

$$S = 2b \int_{-1}^{1} a \sqrt{1 - u^2} du = 2ab \int_{-1}^{1} \sqrt{1 - x^2} dx = ab\pi.$$

401. *If*

$$f(x) = \int_{0}^{x} \sqrt{\cos t} \, dt, \tag{6.4}$$

find the length of the arc of the graph of f from the point where $x = 0$ to the point where $x = \dfrac{\pi}{2}$.

Solution. From (6.4) we get $f'(x) = \sqrt{\cos x}$. Therefore, we have

$$
\begin{aligned}
L &= \int_{0}^{\frac{\pi}{2}} \sqrt{1 + \left(f'(x)\right)^2} dx = \int_{0}^{\frac{\pi}{2}} \sqrt{1 + \cos x} dx \\
&= \int_{0}^{\frac{\pi}{2}} \sqrt{2 \cos^2 \frac{x}{2}} dx = \int_{0}^{\frac{\pi}{2}} \sqrt{2} \cos \frac{x}{2} dx \\
&= 2\sqrt{2} \sin \frac{x}{2} \Big|_{0}^{\frac{\pi}{2}} = 2\sqrt{2} \left(\sin \frac{\pi}{4} - \sin 0 \right) = 2.
\end{aligned}
$$

402. *Show that the area between the graph of the function $f(x) = \dfrac{1}{c} \cosh(cx)$, the x-axis and the ordinates of two points of the graph is numerically $\dfrac{1}{c}$ times the length of the arc between these points.*

Solution. Suppose that (a_1, b_1) and (a_2, b_2) be two points on the graph of f. The area between the graph of f, x-axis, $x = a_1$ and $x = a_2$ is

$$
\begin{aligned}
A &= \int_{a_1}^{a_2} \frac{1}{c} \cosh(cx) = \frac{1}{c^2} \sinh(cx) \Big|_{a_1}^{a_2} \\
&= \frac{1}{c^2} \left(\sinh(ca_2) - \sinh(ca_1) \right).
\end{aligned}
$$

On the other hand, the length of the arc between (a_1, b_1) and (a_2, b_2) is

$$
\begin{aligned}
L &= \int_{a_1}^{a_2} \sqrt{1 + \left(f'(x)\right)^2} dx = \int_{a_1}^{a_2} \sqrt{1 + \sinh^2(cx)} dx \\
&= \int_{a_1}^{a_2} \cosh(cx) dx = \frac{1}{c} \sinh(cx) \Big|_{a_1}^{a_2} \\
&= \frac{1}{c} \left(\sinh(ca_2) - \sinh(ca_1) \right).
\end{aligned}
$$

Fig. 6.12 Curve $f(x) = x^{\frac{2}{3}}$ from the point $(-1, 1)$ to $(8, 4)$

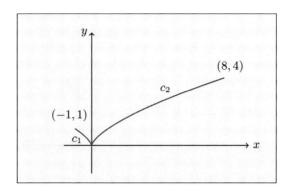

Therefore, $A = \dfrac{1}{c}L$, this completes the proof.

403. *Find the length of the arc of the curve* $f(x) = x^{\frac{2}{3}}$ *from the point* $(-1, 1)$ *to* $(8, 4)$.

Solution. See Fig. 6.12. Note that f is not differentiable at $x = 0$. So, we cannot use the formula directly to find the length of the curve in the specified interval. We compute the length of arc as a sum of the lengths of two arcs. Indeed, we can consider two curves c_1 and c_2, where

$$c_1 : g_1(y) = -y^{\frac{3}{2}}, \text{ for } 0 \le y \le 1,$$
$$c_2 : g_2(y) = y^{\frac{3}{2}}, \text{ for } 0 \le y \le 4.$$

Then, we have $g_1'(y) = -\dfrac{3}{2}\sqrt{y}$ and $g_2'(y) = \dfrac{3}{2}\sqrt{y}$.

Hence, the required length is equal to

$$L = \int_0^1 \sqrt{1 + \left(g_1'(y)\right)^2}\,dy + \int_0^4 \sqrt{1 + \left(g_2'(y)\right)^2}\,dy$$
$$= \int_0^1 \sqrt{1 + \frac{9}{4}y}\,dy + \int_0^4 \sqrt{1 + \frac{9}{4}y}\,dy.$$

To evaluate the above definite integrals, making substitution $u = 1 + \dfrac{9}{4}y$, then $dy = \dfrac{4}{9}du$. Therefore, we obtain

$$L = \int_1^{\frac{13}{4}} \frac{4}{9}\sqrt{u}\,du + \int_1^{10} \frac{4}{9}\sqrt{u}\,du$$
$$= \frac{8}{27}u\sqrt{u}\Big|_1^{\frac{13}{4}} + \frac{8}{27}u\sqrt{u}\Big|_1^{10}$$
$$= \frac{1}{27}\left(13\sqrt{3} + 80\sqrt{10} - 16\right).$$

404. *The integral*

$$E(k) = \int_0^{\frac{\pi}{2}} \sqrt{1 - k^2 \sin^2 t} \, dt \quad (0 < k < 1)$$

defines a non-elementary function of k, known as the complete elliptic integral of the second kind. Show that the ellipse $\dfrac{x^2}{a^2} + \dfrac{y^2}{b^2} = 1$ when $a > b > 0$ is of length

$$L = 4a E\left(\frac{\sqrt{a^2 - b^2}}{a}\right).$$

Solution We can parameterize the points of the ellipse in the first quadrant by $x(t) = a \sin t$ and $y(t) = b \cos t$, where $0 \le t \le \dfrac{\pi}{2}$. Hence, we have

$$L = 4 \int_0^{\frac{\pi}{2}} \sqrt{(x'(t))^2 + (y'(t))^2} \, dt = 4 \int_0^{\frac{\pi}{2}} \sqrt{a^2 \cos^2 t + b^2 \sin^2 t} \, dt. \quad (6.5)$$

We can simplify (6.5) as follows:

$$\begin{aligned} I &= 4 \int_0^{\frac{\pi}{2}} \sqrt{a^2(1 - \sin^2 t) + b^2 \sin^2 t} \, dt \\ &= 4 \int_0^{\frac{\pi}{2}} \sqrt{a^2 + (b^2 - a^2) \sin^2 t} \, dt \\ &= 4a \int_0^{\frac{\pi}{2}} \sqrt{1 + \frac{b^2 - a^2}{a^2} \sin^2 t} \, dt \\ &= 4a E\left(\frac{\sqrt{a^2 - b^2}}{a}\right). \end{aligned}$$

405. *Let f be a differentiable function such that f' is continuous and for every $x \in [0, 1]$, $f(x) \ge 0$. If $f(0) = 0$, $f(1) = 1$ and L is the length of f on $[0, 1]$, prove that*

$$\sqrt{2} \le L < 2.$$

Solution. Since $f(x) \ge 0$, it follows that

$$\frac{\sqrt{2}}{2}(1 + f'(x)) \le \sqrt{1 + (f'(x))^2} \le 1 + f'(x).$$

Hence, we deduce that

$$\frac{\sqrt{2}}{2} \int_0^1 \left(1 + f'(x)\right) dx \leq \int_0^1 \sqrt{1 + \left(f'(x)\right)^2} \, dx \leq \int_0^1 \left(1 + f'(x)\right) dx,$$

and so

$$\frac{\sqrt{2}}{2} \left(x + f(x)\right)\Big|_0^1 \leq L \leq \left(x + f(x)\right)\Big|_0^1.$$

This implies that $\sqrt{2} \leq L \leq 2$. Now, we show that $L \neq 2$. Indeed, if $L = 2$, then $\sqrt{1 + \left(f'(x)\right)^2} = 1 + f'(x)$. Since $f'(x) \geq 0$, it follows that $f'(x) = 0$. Thus, f must be constant, and this is a contradiction with the assumption $f(0) = 0$ and $f(1) = 1$.

406. *A solid of revolution is generated by rotating the graph of a continuous function f around the interval $[0, a]$ on the x-axis. If, for every $a > 0$, the volume is $a^2 + a$, find the function f.*

Solution. The volume of the solid of revolution can be obtained from the following formula

$$V(a) = \pi \int_0^a \left(f(x)\right)^2 dx.$$

Hence, we have

$$\pi \int_0^a \left(f(x)\right)^2 dx = a^2 + a. \tag{6.6}$$

Taking derivative from (6.6) with respect to a we obtain

$$\pi \left(f(a)\right)^2 = 2a + 1.$$

Since a is arbitrary, it follows that $f(x) = \sqrt{\dfrac{2x + 1}{\pi}}$ for every positive number x.

407. *Find the volume of the solid generated by rotating the region bounded by $f(x) = x^2 - 4x + 5$, $x = 1$, $x = 4$ and the x-axis about the x-axis.*

Solution. The points of intersection are $(1, 2)$ and $(4, 5)$. Figure 6.13 shows the region. Also, Fig. 6.14 shows the solid obtained by rotating the region about the x-axis. Then, the volume of the solid of revolution is

$$V = \pi \int_1^4 \left(f(x)\right)^2 dx = \pi \int_1^4 \left(x^2 - 4x + 5\right)^2 dx$$
$$= \pi \int_1^4 \left(x^4 - 8x^3 + 26x^2 - 40x + 25\right) dx$$
$$= \pi \left(\frac{1}{5}x^5 - 2x^4 + \frac{26}{3}x^3 - 20x^2 + 25x\right)\Big|_1^4$$
$$= \frac{78\pi}{5}.$$

Fig. 6.13 Region bounded
by $f(x) = x^2 - 4x + 5$,
$x = 1, x = 4$ and the x-axis

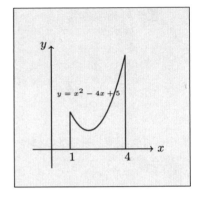

Fig. 6.14 Solid generated by
rotating the region bounded
by $f(x) = x^2 - 4x + 5$,
$x = 1, x = 4$ and the x-axis
about the x-axis

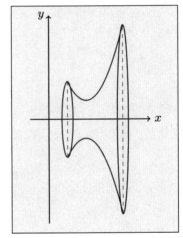

408. *Derive a formula for the volume of a sphere of radius r.*

Solution. We can generate the sphere by revolving the region bounded by the x-axis
and the graph of the function

$$y = \sqrt{r^2 - x^2} \quad (-r \le x \le r)$$

about the x-axis (Fig. 6.15).
    Therefore, we get

$$V = \pi \int_{-r}^{r} y^2 dx = \pi \int_{-r}^{r} (r^2 - x^2) dx$$
$$= \pi \left(r^2 x - \frac{1}{3} x^3 \right) \Big|_{-r}^{r} = \frac{4}{3} \pi r^3.$$

Fig. 6.15 A sphere of radius
r

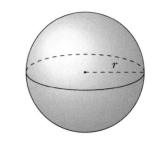

Fig. 6.16 An ellipsoid

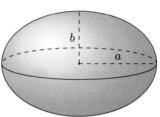

409. *The solid generated by rotating the region inside the ellipse with equation*
$\dfrac{x^2}{a^2} + \dfrac{y^2}{b^2} = 1$ *around the x-axis is called an ellipsoid. Show that the ellipsoid has*
volume $\dfrac{4}{3}\pi ab^2$.

Solution. Rotating the ellipse about the x-axis produces an ellipsoid whose cross
sections are disks with radius $b\sqrt{1 - \dfrac{x^2}{a^2}}$ (Fig. 6.16).
 So, the volume of the ellipsoid is equal to

$$V = \pi \int_{-a}^{a} \left(b\sqrt{1 - \frac{x^2}{a^2}}\right)^2 dx = \pi b^2 \int_{-a}^{a} \left(1 - \frac{x^2}{a^2}\right)dx$$
$$= \pi b^2 \left(x - \frac{1}{3a^2}\right)\Big|_{-a}^{a} = \frac{4}{3}\pi ab^2.$$

410. *A frustum of a cone is the part of the cone that remains after the top of the*
cone is cutoff parallel to the base of the cone. Find the volume of the frustum of a
right circular cone with height h, lower base radius R and top radius r.

Solution. Since the frustum has rotational symmetry, let us set it up so a representative
rectangle is rotated about the x-axis (see Fig. 6.17).
 The equation of the line AB in Fig. 6.18 is

$$y = \frac{r - R}{h}x + R.$$

So, the volume of the solid of revolution, i.e., the volume of the frustum, is

Fig. 6.17 A frustum of a cone

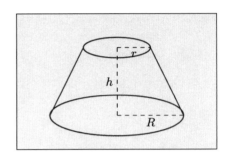

Fig. 6.18 Line AB

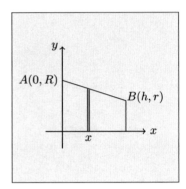

$$V = \int_0^h \pi y^2 dx = \pi \int_0^h \left(\frac{r-R}{h}x + R\right)^2 dx$$

$$= \pi \int_0^h \left(\frac{(r-R)^2}{h^2}x^2 + \frac{2R(r-R)}{h}x + R^2\right)dx$$

$$= \pi \left(\frac{(r-R)^2}{3h^2}x^3 + \frac{R(r-R)}{h}x^2 + R^2 x\right)\Big|_0^h$$

$$= \pi \left(\frac{(r-R)^2}{3}h + R(r-R)h + R^2 h\right)$$

$$= \frac{\pi h}{3}(r^2 + rR + R^2).$$

411. *Find the volume formed when the area between $y = 8 - x^2$ and $y = x^2$ rotating about the y-axis.*

Solution. The points of intersection are $(2, 4)$ and $(-2, 4)$. Figure 6.19 shows the region and the solid obtained by rotating the region about the y-axis.

So, the volume of the solid revolution is

$$V = \pi \int_0^4 y\,dy + \pi \int_4^8 (8 - y)dy$$

$$= \pi\left(\frac{1}{2}y^2\right)\Big|_0^4 + \pi\left(8y - \frac{1}{2}y^2\right)\Big|_4^8$$

$$= 16\pi.$$

Fig. 6.19 Solid formed when the area between $y = 8 - x^2$ and $y = x^2$ rotating about the y-axis

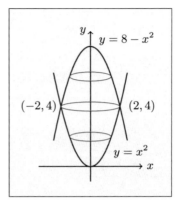

Fig. 6.20 A horn-shaped solid

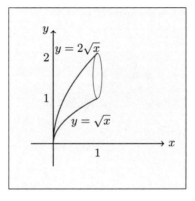

412. *Find the volume of the horn-shaped solid in Fig. 6.20, whose cross sections by planes perpendicular to the x-axis are circular disks. The endpoints of a diameter of each disk lie on the curves $y = \sqrt{x}$ and $y = 2\sqrt{x}$, and the solid extends from its tip at the origin of the xy plane to its cross section at $x = 1$.*

Solution. The area $A(x)$ of the cross section is

$$A(x) = \pi \left(\frac{2\sqrt{x} - \sqrt{x}}{2} \right)^2 = \frac{\pi}{4} x.$$

Therefore, the volume of the horn-shaped solid is

$$V = \int_0^1 A(x)\,dx = \int_0^1 \frac{\pi}{4} x = \frac{\pi}{8}.$$

413. *Find the volume of a pyramid with height h and square base whose sides have length a (see Fig. 6.21).*

Fig. 6.21 A pyramid

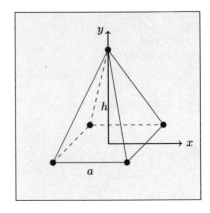

Fig. 6.22 Similar triangles

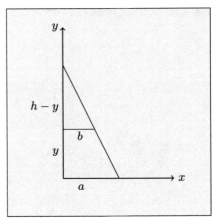

Solution. Each cross section of the pyramid perpendicular to the y-axis is a square. In order to determine the length of the side of the square at y, we consider the triangle in Fig. 6.22, bounded by the y-axis, the x-axis and the line along the side of the pyramid directly above the x-axis. The length of the side of the cross-sectional square at y is $2b$ and the cross-sectional area at y is $A(y) = 4b^2$. On the other hand, by similar triangles we have

$$\frac{h - y}{b} = \frac{h}{a}.$$

This implies that $b = \dfrac{a(h - y)}{h}$.

Consequently, we get

$$A(y) = \frac{4a^2}{h^2}(h - y)^2 = \frac{4a^2}{h^2}(h^2 - 2hy + y^2).$$

Therefore, the volume of the pyramid is equal to

$$V = \int_0^h A(y)dy = \frac{4a^2}{h^2} \int_0^h (h^2 - 2hy + y^2)dy$$
$$= \frac{4a^2}{h^2} \left(h^2 y - hy^2 + \frac{1}{3}y^3 \right)\Big|_0^h = \frac{4}{3}a^2 h.$$

414. *We rotate a region between two curves* $y = f(x)$ *and* $y = g(x)$ *and the lines* $x = a$ *and* $x = b$ *around a line of the form* $y = k$ *to generate a solid. Derive a formula for the volume of the solid.*

Solution. Suppose that $|f(x) - k| \geq |g(x) - k| \geq 0$, for all $x \in [a, b]$. The cross sections of the solid are washers with area

$$A(x) = \pi \left((f(x) - k)^2 - (g(x) - k)^2 \right).$$

Therefore, the volume of the solid is area

$$V = \pi \int_a^b \left((f(x) - k)^2 - (g(x) - k)^2 \right) dx.$$

415. *Find the volume of the solid generated if the region bounded by the curve* $y = \sin^2 x$ *and the x-axis from* $x = 0$ *to* $x = \pi$ *is revolved about the line* $y = 1$.

Solution. We apply the formula of Problem (414). Since $|\sin^2 x - 1| \leq |0 - 1|$, it follows that the volume of the solid is equal to

$$V = \pi \int_0^\pi \left((0 - 1)^2 - (\sin^2 x - 1)^2 \right) dx$$
$$= \pi \int_0^\pi \left(1 - \cos^4 x \right) dx$$
$$= \pi \int_0^\pi \sin^4 x \, dx.$$

According to Problem (374) we have

$$\int_0^\pi \sin^4 x \, dx = \left(-\frac{1}{4} \sin^3 x \cos x - \frac{3}{8} \sin x \cos x + \frac{3}{8}x \right)\Big|_0^\pi = \frac{3}{8}\pi.$$

Therefore, we obtain $V = \frac{3}{8}\pi^2$.

416. *Find the volume of the solid generated by revolving the line* $x = -4$ *the region bounded by the parabolas* $x = y - y^2$ *and* $x = y^2 - 3$.

Solution. The curves of the parabolas intersect at the points $(2, -1)$ and $(-\frac{3}{4}, \frac{3}{2})$. Therefore, the volume of the solid is equal to

$$V = \pi \int_{-1}^{\frac{3}{2}} \left(\left(4 + y - y^2\right)^2 - \left(4 + y^2 - 3\right)^2 \right) dy$$

$$= \pi \int_{-1}^{\frac{3}{2}} \left(-2y^3 - 9y^2 + 8y + 15 \right) dy$$

$$= \pi \left(-\frac{1}{2}y^4 - 3y^3 + 4y^2 + 15y \right) \Big|_{-1}^{\frac{3}{2}} = \frac{875}{32}\pi.$$

417. *Let R be the region between the curve $y = f(x) \geq 0$, the x-axis, the line $x = a$ and the line $x = b$. We rotate the region R around a line of the form $x = k$ to generate a solid. Give a formula for the volume of the solid.*

Solution. If $k \leq a \leq b$, then by using shell method, the volume of the solid is

$$V = 2\pi \int_a^b (x - k)f(x)dx. \tag{6.7}$$

For $a \leq b \leq k$ we obtain the following formula

$$V = 2\pi \int_a^b (k - x)f(x)dx. \tag{6.8}$$

418. *Find the volume of the solid generated by revolving about the line $x = 3$ the region in the first quadrant bounded by $y = 9 - x^2$, $x = 0$ and $y = 0$.*

Solution. Figure 6.23 shows the region. We use shell method by applying Formula (6.8).

So, the volume of the solid is equal to

Fig. 6.23 Solid generated by revolving about the line $x = 3$ the region in the first quadrant bounded by $y = 9 - x^2$, $x = 0$ and $y = 0$

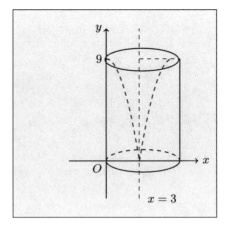

Fig. 6.24 Region bounded by the curve $y = f(x)$ between the points $A(a, f(a))$ and $B(b, f(b))$, the line $y = mx + k$ and perpendicular to the line from A and B

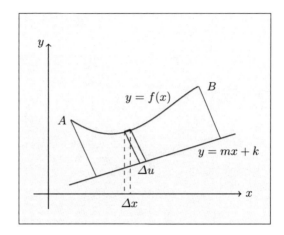

$$V = 2\pi \int_0^3 (3 - x)(9 - x^2)dx = 2\pi \int_0^3 (27 - 9x - 3x^2 + x^3)dx$$
$$= 2\pi \left(27x - \frac{9}{2}x^2 - x^3 + \frac{1}{4}x^4 \right)\Big|_0^3 = \frac{135}{2}\pi.$$

419. *Find a formula for the volume of a solid of revolution when the axis of the rotation is a slanted line (suppose that the lines perpendicular to the axis of the rotation cut off the curve at just one point).*

Solution. Let C be the curve $y = f(x)$ between the points $A(a, f(a))$ and $B(b, f(b))$, and let $\mathcal{R}$ be the region bounded by C, the line $y = mx + k$ (which lies entirely below C) and perpendicular to the line from A and B (Fig. 6.24). We revolve the region $\mathcal{R}$ about $y = mx + k$.

Using the disk method, where the radius of a disk is $r(x)$ and its thickness is Δu, we can write $\Delta V = \pi \left(r(x)\right)^2 \Delta u$. Here, r is the distance between the point $(x, f(x))$ and the line $y = mx + k$. So, we have

$$r(x) = \frac{|f(x) - mx - k|}{\sqrt{m^2 + 1}}.$$

In order to find Δu in terms of Δx, let Δd represent the arc length along $f(x)$ corresponding to Δx. Then, we have $\Delta x = \cos \theta \cdot \Delta d$. Now, to obtain the project Δd onto $y = mx + k$, we find that by rotating everything by $-\tan^{-1} m$. Thus, $\Delta u = \Delta d \cos \left(\theta - \tan^{-1} m\right)$. Hence, we deduce that

$$\Delta u = \frac{\cos \left(\theta - \tan^{-1} m\right)}{\cos \theta} \Delta x$$
$$= \frac{\cos \theta \cos \left(\tan^{-1} m\right) + \sin \theta \sin \left(\tan^{-1} m\right)}{\cos \theta} \Delta x.$$

Now, we use the following simple formulas:

$$\cos\left(\tan^{-1}m\right) = \frac{1}{\sqrt{m^2+1}} \text{ and } \sin\left(\tan^{-1}m\right) = \frac{m}{\sqrt{m^2+1}}.$$

Then, we get

$$\Delta u = \left(\frac{1}{\sqrt{m^2+1}} + \frac{m}{\sqrt{m^2+1}}\tan\theta\right)\Delta x$$

$$= \frac{1}{\sqrt{m^2+1}}\left(1 + m\tan\theta\right)\Delta x.$$

Hence,

$$V = \pi\int_a^b r^2 du$$

$$= \pi\int_a^b \left(\frac{|f(x) - mx - k|}{\sqrt{m^2+1}}\right)^2\left(\frac{1}{\sqrt{m^2+1}}\left(1 + mf'(x)\right)\right)dx.$$

Therefore, the volume of the solid is equal to

$$V = \frac{\pi}{(m^2+1)^{\frac{3}{2}}}\int_a^b \left(f(x) - mx - k\right)^2\left(1 + mf'(x)\right)dx. \qquad (6.9)$$

420. *(1) Find the volume of the solid obtained by rotating $f(x) = x^2$ about the line $y = x - 1$ on $[0, 1]$.*

(2) Find the volume of the solid generated by revolving the region between the curves $y = \sqrt{x}$ and $y = x^2$ about the line $y = x$.

Solution. (1) By using Formula (6.9) the volume of the solid of rotation is equal to

$$V = \frac{\pi}{2\sqrt{2}}\int_0^1 \left(x^2 - x + 1\right)^2(1 + 2x)dx$$

$$= \frac{\pi}{2\sqrt{2}}\int_0^1 \left(2x^5 - 3x^4 + 4x^3 - x^2 + 1\right)dx$$

$$= \frac{\pi}{2\sqrt{2}}\left(\frac{1}{3}x^6 - \frac{3}{5}x^5 + x^4 - \frac{1}{3}x^3 + x\right)\Big|_0^1 = \frac{7\pi}{10\sqrt{2}}.$$

(2) Since the graphs of $y = \sqrt{x}$ and $y = x^2$ are symmetric with respect to $y = x$, it is enough we consider $y = \sqrt{x}$ to obtain the volume of the solid. So, by using Formula (6.9) the volume of the solid of rotation is equal to

$$V = \frac{\pi}{2\sqrt{2}} \int_0^1 (\sqrt{x} - x)^2 \left(1 + \frac{1}{2\sqrt{x}}\right) dx$$

$$= \frac{\pi}{2\sqrt{2}} \int_0^1 \left(x + x^2 - 2x^{\frac{3}{2}}\right)\left(1 + \frac{1}{2}x^{-\frac{1}{2}}\right) dx$$

$$= \frac{\pi}{2\sqrt{2}} \int_0^1 \left(x^2 - \frac{3}{2}x^{\frac{3}{2}} + \frac{1}{2}x^{\frac{1}{2}}\right) dx$$

$$= \frac{\pi}{2\sqrt{2}} \left(\frac{1}{3}x^3 - \frac{3}{5}x^{\frac{5}{2}} + \frac{1}{3}x^{\frac{3}{2}}\right)\Big|_0^1 = \frac{\pi}{30\sqrt{2}}.$$

421. *Suppose the region under the curve*

$$y = a \cosh \frac{x}{a} \quad (0 \le x \le b, \; a > 0)$$

is revolved about the x-axis. What is the ratio of the volume V of the resulting solid of revolution to its lateral surface area S.

Solution. See Fig. 6.25. The volume of the solid of revolution is equal to

$$V = \pi \int_0^b \left(f(x)\right)^2 dx = \pi a^2 \int_0^b \cosh^2 \frac{x}{a} dx = \pi a^2 B.$$

On the other hand, the area of the surface of revolution is equal to

$$S = 2\pi \int_0^b f(x)\sqrt{1 + \left(f'(x)\right)^2}\,dx = 2\pi \int_0^b a \cosh \frac{x}{a}\sqrt{1 + \sinh^2 \frac{x}{a}}\,dx$$

$$= 2\pi a \int_0^b \cosh^2 \frac{x}{a}\,dx = 2\pi a B.$$

Fig. 6.25 Solid of revolution

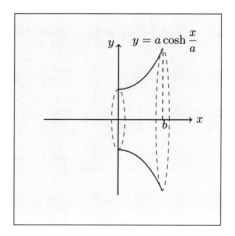

Fig. 6.26 Solid by the
revolution of an astroid about
the x-axis

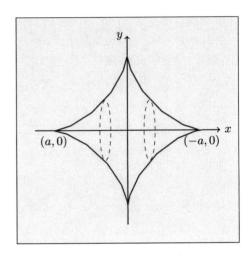

Therefore, we deduce that $\dfrac{V}{S} = \dfrac{1}{2}a$.

422. *Determine the area of the surface of the solid by the revolution of the astroid* $x = a \cos^3 t$ *and* $y - a \sin^3 t$ *about the x-axis.*

Solution. See Fig. 6.26. We have

$$\frac{dx}{dt} = -3a \cos^2 t \sin t \text{ and } \frac{dy}{dt} = 3a \sin^2 t \cos t.$$

By the symmetry, it is enough to calculate the surface of revolution for $x \in [0, a]$. So, the area of the surface of revolution is equal to

$$S = 2(2\pi) \int_0^{\frac{\pi}{2}} ay(t) \sqrt{\left(\frac{dx}{dt}\right)^2 + \left(\frac{dy}{dt}\right)^2} dt$$

$$= 4\pi \int_0^{\frac{\pi}{2}} a \cos^3 t \sqrt{\left(-3a \cos^2 t \sin t\right)^2 + \left(3a \sin^2 t \cos t\right)^2} dt$$

$$= 4\pi \int_0^{\frac{\pi}{2}} a \cos^3 t \sqrt{9a^2 \sin^2 t \cos^2 t} \, dt$$

$$= 12\pi a^2 \int_0^{\frac{\pi}{2}} \cos^4 t \sin t \, dt$$

$$= 12\pi a^2 \left(\frac{1}{5} \sin^5 t\right)\Big|_0^{\frac{\pi}{2}} = \frac{12\pi a^2}{5}.$$

423. *Prove that the lateral surface area of a right circular cone of radius r and height h is* $\pi r \sqrt{h^2 + r^2}$.

Solution. Consider the line segment by

$$\frac{x}{r} + \frac{y}{h} = 1,$$

where $x \in [0, r]$. The surface area of the cone of radius r and height h generated by revolving this line segment about y-axis is equal to

$$\begin{aligned} S &= 2\pi \int_0^h r\left(1 - \frac{y}{h}\right)\sqrt{1 + \left(\frac{r}{h}\right)^2}\,dy \\ &= 2\pi r\sqrt{1 + \left(\frac{r}{h}\right)^2}\int_0^h \left(1 - \frac{y}{h}\right)dy \\ &= 2\pi r\sqrt{1 + \left(\frac{r}{h}\right)^2}\left(y - \frac{1}{2h}y^2\right)\Big|_0^h \\ &= 2\pi r\sqrt{1 + \left(\frac{r}{h}\right)^2}\left(h - \frac{1}{2}h\right) \\ &= \pi r\sqrt{h^2 + r^2}. \end{aligned}$$

424. *Find a formula for the area of a surface of revolution when the axis of the rotation is a slanted line.*

Solution. Consider Fig. 6.24. According to Problem (419), we have

$$r(x) = \frac{f(x) - mx - k}{\sqrt{m^2 + 1}}.$$

This acts as the radius rotation around the line $y = mx + b$. A small arc length along the curve is

$$\Delta\sigma = \sqrt{(\Delta x)^2 + (\Delta y)^2} = \sqrt{1 + \left(\frac{\Delta y}{\Delta x}\right)^2}\,\Delta x$$

Hence, we conclude that $\Delta S = 2\pi r(x)\Delta\sigma$, and so we get

$$S = 2\pi \int_a^b \frac{f(x) - mx - k}{\sqrt{m^2 + 1}}\sqrt{1 + (f'(x))^2}\,dx.$$

Therefore, the area of a surface of revolution is equal to

$$S = \frac{2\pi}{\sqrt{m^2 + 1}}\int_a^b (f(x) - mx - k)\sqrt{1 + (f'(x))^2}\,dx. \qquad (6.10)$$

425. *Consider the surface generated by revolving the arc of the unit circle $x^2 + y^2 = 1$ in the first quadrant about the line $x + y = 1$. What is the area of surface?*

Solution. See Fig. 6.27. By using Formula (6.10) the area of the surface of rotation is equal to

Fig. 6.27 Surface generated by revolving the arc of the unit circle $x^2 + y^2 = 1$ in the first quadrant about the line $x + y = 1$

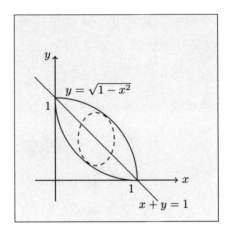

Fig. 6.28 A tank is full of water

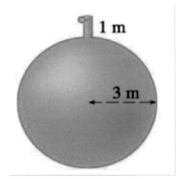

$$S = \frac{2\pi}{\sqrt{2}} \int_0^1 \left(\sqrt{1-x^2} + x - 1\right)\sqrt{1 + \frac{x^2}{1-x^2}}\,dx$$

$$= \frac{2\pi}{\sqrt{2}} \int_0^1 \left(\sqrt{1-x^2} + x - 1\right)\sqrt{\frac{1}{1-x^2}}\,dx$$

$$= \frac{2\pi}{\sqrt{2}} \int_0^1 \left(1 + \frac{x}{\sqrt{1-x^2}} - \frac{1}{\sqrt{1-x^2}}\right)dx$$

$$= \frac{2\pi}{\sqrt{2}} \left(x - \sqrt{1-x^2} - \sin^{-1} x\right)\Big|_0^1$$

$$= \frac{2\pi}{\sqrt{2}} \left(2 - \frac{\pi}{2}\right) = \frac{\pi(4-\pi)}{\sqrt{2}}.$$

426. *A tank is full of water (see Fig. 6.28). Find the work required to pump the water out of the spout.*

Solution. We consider a cross section of the tank (Fig. 6.29) to find the radius of the representative slice. We have $(3 - y)^2 + r^2 = 9$, and so $r = \sqrt{6y - y^2}$.

Fig. 6.29 Cross section of
the tank to find the radius of
the representative slice

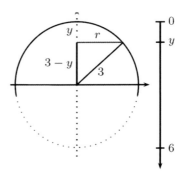

Fig. 6.30 A representative
slice

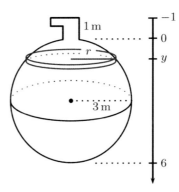

In Fig. 6.30, a representative slice is shown. We show the thickness of the representative slice by Δy.

Therefore, we have

$$\text{Volume}_{\text{slice}} = \pi(6y - y^2)\Delta y.$$

The density of water is $1000 \, \text{kg/m}^3$. Hence, we get

$$\text{Mass}_{\text{slice}} = 1000\pi(6y - y^2)\Delta y.$$

From Newton's second law, $f = ma$, where $a = 9.8 \, \text{m/s}^2$ is the acceleration due to gravity. Thus, the force on the slice is given by

$$\text{Force}_{\text{slice}} = (9.8)(1000)\pi(6y - y^2)\Delta y.$$

As the water is pumped up and out of the tank's spout, each slice moves a distance of $y - (-1) = y + 1m$, so the work done on each slice is

$$\text{Work}_{\text{slice}} = (9.8)(1000)\pi(y + 1)(6y - y^2)\Delta y.$$

Therefore, we conclude that

$$W = \lim_{n \to \infty} \sum_{i=1}^{n} (9.8)(1000)\pi (y + 1)(6y - y^2)dy$$

$$= \int_0^6 9800\pi (y + 1)(6y - y^2)dy$$

$$= 9800\pi \int_0^6 (-y^3 + 5y^2 + 6y)dy \approx 44 \times 10^6.$$

6.3 Exercises

Easier Exercises

Find the area of the given region:

1. The region bounded by $y = 12 - x - x^2$, the x-axis, and the lines $x = -3$ and $x = 2$.
2. The region bounded by $y = x^3 + x$, the x-axis, and the lines $x = -2$ and $x = 1$.
3. The region bounded by $x^4 - 4x^3 + 4x^2$ and the x-axis.
4. The region bounded by $x = y^3$ and $x = y^2$.
5. The region bounded by $x = 3y - y^2$ and $x + y = 3$.
6. The region bounded by $y = \cos(\frac{\pi x}{3})$ and $y = 1 - x^2$ (in the first quadrant).
7. The region bounded by $y = \sin(\frac{\pi x}{3})$ and $y = x$ (in the first quadrant).
8. The region bounded by $y = \sqrt{x}$ and $y = \sqrt{x + 1}$, $(0 \le x \le 4)$.
9. The region bounded by $y = \sin x \cos x$ and $\sin x$, $(0 \le x \le \pi)$.
10. The region bounded by $y = x^{\frac{3}{2}}$ and $y = x^{\frac{2}{3}}$.
11. The region bounded by $y = \sin^{-1} x$, the x-axis and the line $x = 1$.
12. The region bounded by $y = \sin^{-1} x$, the y-axis and the line $y = \frac{\pi}{2}$.
13. The region bounded by $y = \cos^{-1} x$ and the coordinate axes.
14. Prove that the area bounded by the ellipse $\dfrac{x^2}{a^2} + \dfrac{y^2}{b^2} = 1$ is πab.

Compute the arc length of the graph of the given function:

15. $f(x) = \frac{4}{3}(x^2 + 1)^{\frac{3}{2}}$ on $[1, 4]$,

16. $f(x) = \frac{x^3}{6} + \frac{1}{2x}$ on $[1, 3]$,

17. $f(x) = \ln(\cos x)$ on $[0, \frac{\pi}{3}]$,

18. $f(x) = \ln(\sec x)$ on $[0, \frac{\pi}{4}]$,

19. $f(x) = \int_0^x \sqrt{\cos 2t} \, dt$ on $[0, \frac{\pi}{4}]$,

20. $f(x) = \int_0^x \sqrt{\cosh t} \, dt$ on $[0, 2]$.

Find the arc length of the given parametric curve:

21. $x = \cos t + t \sin t$ and $y = \sin t - t \cos t$, $(0 \le t \le 10)$,
22. $x = (t^2 - 2) \sin t + 2t \cos t$ and $y = (2 - t^2) \cos t + 2t \sin t$, $(0 \le t \le \pi)$,
23. $x = e^{-t} \cos t$ and $y = e^{-t} \sin t$, $(a \le t \le b)$.

Find the arc length of the graph of each equation between the indicated points:

24. $x = \frac{2}{3}(y - 5)^{\frac{3}{2}}$ from $(0, 5)$ to $(\frac{2}{3}, 6)$,

25. $12xy = 4x^4 + 3$ from $(1, \frac{7}{12})$ to $(3, \frac{109}{12})$,

26. $x = \frac{y^3}{3} + \frac{1}{4y}$ from $(\frac{7}{12}, 1)$ to $(\frac{67}{24}, 2)$.

27. Find a curve through the origin whose length is

$$\int_0^4 \sqrt{1 + \frac{1}{4x}}\, dx.$$

28. Find the volume of the solid generated by revolving the region bounded by $y = 3x^4$ and the line $x = 1$ and $x = -1$ about:

   (a) the x-axis,   (c) the line $x = 1$,
   (b) the x-axis,   (d) the line $y = 3$.

29. Find the volume of the solid generated by revolving the region bounded by $y = \frac{4}{x^3}$ and the line $x = 1$ and $y = \frac{1}{2}$ about:

   (a) the x-axis,   (c) the line $x = 2$,
   (b) the x-axis,   (d) the line $y = 4$.

30. The region bounded by the y-axis and the curves $y = \sin x$ and $y = \cos x$ for $0 \le x \le \frac{\pi}{4}$ is revolved about the x-axis. Find the volume of the solid of revolution generated.

31. Find the volume of the solid generated by revolving about the line $x = -4$ the region bounded by that line and the parabola $x = 4 + 6y - 2y^2$.

Find the area of the surface generated by revolving the given curve about the specified axis

32. $x = 2\cos t + \cos 2t$, $y = 2\sin t - \sin 2t$ $\,(\frac{2\pi}{3} \le t \le \frac{4\pi}{3})$, y-axis,

33. $y = \sqrt{x^2 + 2}$ $\,(0 \le x \le 2)$, x-axis,

34. $y = |x - 1| + |x|$ $\,(-1 \le x \le 2)$, the line $y = 2$,

35. $x = 2y^{\frac{1}{4}} - \frac{2}{7}y^{\frac{7}{4}}$ $\,(0 \le y \le 1)$, x-axis.

36. A particle moves along a straight line with acceleration given by $a(t) = 1 - \sin(\pi t)$. Suppose that when $t = 0$, $x(t) = v(t) = 0$. Find $x(t)$ and $v(t)$.

37. A particle moves along a straight line with acceleration given by $a(t) = -\cos t$, and $x(0) = 1$ and $v(0) = 0$. Find the maximum distance the particle travels from zero and find its maximum speed.

38. An article is moved from $x = a$ to $x = b$. The force used at position x is $f(x)$. Compute the total amount of work done.

   (a) $f(x) = x\sqrt{1 - x^2}$, $a = 0$ to $b = 1$,
   (b) $f(x) = x^2 e^{-x}$, $a = 0$ to $b = 3$.

Harder Exercises

39. Find the area of the region bounded by $x^2 + y^2 - 2x + 4y - 11 = 0$ and $y = -x^2 + 2x + 1 - 2\sqrt{3}$.

40. Find the area of the region cut from the first quadrant by the curve

$$\sqrt{x} + \sqrt{y} = \sqrt{a}.$$

41. Find the area between the curves $y = 2\dfrac{\log_2 x}{x}$ and $y = 2\dfrac{\log_4 x}{x}$ and x-axis from $x = 1$ to $x = e$. What is the ratio of the longer to the smaller?

42. Find $f(\dfrac{\pi}{2})$ from the following information:

   (a) f is positive and continuous.
   (b) The area under the graph of $y = f(x)$ from $x = 0$ to $x = a$ is

$$\frac{a^2}{2} + \frac{a}{2}\sin a + \frac{\pi}{2}\cos a.$$

43. Let f be a function with continuous first derivative, and let Δs denote the arc length of the graph of f between the point $P(a, f(a))$ and the point $Q(a + \Delta x, f(a + \Delta x))$. If Δl denotes the length of the straight line segment joining the point P and the point Q, prove that

$$\lim_{\Delta x \to 0} \frac{\Delta l}{\Delta s} = 1.$$

44. Let α be a curve in $\mathbb{R}^2$, defined on $[a, b]$ and let f be a continuous one-to-one function from $[c, d]$ onto $[a, b]$ such that $f(c) = a$. Define $\beta(t) = \alpha(f(t))$. Prove that β is an arc, a closed curve or a rectifiable curve if and only if the same is true of α. Prove that α and β have the same length.

45. Find the volume of the solid S whose base is the circular disk $x^2 + y^2 \leq 9$ if every cross section of S by a plane perpendicular to the x-axis is

   (a) A square.
   (b) A semicircle with its diameter in the xy plane.
   (c) An equilateral triangle.
   (d) An isosceles right triangle with its hypotenuse in the xy plane.

46. Find the volume of

    (a) An *oblique circular cylinder* of height h and base radius r.
    (b) An *oblique circular cone* of height h and base radius r.

47. A *torus* (see Fig. 6.31) is essentially the shape of an inner tube with tube radius r for which the radius of the circle in the center of the tube is $R > r$. Derive the formula for the volume of a torus of main radius R and tube radius r.

48. A hole of radius r is bored through the middle of a cylinder of radius $R > r$ at right angles to the axis of the cylinder. Set up an integral for the volume cut out and evaluate it.

49. Find the volume common to two spheres, each with radius , if the center of each sphere lies on the surface of the other sphere.

50. A hole of radius r has been drilled through a sphere of radius $R > r$.

    (a) Find the volume of the remaining part of the sphere.
    (b) Find the volume of the material that was drilled out.

51. Find the volume common to two cylinders that have the same radius and whose axes intersect at a right angle (see Fig. 6.32).

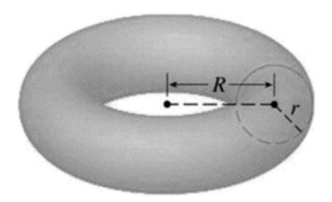

Fig. 6.31 A torus

Fig. 6.32 Intersection of two cylinders

Fig. 6.33 A water tank in
the form of an inverted right
circular cone

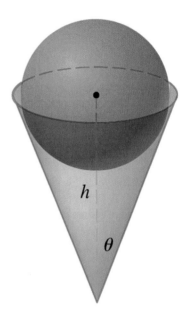

h

θ

52. A wedge is cut from a solid in the shape of a right circular cylinder with a radius
    of r cm by a plane through a diameter of the base and inclined to the plane of
    the base at an angle of measurement $45°$. Find the volume of wedge.

53. A paper drinking cup filled with water has the shape of a cone with height h and
    semivertical angle θ (see Fig. 6.33). A ball is placed carefully in the cup, thereby
    displacing some of the water and making it overflow. What is the radius of the
    ball that causes the greatest volume of water to spill out of the cup?

54. The parabola $y = x^2$ divides the triangle with vertices $(0, 0)$, $(2, 0)$ and $(0, 2)$
    into two regions, one above the parabola and the other below the parabola. Find
    the volume of the solid generated by revolving each region about

    (a) The x-axis.
    (b) The y-axis.

55. Find the volume of the solid generated by revolving the triangle with vertices
    $(1, 0)$, $(4, 1)$ and $(2, 3)$ about

    (a) The x-axis.
    (b) The y-axis.

56. Suppose that a region R has area A and lies above the x-axis. When R is rotated
    about the x-axis, it sweeps out a solid with volume V_1. When it is rotated about
    the line $y = -k$ (where k is a positive number), it sweeps out a solid with volume
    V_2. Express V_2 in terms of V_1, k and A.

57. An infinite solid (called *Torricelli's trumpet*) with finite volume enclosed by a
    surface with infinite surface area (see Fig. 6.34).

Fig. 6.34 Torricelli's trumpet

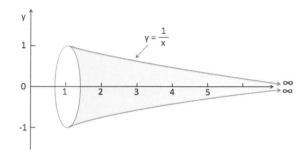

For $a > 1$, consider the funnel or trumpet formed by revolving the curve $y = \dfrac{1}{x}$, $1 \le x \le a$, about the x-axis. Let $V(a)$ and $S(a)$ denote the volume and the surface area of the funnel. Show that

$$\lim_{a \to \infty} V(a) = \pi \quad \text{and} \quad \lim_{a \to \infty} S(a) = \infty.$$

58. A certain oil tank is shaped like a right circular cylinder that is lying on its side. The tank has a radius of r and a length l.

   (a) Find the amount of work that it takes to pump out the whole contents through the top of the tank.
   (b) Find the amount of work that it takes to empty the tank through the top if the tank is half full.
   (c) Find the amount of work that it takes to empty the tank through the top if the tank is three quarters full.

59. A tank full of water in the form of a rectangular parallelepiped 2 m deep, 3 m wide and 5 m long. Find the work required to pump the water in the tank up to a level 1 m above the surface of the tank.

Fig. 6.35 A paper drinking cup and a ball is placed in the cup

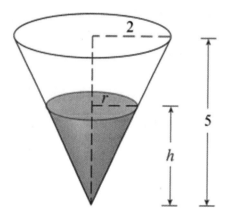

60. A water tank in the form of an inverted right circular cone has a height of 5 m and a base radius of 2 m and is filled with water to a depth of h meters (see Fig. 6.35). Determine the amount of work needed to pump all of the water to the top of the tank.

61. Consider the infinitely deep well formed by rotating the region bounded by the curve $y = \ln x$, the y-axis and the x-axis in the fourth quadrant. It is filled with water (density $D = 1000 \, \text{kg/m}^3$. How much work W is done in emptying the well (raising water to ground level $H = 0$). The formula for work is

$$W = D \int_a^b (H - y)\pi \left(f(y)\right)^2 dy,$$

where a and b are the depths of the bottom and top of the well and $x = f(y)$ represents the equation of the curve as a function of x.

62. A heavy rope, 15 m long, weighs 0.75 kg/m and hangs over the edge of a building 40 m high.

(a) How much work is done in pulling the rope to the top of the building?
(b) How much work is done in pulling half the rope to the top of the building?

Chapter 7
Sequences and Series

7.1 Basic Concepts and Theorems

Sequences

A sequence is a function whose domain is a set of the form $\{n \in \mathbb{Z} \ : \ n \geq m\}$; m is usually 1 or 0. Thus, a sequence is a function that has a specified value for each integer $n \geq m$. It is customary to denote a sequence by a latter such as a and to denote its value at n as a_n rather than $a(n)$. It is often convenient to write the sequence as $\{a_n\}_{n=m}^{\infty}$. If $m = 1$ we may write $\{a_n\}_{n \in \mathbb{N}}$. We study the sequences whose range values are real numbers; i.e., each a_n represent a real number. Sometimes we write $\{a_n\}$ when the domain is understood. A sequence $\{a_n\}$ of real numbers is said to be *convergent* to a real number L provided that:

For every $\epsilon > 0$ there exists a number N such that $n > N$ implies $|a_n - L| < \epsilon$.

If $\{a_n\}$ converges to L we write $\lim_{n \to \infty} a_n = L$ or $a_n \to L$ as $n \to \infty$. The number L is called the *limit of the sequence* $\{a_n\}$. A sequence that does not converge to any real number is said to be *divergent*. There are limit theorems for sequences that are analogous to limit theorem for functions. For instance, if $\{a_n\}$ and $\{b_n\}$ are convergent sequences and c is a constant, then

(1) $\lim_{n \to \infty} ca_n = c \lim_{n \to \infty} a_n$,

(2) $\lim_{n \to \infty} (a_n \pm b_n) = \lim_{n \to \infty} a_n \pm \lim_{n \to \infty} b_n$,

(3) $\lim_{n \to \infty} a_n b_n = \lim_{n \to \infty} a_n \cdot \lim_{n \to \infty} b_n$

(4) $\lim_{n \to \infty} \dfrac{a_n}{b_n} = \dfrac{\lim_{n \to \infty} a_n}{\lim_{n \to \infty} b_n}$ if $\lim_{n \to \infty} b_n \neq 0$.

© The Author(s), under exclusive license to Springer Nature Singapore Pte Ltd. 2020
B. Davvaz, *Examples and Problems in Advanced Calculus: Real-Valued Functions*,
https://doi.org/10.1007/978-981-15-9569-1_7

Sandwich theorem for sequences: Let $\{a_n\}$, $\{b_n\}$ and $\{c_n\}$ be sequences of real numbers. If $a_n \leq b_n \leq c_n$ holds for all n beyond some index m and if $\lim\limits_{n\to\infty} a_n = \lim\limits_{n\to\infty} c_n = L$, then $\lim\limits_{n\to\infty} b_n = L$.

The continuous function theorem for sequences: Let $\{a_n\}$ be a sequence of real numbers. If $\lim\limits_{n\to\infty} a_n = L$ and if f is a function that is continuous at L and defined at all a_n, then $\lim\limits_{n\to\infty} f(a_n) = f(L)$.

The following theorem formalizes the connection between $\lim\limits_{n\to\infty} a_n$ and $\lim\limits_{x\to\infty} f(x)$. It enables us to use L'Hospital's rule to find the limits of some sequences.

Suppose that $f(x)$ is a function defined for all $x \geq m$ and $\{a_n\}$ is the sequence of real numbers $a_n = f(n)$ for all $n \geq m$. If $\lim\limits_{x\to\infty} f(x) = L$, then $\lim\limits_{n\to\infty} a_n = L$.

A sequence $\{a_n\}$ is said to be

(1) *increasing* if $a_n \leq a_{n+1}$ for all n,
(2) *decreasing* if $a_n \geq a_{n+1}$ for all n.

If $a_n < a_{n+1}$ for all n, the sequence is *strictly increasing*. If $a_n > a_{n+1}$ for all n, the sequence is *strictly decreasing*. By a *monotone sequence* is meant a sequence which is either increasing or decreasing. A sequence $\{a_n\}$ is said to be *bounded* if there is a number $M > 0$ such that $|a_n| \leq M$. The following are two important theorems:

(1) Every convergent sequence is bounded.
(2) Every bounded monotonic sequence is convergent.

Given a sequence $\{a_n\}$, consider a sequence $\{n_k\}$ of positive integers such that $n_1 < n_2 < n_3 < \dots$. Then, the sequence $\{a_{n_k}\}$ is called a *subsequence* of $\{a_n\}$. If $\{a_{n_k}\}$ converges, its limit is called a *subsequential limit* of $\{a_n\}$. It is clear that $\{a_n\}$ converges to L if and only if every subsequence of $\{a_n\}$ converges to L.

Infinite Series

Let $\{a_n\}$ be a sequence. Then, the expression $a_1 + a_2 + \dots + a_n + \dots$ or more concisely

$$\sum_{n=1}^{\infty} a_n \tag{7.1}$$

is called an *infinite series*, or just a *series*. Sometimes we show briefly a series by $\sum a_n$. The number a_n is called the *nth term* or *general term*. The sum of the first n terms of the series (7.1),

$$S_n = a_1 + a_2 + \dots + a_n$$

is called the *nth partial sum* of the series. Let $\{s_n\}$ be the sequence of partial sums of the series $\sum_{n=1}^{\infty} a_n$. If $\lim_{n\to\infty} s_n$ exists and is equal to a real number s, the given series is said to be *convergent* and s is the sum of the given series. If $\lim_{n\to\infty} s_n$ does not exist, the series is *divergent* and the series does not have a sum. Hence, a series is convergent if and only if the corresponding sequence of partial sums is convergent.

Necessary condition for convergence of a series: If $\sum a_n$ is convergent, then $\lim_{n\to\infty} a_n = 0$. Equivalently, the series $\sum a_n$ is divergent if $\lim_{n\to\infty} a_n \neq 0$.

If $\{s_n\}$ is the sequence of partial sums for a given series $\sum a_n$, then the series $\sum a_n$ converges if and only if for every $\epsilon > 0$ there exists a number N such that $|s_i - s_j| < \epsilon$ whenever $i, j > N$.

If $\sum a_n$ and $\sum b_n$ are two series, differing at most in their first m terms, i.e., $a_k = b_k$ for all $k > m$, then either both series converge or both series diverge.

If $\sum a_n = A$ and $\sum b_n = B$ are convergent series, then $\sum(a_n \pm b_n) = A \pm B$ and $\sum c a_n = c A$, for each real number c.

A *geometric series* is a series of the form

$$\sum_{n=0}^{\infty} a r^n.$$

The *n*th partial sum of this series is

$$s_n = a(1 + r + r^2 + \cdots + r^{n-1})$$
$$= a \frac{1 - r^n}{1 - r} \text{ if } r \neq 1.$$

The geometric series converges to the sum $\dfrac{a}{1 - r}$ if $|r| < 1$, and the geometric series diverges if $|r| \geq 1$.

Infinite Series of Positive Terms

An infinite series is said to be *positive* or *non-negative* if all terms are non-negative.

Convergence criterion for non-negative series: A non-negative series $\sum a_n$ is convergent if and only if the sequence of partial sums has an *upper bound*, that is, if there is a number $M > 0$ such that $s_n = a_1 + a_2 + \cdots + a_n \leq M$, for all n.

Comparison test for series: Let $\sum a_n$ and $\sum b_n$ be two non-negative series such that $a_n \leq b_n$ for all sufficiently large n.

(1) If $\sum b_n$ converges, so does $\sum a_n$;
(2) If $\sum a_n$ diverges, so does $\sum b_n$.

Limit comparison test for series: Let $\sum a_n$ and $\sum b_n$ be two non-negative series such that $b_n \neq 0$, for all n, and

$$\lim_{n \to \infty} \frac{a_n}{b_n} = L,$$

where the case $L = \infty$ is allowed.

(1) If $0 \leq L < \infty$ and $\sum b_n$ is convergent, then $\sum a_n$ is convergent too;
(2) If $L > 0$ or $L = \infty$ and $\sum b_n$ is divergent, then $\sum a_n$ is divergent too;
(3) In particular, if L is a positive real number, the two series $\sum a_n$ and $\sum b_n$ are either both convergent or both divergent.

If $\sum a_n$ is a given convergent series of positive terms, it terms can be grouped in any manner or the order of the terms can be rearranged, and resulting series also will be convergent and will have the same sum as the given series.

Integral test: Let f be a continuous, positive and decreasing function on $[1, \infty)$ and let $a_n = f(n)$ for all $n \in \mathbb{N}$. Then, the series

$$\sum_{n=1}^{\infty} a_n$$

and the improper integral

$$\int_1^{\infty} f(x)dx$$

are either both convergent or both divergent.

A series of the form

$$\sum_{n=1}^{\infty} \frac{1}{n^p} = \frac{1}{1^p} + \frac{1}{2^p} + \frac{1}{3^p} + \cdots$$

is a *p-series*, where p is a constant. For $p = 1$, the series

$$\sum_{n=1}^{\infty} \frac{1}{n}$$

known as the *harmonic series*. The integral test is convenient for establishing the convergence or divergence of p-series.

Convergence of p-series: The p-series converges if $p > 1$, and diverges if $p \leq 1$.

Alternating Series

In continue infinite series having both positive and negative terms are considered. The simplest such series is an *alternating series*, whose terms alternate in sign. Alternating series occur in two ways: Either the odd terms are negative or the even terms are negative.

Alternating series test: Let $a_n > 0$, for all n. The alternating series

$$\sum_{n=1}^{\infty}(-1)^n a_n \quad \text{and} \quad \sum_{n=1}^{\infty}(-1)^{n+1} a_n$$

converge if the following two conditions hold:

(1) $\lim_{n \to \infty} a_n = 0$,

(2) $a_{n+1} \le a_n$ for all n.

The second condition in the alternating series test can be modified to require only that $0 < a_{n+1} \le a_n$ for all n greater than some integer m. For a convergent alternating series, the partial sum s_n can be a useful approximation for the sum s of the series.

Alternating series reminder: If a convergent alternating series satisfies the condition $a_{n+1} \le a_n$, then the absolute value of the remainder r_k involved in approximating the sum s by s_k is less than (or equal to) the first neglected term, that is, $|s - s_k| = |r_k| < a_{k+1}$.

Absolute and Conditional Convergence

A series

$$\sum_{n=1}^{\infty} a_n$$

is said to be *absolutely convergent* if the related series

$$\sum_{n=1}^{\infty} |a_n| = |a_1| + |a_2| + \ldots |a_n| + \ldots$$

is convergent. A series that is convergent, but not absolutely convergent, is said to be *conditionally convergent*.

If $\sum a_n$ is absolutely convergent, then it is convergent and

$$\left|\sum_{n=1}^{\infty} a_n\right| \leq \sum_{n=1}^{\infty} |a_n|.$$

If a series is absolutely convergent, its terms can be rearranged arbitrary without changing the sum of the series. If we rearrange the terms of a conditionally convergent series, we get different results.

The Ratio and Root Tests

Now, we present two further convergent tests.

Ratio test: Let $\sum a_n$ be a series with non-zero terms and

$$\lim_{n \to \infty} \left|\frac{a_{n+1}}{a_n}\right| = L$$

exists, where the case $L = \infty$ allowed.

(1) If $0 \leq L < 1$, then the series is absolutely convergent;
(2) If $L > 1$ or $L = \infty$, then the series is divergent;
(3) When $L = 1$, the test is inconclusive.

The next test for convergence or divergence of series works especially well for series involving nth powers.

Root test: Let $\sum a_n$ be a series and

$$\lim_{n \to \infty} \sqrt[n]{|a_n|} = L$$

exists, where the case $L = \infty$ allowed.

(1) If $0 \leq L < 1$, then the series is absolutely convergent;
(2) If $L > 1$ or $L = \infty$, then the series is divergent;
(3) When $L = 1$, the test is inconclusive.

Power Series

Let x be an independent variable, and let $\{a_n\}$ be any sequence of real numbers. Then, an infinite series of the form

$$\sum_{n=0}^{\infty} a_n x^n = a_0 + a_1 x + a_2 x^2 + \cdots + a_n x^n + \cdots$$

is called a *power series* and the numbers $a_0, a_1, a_2, \ldots, a_n, \ldots$ are called the *coefficients of the series*. More generally, an infinite series of the form

$$\sum_{n=0}^{\infty} a_n(x-c)^n = a_0 + a_1(x-c) + a_2(x-c)^2 + \cdots + a_n(x-c)^n + \ldots$$

is called a *power series centered at c*, where c is a constant. In considering a power series we ask: For what values of x does the power series converges? For each value of x for which the power series converges, the series represents the number that is the sum of the series. Therefore, a power series defines a function. The function f, with function value

$$f(x) = \sum_{n=0}^{\infty} a_n(x-c)^n = a_0 + a_1(x-c) + a_2(x-c)^2 + \cdots + a_n(x-c)^n + \ldots$$

has as its domain all values of x for which the power series converges.

Convergence of a power series: For a power series centered at c precisely one of the following is true:

(1) The series converges only at c.
(2) There exists a real number $R > 0$ such that the series converges absolutely for $|x - c| < R$ and diverges for $|x - c| > R$.
(3) The series converges absolutely for all x.

The number R is the *radius of convergence of the power series*. If the series converges only at c the radius of convergence is $R = 0$ and if the series converges for all x the radius of convergence is $R = \infty$. The set of all values of x for which the power series converges is the *interval of convergence of the power series*.

Usually, we use the ratio test or root test to find the interval where the series converges absolutely.

Operations on Power Series

On the intersection of their intervals of convergence, two power series can be added and subtracted term by term.

The series multiplication theorem for power series: If

$$f(x) = \sum_{n=0}^{\infty} a_n x^n \quad \text{and} \quad g(x) = \sum_{n=0}^{\infty} b_n x^n$$

converges absolutely for $|x| < R$ and

$$c_n = a_0 b_n + a_1 b_{n-1} + a_2 b_{n-2} + \cdots + a_{n-1} b_1 + a_n b_0,$$

then $\sum_{n=0}^{\infty} c_n x^n$ converges absolutely to $f(x)g(x)$ for $|x| < R$.

We can also substitute a function $f(x)$ for x in a convergent power series. Indeed, if $\sum a_n x^n$ converges absolutely for $|x| < R$, then $\sum a_n \big(f(x)\big)^n$ converges absolutely for any continuous function f on $|f(x)| < R$.

Terms by terms differentiation of a power series: If the function f given by

$$f(x) = \sum_{n=0}^{\infty} a_n (x - c)^n$$

has a radius of convergence $R > 0$, then f is differentiable on the interval $(c - R, c + R)$ and

$$f'(x) = \sum_{n=1}^{\infty} n a_n (x - c)^{n-1}.$$

The radius of convergence of the series obtained by differentiating is the same as that of the original power series.

It is also true that a power series can be integrated term by term throughout its interval of convergence.

Terms by terms integration of a power series: If the function f given by

$$f(x) = \sum_{n=0}^{\infty} a_n (x - c)^n$$

has a radius of convergence $R > 0$, then f is integrable and for each $x \in (c - R, c + R)$ we have

$$\int f(x)dx = \sum_{n=1}^{\infty} \frac{a_n}{n + 1} (x - c)^{n+1} + C.$$

Taylor and Maclaurin Series

Let f be a function with derivatives of all orders throughout some interval containing a as an interior point. Then the *Taylor series* generated by f at c is

$$\sum_{k=0}^{\infty} \frac{f^{(k)}(c)}{k!}(x-c)^k = f(c) + \frac{f'(c)}{1!}(x-c) + \cdots + \frac{f^{(n)}(c)}{n!}(x-c)^n + \ldots$$

Moreover, if $c = 0$ then the series is the *Maclaurin series* for f.

Let f have derivatives of all orders in an open interval I centered at c. The Taylor series for f may fail to converge for some x in I. Or, even if it is convergent, it may fail to have $f(x)$ as its sum. Nevertheless, Taylor's theorem in Chap. 3 tells us that for each n,

$$f(x) = f(c) + \frac{f'(c)}{1!}(x-c) + \cdots + \frac{f^{(n)}(c)}{n!}(x-c)^n + R_n(x),$$

where

$$R_n(x) = \frac{f^{(n+1)}(\xi)}{(n+1)!}(x-c)^{n+1}.$$

Note that in this remainder formula the particular value of ξ that makes the remainder formula true depends on the values of x and n.

Convergence of Taylor series: If $\lim_{n\to\infty} R_n(x) = 0$ for all x in the interval I, then the Taylor series for f converges and equals $f(x)$,

$$f(x) = \sum_{k=0}^{\infty} \frac{f^{(k)}(c)}{k!}(x-c)^k.$$

If the reader is interested to see the proof of theorems that are presented in this chapter, we refer him/her to [1–16].

7.2 Problems

427. *By the definition of limit, prove that*

$$\lim_{n\to\infty} \frac{2n+1}{5n-3} = \frac{2}{5}.$$

Solution. For each $\epsilon > 0$, we need to decide how big n must be to guarantee that

$$\left| \frac{2n+1}{5n-3} - \frac{2}{5} \right| < \epsilon \text{ or } \left| \frac{11}{5(5n-3)} \right| < \epsilon,$$

or equivalently $\dfrac{11}{25\epsilon} + \dfrac{3}{5} < n$. Our steps are reversible, so we put $N = \dfrac{11}{25\epsilon} + \dfrac{3}{5}$. Thus, for each $n > N$, we have $5n > \dfrac{11}{5\epsilon} + 3$. Hence, $5n - 3 > \dfrac{11}{5\epsilon}$ and so $\left| \dfrac{2n+1}{5n-3} - \dfrac{2}{3} \right| = \dfrac{11}{5(5n-3)} < \epsilon$. This completes the proof.

428. *By the definition of limit, prove that*

$$\lim_{n \to \infty} \frac{\sqrt{n^2 + 1}}{n!} = 0.$$

Solution. We have

$$\frac{\sqrt{n^2 + 1}}{n!} \leq \frac{\sqrt{2n^2}}{n!} = \frac{\sqrt{2}n}{n!} = \frac{\sqrt{2}}{(n-1)!} \leq \frac{\sqrt{2}}{n-1}.$$

Now, let $\epsilon > 0$ be arbitrary. Since $\dfrac{\sqrt{2}}{n-1} < \epsilon$ if and only if $n > \dfrac{\sqrt{2}}{\epsilon} + 1$, we pick $N = \dfrac{\sqrt{2}}{\epsilon} + 1$. Then, we obtain

$$0 \leq \frac{\sqrt{n^2 + 1}}{n!} \leq \frac{\sqrt{2}}{n-1} < \frac{\sqrt{2}}{N-1} = \epsilon, \text{ whenever } n > N.$$

This completes the proof.

429. *Show that the sequence* $\{(-1)^n + 1\}$ *is divergent.*

Solution. Suppose that the sequence is convergent and $(-1)^n + 1 \to L$ as $n \to \infty$. Let $\epsilon = \dfrac{1}{2}$. Then, there is $N \in \mathbb{N}$ such that

$$|(-1)^n + 1 - L| < \frac{1}{2}, \text{ whenever } n > N.$$

If $n > N$ is even, then $|2 - L| < \dfrac{1}{2}$. If $n > N$ is odd, then $|L| < \dfrac{1}{2}$. So, we get

$$2 = |2 - L + L| \leq |2 - L| + |L| < \frac{1}{2} + \frac{1}{2} = 1.$$

This is a contradiction.

430. *Let $\{x_n\}$ and $\{y_n\}$ be two sequences in $\mathbb{R}$ and let $\{z_n\}$ be the sequence in $\mathbb{R}$ defined by*

$$z_1 = x_1, \qquad z_2 = y_1,$$
$$z_3 = x_2, \qquad z_4 = y_2,$$
$$\vdots \qquad \qquad \vdots$$
$$z_{2n-1} = x_n, \qquad z_{2n} = y_n,$$
$$\vdots \qquad \qquad \vdots$$

Is it true that $\{z_n\}$ is convergent if and only if $\{x_n\}$ and $\{y_n\}$ are convergent and $\lim\limits_{n\to\infty} x_n = \lim\limits_{n\to\infty} y_n$.

Solution. Let $z_n \to L$ as $n \to \infty$. Then, every subsequence of $\{z_n\}$ is convergent. Since $\{z_{2n-1}\}$ and $\{z_{2n}\}$ are two subsequence of $\{z_n\}$, it follows that the sequences $\{x_n\}$ and $\{y_n\}$ are convergent and $\lim\limits_{n\to\infty} x_n = \lim\limits_{n\to\infty} y_n = \lim\limits_{n\to\infty} z_n = L$.

Now, suppose that the sequences $\{x_n\}$ and $\{y_n\}$ are convergent and $\lim\limits_{n\to\infty} x_n = \lim\limits_{n\to\infty} y_n = L$. For $\epsilon > 0$, there exist $N_1, N_2 \in \mathbb{N}$ such that

$$|x_n - L| < \epsilon, \text{ whenever } n \geq N_1,$$
$$|y_n - L| < \epsilon, \text{ whenever } n \geq N_2.$$

If $N = \max\{N_1 N_2\}$, then

$$|z_{2n-1} - L| < \epsilon \text{ and } |z_{2n} - L| < \epsilon, \text{ whenever } n \geq N,$$

or equivalently, $|z_n - L| < \epsilon$, whenever $n \geq N$. Therefore, $z_n \to L$ as $n \to \infty$.

431. *Is it true Problem (430) if we remove the condition* $\lim\limits_{n\to\infty} x_n = \lim\limits_{n\to\infty} y_n$.

Solution. No. For example, we can consider the sequence $\{(-1)^n\}$. This sequence is divergent while the sequences $\{(-1)^{2n-1}\}$ and $\{(-1)^{2n}\}$ are constant and $\lim\limits_{n\to\infty} \{(-1)^{2n-1}\} = -1$ and $\lim\limits_{n\to\infty} \{(-1)^{2n}\} = 1$.

432. *Suppose that $\{a_n\}$ is a sequence and the arithmetic mean sequence $\{\sigma_n\}$ is defined as follows:*
$$\sigma_n = \frac{a_1 + a_2 + \cdots + a_n}{n}.$$

If $\lim\limits_{n\to\infty} a_n = a$, *prove that* $\lim\limits_{n\to\infty} \sigma_n = a$.

Solution. If $a = 0$, then for $\epsilon > 0$, there is $N_1 \in \mathbb{N}$ such that $|a_n| < \dfrac{\epsilon}{2}$, for all $n \geq N_1$. Let $M = \max\{|a_1|, |a_2|, \ldots, |a_{N_1-1}|\}$. Then, for $n \geq N_1$ we obtain

$$|\sigma_n| \leq \frac{1}{n}\left((|a_1|, |a_2|, \ldots, |a_{N_1-1}|) + (|a_{N_1}| + \cdots + |a_n|)\right)$$

$$\leq \frac{(N_1 - 1)M + (n - N_1 + 1)\dfrac{\epsilon}{2}}{n}$$

$$\leq \frac{(N_1 - 1)M}{n} + \frac{\epsilon}{2}.$$

Now, we pick $N_2 \in \mathbb{N}$ such that $2(N_1 - 1)M < \epsilon N_2$. Hence, $\dfrac{(N_1 - 1)M}{n} < \dfrac{\epsilon}{2}$, for all $n \geq N_2$. If $N = \max\{N_1, N_2\}$, then $|\sigma_n| < \epsilon$, for all $n \geq N$. Therefore, we conclude that $\lim_{n \to \infty} \sigma_n = 0$.

If $a \neq 0$, then $\lim_{n \to \infty} (a_n - a) = 0$. By the above argument we have

$$\lim_{n \to \infty} \frac{(a_1 - a) + (a_2 - a) + \ldots (a_n - a)}{n} = 0,$$

which implies that $\lim_{n \to \infty} \sigma_n = a$.

433. *Show that the converse of Problem (432) is not, generally, true.*

Solution. We consider the sequence $a_n = (-1)^{n-1} + 1$, for all $n \in \mathbb{N}$. Then, its arithmetic mean sequence is

$$\sigma_n = \begin{cases} \dfrac{1}{2} & \text{if } n \text{ is even} \\ \dfrac{n + 1}{2n} & \text{if } n \text{ is odd,} \end{cases}$$

and so $\sigma_n \to \dfrac{1}{2}$ as $n \to \infty$, while $\{a_n\}$ is divergent.

434. *Let $\{a_n\}$ and $\{b_n\}$ be two convergent sequences such that $\lim_{n \to \infty} a_n = a$ and $\lim_{n \to \infty} b_n = b$. Prove that*

$$\lim_{n \to \infty} \frac{a_1 b_n + a_2 b_{n-1} + \cdots + a_n b_1}{n} = ab.$$

Solution. We can write

$$\frac{a_1 b_n + a_2 b_{n-1} + \cdots + a_n b_1}{n}$$

$$= a \frac{b_1 + b_2 + \cdots + b_n}{n} + \frac{(a_1 - a)b_n + (a_2 - a)b_{n-1} + \cdots + (a_n - a)b_1}{n} \quad (7.2)$$

By Problem (432), the limit of the first term in the right side of Eq. (7.2) is equal to ab. So, it is enough to prove that the limit of the second term is equal to 0. Since

$\{b_n\}$ is convergent, it follows that there exists a number M such that $|b_n| \leq M$, for all $n \in \mathbb{N}$. Hence, we have

$$\frac{|(a_1 - a)b_n + \cdots + (a_n - a)b_1|}{n} \leq M \frac{|a_1 - a| + \cdots + |a_n - a|}{n}.$$

Since $|a_n - a| \to 0$ as $n \to \infty$, by Problem (432), we conclude that

$$\frac{|a_1 - a| + |a_n - a| + \cdots + |a_n - a|}{n} = 0.$$

This completes the proof of our statement.

435. *If $x > 0$, find $L = \lim\limits_{n \to \infty} n \ln(\dfrac{nx + 1}{nx})$.*

Solution. We have

$$\begin{aligned}
L &= \lim_{n \to \infty} \frac{\ln(nx + 1) - \ln(nx)}{\dfrac{1}{n}} \\
&= \lim_{h \to 0} \frac{\ln(\dfrac{x}{h} + 1) - \ln(\dfrac{x}{h})}{h} \\
&= \lim_{h \to 0} \frac{\ln(x + h) - \ln h - \ln x + \ln h}{h} \\
&= \lim_{h \to 0} \frac{\ln(x + h) - \ln x}{h} \\
&= (\ln x)' = \frac{1}{x}.
\end{aligned}$$

436. *Find $\lim\limits_{n \to \infty} (\sqrt{n^2 + n} - n)$ if it exists.*

Solution. Suppose that $x_n = \sqrt{n^2 + n} - n$. Then, we have

$$\begin{aligned}
x_n &= \frac{n^2 + n - n^2}{\sqrt{n^2 + n} + n} = \frac{n}{\sqrt{n^2 + n} + n} \leq \frac{n}{n + n} = \frac{1}{2}, \\
x_n &= \frac{n}{\sqrt{n^2 + n} + n} \geq \frac{n}{\sqrt{n^2 + 2n + 1} + n} = \frac{n}{n + 1 + n} = \frac{n}{2n + 1}.
\end{aligned}$$

Therefore, $\dfrac{n}{2n + 1} \leq x_n \leq \dfrac{1}{2}$. Now, by Sandwich theorem, we conclude that $\lim\limits_{n \to \infty} (\sqrt{n^2 + n} - n) = \dfrac{1}{2}$.

437. *Prove that*

$$\lim_{n \to \infty} \frac{[x] + [2x] + \cdots + [nx]}{n^2} = \frac{x}{2}.$$

Solution. Clearly, we have

$$x - 1 < [x] \leq x, \; 2x - 1 < [2x] \leq 2x, \; \ldots \; nx - 1 < [nx] \leq nx.$$

So,

$$(x - 1) + (2x - 1) + \cdots + (nx - 1) < [x] + [2x] + \cdots + [nx] \leq x + 2x + \cdots + nx.$$

This implies that

$$(1 + 2 + \cdots + n)x - n < [x] + [2x] + \cdots + [nx] \leq (1 + 2 + \cdots + n)x.$$

Hence, we obtain

$$\frac{n(n + 1)}{2}x - n < [x] + [2x] + \cdots + [nx] \leq \frac{n(n + 1)}{2}x.$$

Therefore,

$$\frac{n(n + 1)}{2n^2}x - \frac{1}{n} < \frac{[x] + [2x] + \cdots + [nx]}{n^2} \leq \frac{n(n + 1)}{2n^2}x.$$

Since $\lim\limits_{n \to \infty} \left(\frac{n(n + 1)}{2n^2}x - \frac{1}{n} \right) = \lim\limits_{n \to \infty} \frac{n(n + 1)}{2n^2}x = \frac{x}{2}$, by Sandwich's theorem the proof completes.

438. *Let a be any real number greater than 1 and p is a positive real number. Show that*

$$\lim_{n \to \infty} \frac{n^p}{a^n} = 0.$$

Solution. We can write $a = 1 + \delta$, where $\delta > 0$. Let k be a positive number such that $k > p$. For $n > 2k$, we have

$$a^n = (1 + \delta)^n = 1 + \binom{n}{1}\delta + \binom{n}{2}\delta^2 + \cdots + \binom{n}{k}\delta^k + \cdots + \delta^n$$

$$> \binom{n}{k}\delta^k$$

$$= \frac{n(n - 1) \ldots (n - k + 1)}{k!}\delta^k$$

$$> \frac{n^k \delta^k}{2^k k!}.$$

Therefore, we obtain

$$0 < \frac{n^p}{a^n} < \frac{2^k k!}{\delta^k n^{k-p}} \quad \text{(for } n > 2k\text{)}.$$

Since $k - p > 0$, it follows that $\dfrac{1}{n^{k-p}} \to 0$ as $n \to \infty$. Now, if we apply Sandwich's

theorem, then we obtain $\lim\limits_{n\to\infty} \dfrac{n^p}{a^n} = 0$.

439. *Let* $\{a_n\}_{n=1}^{\infty}$ *be a sequence such that* $\lim\limits_{n\to\infty} a_n = \infty$ *and let* $r > 1$. *Show that*

$$\lim_{n\to\infty} \frac{a_n}{r^{a_n}} = 0.$$

Solution. Since $a_n \to \infty$ as $n \to \infty$, it follows that for each large enough positive integer n, we can find $k \in \mathbb{N}$ such that $k \le a_n < k + 1$. Furthermore $k \to \infty$ as $n \to \infty$. Hence, $r^k \le r^{a_n} < r^{k+1}$. So, we have

$$\frac{k}{r^{k+1}} \le \frac{a_n}{r^{a_n}} \le \frac{k+1}{r^k}.$$

Now, by Sandwich's theorem, we conclude that $\dfrac{a_n}{r^{a_n}} \to 0$ as $n \to \infty$.

440. *Let*

$$a_n = k + \frac{d_1}{10} + \frac{d_2}{10^2} + \cdots + \frac{d_n}{10^n} \quad \text{(decimal expansion)},$$

where k *is a non-negative integer and each* d_i *belongs to* $\{0, 1, \ldots, 9\}$. *Prove that* $a_n < k + 1$, *i.e.,* $\{a_n\}$ *is bounded.*

Solution. Let

$$A_n = \frac{9}{10} + \frac{9}{10^2} + \cdots + \frac{9}{10^n}.$$

Then, we obtain

$$A_n = \frac{9}{10} \sum_{k=0}^{n-1} \left(\frac{1}{10}\right)^k = \frac{9}{10} \cdot \frac{1 - \left(\dfrac{1}{10}\right)^n}{1 - \dfrac{1}{10}} = 1 - \left(\frac{1}{10}\right)^n.$$

Now, we have

$$a_n = k + \frac{d_1}{10} + \frac{d_2}{10^2} + \cdots + \frac{d_n}{10^n}$$

$$\le k + \frac{9}{10} + \frac{9}{10^2} + \cdots + \frac{9}{10^n}$$

$$= k + 1 - \frac{1}{10^n} < k + 1.$$

441. *Let* $a_1 = 1$ *and* $a_{n+1} = \dfrac{a_n^2 + 2}{2a_n}$, *for all* $n \ge 1$. *Suppose that* $\{a_n\}$ *converges and find its limit.*

Solution. Let $a = \lim_{n \to \infty} a_n$. Then,

$$\lim_{n \to \infty} a_{n+1} = \lim_{n \to \infty} \frac{a_n^2 + 2}{2a_n},$$

which implies that $a = \frac{a^2 + 2}{2a}$ or $a = \pm\sqrt{2}$. Consequently, $a = \sqrt{2}$, thanks to the positivity of each a_n, which can be proved by induction on n.

442. *Let c be any positive number. Show that the sequence*

$$a_1 = \sqrt{c}, \ a_2 = \sqrt{c + \sqrt{c}}, \ a_3 = \sqrt{c + \sqrt{c + \sqrt{c}}}, \dots$$

is convergent and then find its limit.

Solution. We can write the sequence by the recursion formula $a_n = \sqrt{c + a_{n-1}}$ for $n \geq 2$. Clearly, $\{a_n\}$ is an increasing sequence and by mathematical induction we can see that $a_n \leq \sqrt{c} + 1$, i.e., $\{a_n\}$ is bounded. Since $\{a_n\}$ is increasing and bounded, it follows that its limit exists. Suppose that $a_n \to L$ as $n \to \infty$. Since $a_n = \sqrt{c + a_{n-1}}$, it follows that $a_n^2 = c + a_{n-1}$. So,

$$\lim_{n \to \infty} a_n^2 = c + \lim_{n \to \infty} a_{n-1},$$

which implies that $L^2 = c + L$ or $L^2 - L - c = 0$. Consequently, we get $L = \frac{1 \pm \sqrt{1 + 4c}}{2}$. Since $\frac{1 - \sqrt{1 + 4c}}{2} < 0$, we conclude that $L = \frac{1 + \sqrt{1 + 4c}}{2}$.

443. *If $0 < b \leq a$ and $x_n = (a^n + b^n)^{\frac{1}{n}}$, show that $\lim_{n \to \infty} x_n = a$.*

Solution. We have

$$a = (a^n)^{\frac{1}{n}} < (a^n + b^n)^{\frac{1}{n}} \leq (a^n + a^n)^{\frac{1}{n}} = \sqrt[n]{2}a.$$

Since $a < x_n < \sqrt[n]{2}a$ and $\sqrt[n]{2} \to 1$ as $n \to \infty$, it follows that $x_n \to a$ as $n \to \infty$.

444. *Let $a_1 = 1$ and $a_{n+1} = \frac{n}{n+1}a_n^2$, for $n > 1$.*

(1) Find a_2, a_3 and a_4.
(2) Show that $\lim_{n \to \infty} a_n$ exists.
(3) Prove that $\lim_{n \to \infty} a_n = 0$.

Solution. (1) It is clear that

$$a_2 = \frac{1}{2}a_1^2 = \frac{1}{2},$$

$$a_3 = \frac{2}{3}a_2^2 = \frac{1}{6},$$

$$a_4 = \frac{3}{4}a_3^2 = \frac{1}{48}.$$

(2) We prove that

$$0 < a_{n+1} < a_n < 1, \text{ for all } n \geq 1. \tag{7.3}$$

This is obvious from part (1) for $n = 1, 2, 3$. Suppose that (7.3) holds for $n - 1$. Then, $a_n < 1$ and so

$$a_{n+1} = \frac{n}{n+1}a_n^2 = \left(\frac{n}{n+1}a_n\right)a_n < a_n.$$

Since $a_n > 0$, it follows that $a_{n+1} > 0$. So, $0 < a_{n+1} < a_n < 1$. Therefore, (7.3) holds by induction. Now, according to (7.3), $\{a_n\}$ is a bounded monotone sequence, and so is convergent.

(3) Suppose that $a_n \to a$ as $n \to \infty$. Then, we obtain

$$a = \lim_{n \to \infty} a_{n+1} = \lim_{n \to \infty} \frac{n}{n+1}a_n^2 = a^2.$$

Hence, $a = 1$ or $a = 0$. But $a = 1$ is impossible since $a_n \leq \frac{1}{2}$ for $n \geq 2$. Thus, $a = 0$.

445. *Let a_1 and a_2 be two arbitrary numbers and $a_{n+2} = \dfrac{a_{n+1} + a_n}{2}$, for all $n \geq 1$. If $\lim\limits_{n \to \infty} a_n = L$, show that $L = \dfrac{a_1 + 2a_2}{3}$.*

Solution. We can write

$$a_{n+2} - a_{n+1} = \frac{1}{2}(a_n - a_{n+1}), \tag{7.4}$$

for all $n \geq 1$. By (7.4), we have

$$a_3 - a_2 = \frac{1}{2}(a_1 - a_2),$$

$$a_4 - a_3 = \frac{1}{2}(a_2 - a_3),$$

$$\vdots$$

$$a_n - a_{n-1} = \frac{1}{2}(a_{n-2} - a_{n-1}).$$

At this point, we add the above equalities to each other. We obtain $a_n - a_2 = \frac{1}{2}(a_1 - a_{n-1})$. Hence, $a_n + \frac{1}{2}a_{n-1} = \frac{1}{2}a_1 + a_2$. Therefore,

$$\lim_{n \to \infty} \left(a_n + \frac{1}{2}a_{n-1} \right) = \frac{1}{2}a_1 + a_2,$$

which implies that $L + \frac{1}{2}L = \frac{1}{2}a_1 + a_2$ or $L = \dfrac{a_1 + 2a_2}{3}$.

446. *Let a_0, b_0 be two positive numbers with $a_0 > b_0$. Define*

$$a_{n+1} = \frac{a_n + b_n}{2} \quad and \quad b_{n+1} = \sqrt{a_n b_n} \quad for \ n = 0, 1, 2, \dots.$$

Show that the limits of $\{a_n\}$ and $\{b_n\}$ exist and are equal.

Solution. We observe that

$$a_1 - b_1 = \frac{a_0 + b_0}{2} - \sqrt{a_0 b_0} = \frac{1}{2}\left(\sqrt{a_0} - \sqrt{b_0} \right)^2 > 0,$$
$$a_0 - a_1 = a_0 - \frac{a_0 + b_0}{2} = \frac{a_0 - b_0}{2} > 0,$$
$$b_0 - b_1 = b_0 - \sqrt{a_0 b_0} = \sqrt{b_0}\left(\sqrt{b_0} - \sqrt{a_0} \right) < 0.$$

So, we conclude that $a_0 > a_1 > b_1 > b_0$. Similarly, we obtain $a_1 > a_2 > b_2 > b_1$. If we repeat the above process, for every n we get

$$a_n > a_{n+1} > b_{n+1} > b_n.$$

Hence the sequence $\{a_n\}$ is decreasing and the sequence $\{b_n\}$ is increasing. Moreover, for every n, we have $a_n > b_0$ and $b_n < a_0$. These mean that the sequences $\{a_n\}$ and $\{b_n\}$ are monotone and bounded. Therefore, the sequences $\{a_n\}$ and $\{b_n\}$ are convergent. Suppose that

$$\lim_{n \to \infty} a_n = a \text{ and } \lim_{n \to \infty} b_n = b.$$

We show that $a = b$. Indeed, we have

$$\lim_{n \to \infty} a_{n+1} = \lim_{n \to \infty} \frac{a_n + b_n}{2}.$$

This yields $a = \frac{1}{2}a + \frac{1}{2}b$, and consequently $a = b$.

447. *Let $f : [0, 1] \to [0, 1]$ be a continuous function and differentiable in $(0, 1)$. Let $|f'(x)| \leq M < 1$, for all $x \in (0, 1)$ and some constant M.*

(1) Prove that the equation $f(x) = x$ has a unique solution z in $[0, 1]$.
(2) Let $a \in [0, 1]$ be arbitrary and $\{x_n\}$ be the sequence defined as follows:

$$x_0 = a, \quad x_1 = f(x_0), \quad \ldots, \quad x_{n+1} = f(x_n), \quad \ldots.$$

Prove that $\lim\limits_{n \to \infty} x_n = z$.

Solution. (1) In Problem (101), we proved that the equation $f(x) = x$ has a solution, named z. Now, if $f(z_1) = z_1$, then by the mean value theorem there is c between z and z_1 such that $f(z_1) - f(z) = f'(c)(z_1 - z)$. This implies that $|z_1 - z| \leq M|z_1 - z|$. Since $0 \leq M < 1$, it follows that $z_1 = z$.

(2) For each n, by the mean value theorem, there is c_n between z and x_n such that $f(x_n) - f(z) = f'(c_n)(x_n - z)$. Hence, we get

$$|x_{n+1} - z| = |f(x_n) - f(z)| = |f'(c_n)| \cdot |x_n - z| \leq M|x_n - z|.$$

By induction, we observe that for each positive integer n,

$$|x_n - z| \leq M|x_{n-1} - z| \leq M^2|x_{n-2} - z| \leq \cdots \leq M^n|a - z|.$$

Since $0 \leq M < 1$, it follows that $M^n \to 0$ as $n \to \infty$. Therefore, we conclude that $x_n \to z$ as $n \to \infty$.

448. *Find a continuous function $f : (0, 1) \to \mathbb{R}$ and a convergent sequence $\{a_n\}$ such that $\{f(a_n)\}$ is not convergent.*

Solution. We consider $f : (0, 1) \to \mathbb{R}$ defined by $f(x) = \dfrac{1}{x}$ and we set $a_n = \dfrac{1}{n}$. Clearly, f is a continuous function and $a_n \to 0$ as $n \to \infty$ but $\{f(a_n)\} = \{n\}$ is a divergent sequence.

449. *Let*

$$a_n = \left(\sum_{k=1}^{n} \frac{1}{k} \right) - \ln n.$$

(1) Show that $\{a_n\}$ is a decreasing sequence.
(2) Show that $0 < a_n < 1$, for all $n \geq 2$.
(3) Observe that $\lim\limits_{n \to \infty} a_n$ exists.

Solution. (1) We have

$$a_n - a_{n+1} = \left(\sum_{k=1}^{n} \frac{1}{k} \right) - \ln n - \left(\sum_{k=1}^{n+1} \frac{1}{k} \right) + \ln(n + 1)$$

$$= -\frac{1}{n+1} + \ln(n+1) - \ln n$$

$$= -\frac{1}{n+1} + \int_{n}^{n+1} \frac{1}{t} dt.$$

Since $\dfrac{1}{n+1} < \dfrac{1}{t}$ for all $t \in [n, n+1]$ and then

$$\frac{1}{n+1} = \int_{n}^{n+1} \frac{1}{n+1} dt < \int_{n}^{n+1} \frac{1}{t} dt,$$

it follows that $0 < a_n - a_{n+1}$ or $a_{n+1} < a_n$.

(2) From (1), we conclude that $a_n \leq a_1$, for all $n \geq 2$. Moreover, we have

$$0 < \sum_{k=1}^{n} \left(\frac{1}{k} - \int_{k}^{k+1} \frac{1}{t} dt \right) = \left(\sum_{k=1}^{n} \frac{1}{k} \right) - \int_{1}^{n+1} \frac{1}{t} dt$$

$$\leq \left(\sum_{k=1}^{n} \frac{1}{k} \right) - \int_{1}^{n} \frac{1}{t} dt = a_n.$$

Hence, $0 < a_n < 1$, for all n.

(3) Since $\{a_n\}$ is a bounded monotone sequence, we conclude that $\{a_n\}$ is convergent.

450. *Determine the sequence*

$$x_n = \frac{a^n - b^n}{a^n + b^n}, \quad \text{for all } n \in \mathbb{N},$$

is convergent or divergent, where a and b are real numbers such that $|a| \neq |b|$.

Solution. We can consider two cases, either $|a| < |b|$ or $|b| < |a|$.

If $|a| < |b|$, then we set $r = \dfrac{a}{b}$. Hence, we obtain

$$x_n = \frac{r^n - 1}{r^n + 1}.$$

Since $|r| < 1$, it follows that $\lim_{n \to \infty} r^n = 0$. This implies that $\lim_{n \to \infty} x_n = -1$.

Fig. 7.1 k_n congruent circular disks occupying n rows are inscribed in an equilateral triangle

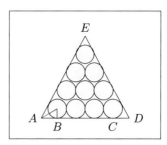

If $|b| < |a|$, then we can write

$$\frac{a^n - b^n}{a^n + b^n} = -\frac{b^n - a^n}{b^n + a^n},$$

and similar to the above argument, we get $\lim_{n\to\infty} x_n = 1$.

451. *Suppose that k_n congruent circular disks occupying n rows are inscribed in an equilateral triangle in the way illustrated in Fig. 7.1, so that*

$$k_1 = 1, \ k_2 = 1 + 2 = 3, \ k_3 = 1 + 2 + 3 = 6, \ \ldots$$

Let S be the area of triangle and S_n the total area of the k_n disks. Show that

$$\lim_{n\to\infty} \frac{S_n}{S} = \frac{\pi}{2\sqrt{3}}.$$

Solution. We know that the angle A is $\frac{\pi}{3}$. So, if r is the radius of each circle, then $\tan\frac{\pi}{6} = \frac{r}{AB}$. Hence, we get $\frac{1}{\sqrt{3}} = \frac{r}{AB}$, or equivalently,

$$\overline{AB} = \sqrt{3}r.$$

Now, we have

$$\overline{BC} = r + \underbrace{2r + \cdots + 2r}_{n-2 \text{ times}} + r = 2(n-1)r,$$

where n is the number of circles in the last row, and

$$\overline{AD} = \overline{AB} + \overline{BC} + \overline{CD} = \sqrt{3}r + 2(n-1)r + \sqrt{3}r = 2(\sqrt{3} + n - 1)r.$$

Since $k_n = k_{n-1} + n$, it follows that $k_n = 1 + 2 + \cdots + n = \frac{n(n+1)}{2}$. The area of each circle is πr^2. So, the total area of the k_n disks is

$$S_n = \pi r^2 k_n = \frac{\pi r^2 n(n+1)}{2}.$$

On the other hand, The area of an equilateral triangle can be found using the formula $\frac{\sqrt{3}}{4}a^2$, where a is the length of one side of the triangle. So, the area of triangle ADE is

$$S = \frac{\sqrt{3}}{4}\overline{AD}^2 = \frac{4\sqrt{3}(\sqrt{3}+n-1)^2 r^2}{4} = \sqrt{3}(\sqrt{3}+n-1)^2 r^2.$$

Therefore, we find that

$$\lim_{n\to\infty}\frac{S_n}{S} = \lim_{n\to\infty}\frac{\pi r^2 n(n+1)}{2\sqrt{3}(\sqrt{3}+n-1)^2 r^2} = \frac{\pi}{2\sqrt{3}}.$$

452. *Let $I_n = [a_n, b_n]$ and $I_{n+1} \subseteq I_n$, for all $n \in \mathbb{N}$. Show that*

$$\bigcap_{n=1}^{\infty} I_n \neq \emptyset.$$

Solution. The sequence $\{a_n\}$ is bounded above by b_1 and it is increasing, so it is convergent. Let $a_n \to x$ as $n \to \infty$. If m and n are arbitrary positive integers, then $a_n \le a_{m+n} \le b_{m+n} \le b_m$. So, $x \le b_m$ for all m. On the other hand, it is clear that $a_m \le x$. Therefore, we conclude that $x \in I_m$, for all $m \in \mathbb{N}$.

453. *Show that every bounded sequence has a convergent subsequence.*

Solution. Suppose that $\{x_n\}$ is a bounded sequence. Then, there exists an interval $[a_1, b_1]$ such that for each n, $x_n \in [a_1, b_1]$. One of the intervals $[a_1, \frac{a_1+b_1}{2}]$ or $[\frac{a_1+b_1}{2}, b_1]$ contains infinitely many terms of $\{x_n\}$. If $[a_1, \frac{a_1+b_1}{2}]$ contains infinitely many terms of $\{x_n\}$, we put $[a_2, b_2] = [a_1, \frac{a_1+b_1}{2}]$, in otherwise, we put $[a_2, b_2] = [\frac{a_1+b_1}{2}, b_1]$. Again, one of the intervals $[a_2, \frac{a_2+b_2}{2}]$ or $[\frac{a_2+b_2}{2}, b_2]$ contains infinitely many terms of $\{x_n\}$. If $[a_2, \frac{a_2+b_2}{2}]$ contains infinitely many terms of $\{x_n\}$, we put $[a_3, b_3] = [a_2, \frac{a_2+b_2}{2}]$, in otherwise, we put $[a_3, b_3] = [\frac{a_2+b_2}{2}, b_2]$. If we continue this construction, then we get a sequence of intervals $\{[a_n, b_n]\}$ such that for all n,

(1) $[a_n, b_n]$ contains infinitely many terms of $\{x_n\}$,
(2) $[a_{n+1}, b_{n+1}] \subseteq [a_n, b_n]$,

(3) $b_{n+1} - a_{n+1} = \dfrac{b_n - a_n}{2}$.

By (3), $b_{n+1} - a_{n+1} = \dfrac{b_1 - a_1}{2^n}$ which tends to zero, as $n \to \infty$. So by Problem (452), the intersection of all the intervals $[a_n, b_n]$ is a singleton x. Since $[a_1, b_1]$ contains infinitely many terms of $\{x_n\}$, it follows that there exists i_1 such that $x_{i_1} \in [a_1, b_1]$. Since $[a_1, b_1]$ contains infinitely many terms of $\{x_n\}$, it follows that there exists i_2 with $i_1 < i_2$ such that $x_{i_2} \in [a_2, b_2]$. If we continue this process, we obtain a sequence $\{x_{i_n}\}$ such that $x_{i_n} \in [a_n, b_n]$. Now, the sequence $\{x_{i_n}\}$ is a subsequence of $\{x_n\}$. Since $a_n \leq x_{i_n} \leq b_n$ and $\lim\limits_{n \to \infty} a_n = \lim\limits_{n \to \infty} b_n = x$, we conclude that $x_{i_n} \to x$ as $n \to \infty$.

454. *A sequence $\{x_n\}$ of real numbers is said to be Cauchy sequence if for every $\epsilon > 0$ there is an integer N such that $|x_n - x_m| < \epsilon$ if $n > N$ and $m > N$. Prove that*

(1) Every convergent sequence is a Cauchy sequence.
(2) Every Cauchy sequence is bounded.
(3) Every Cauchy sequence is convergent.

Solution. (1) Let $\{x_n\}$ be a convergent sequence and $x_n \to x$ as $n \to \infty$. Let $\epsilon > 0$ be arbitrary. Then, there is $N \in \mathbb{N}$ such that $|x_n - x| < \epsilon$, whenever $n > N$. Thus, for all $m, n > N$, we get

$$|x_n - x_m| \leq |x_n - x| + |x - x_m| < \frac{\epsilon}{2} + \frac{\epsilon}{2} = \epsilon.$$

Therefore, $\{x_n\}$ is a Cauchy sequence.

(2) Let $\{x_n\}$ be a Cauchy sequence. If $\epsilon = 1$, then there exists $N \in \mathbb{N}$ such that $|x_n - x_m| < 1$, whenever $m, n \geq N$. Hence, $|x_n| \leq |x_n - x_N| + |x_N| < 1 + |x_N|$. Suppose that

$$M = \max\{|x_1|, \ldots, |x_{N-1}|, 1 + |x_N|\}.$$

Then, $|x_n| \leq M$, for all n. Therefore, $\{x_n\}$ is bounded.

(3) Let $\{x_n\}$ be a Cauchy sequence. By (2), $\{x_n\}$ is bounded. Thus, according to Problem (453), $\{x_n\}$ has a convergent subsequence $\{x_{i_n}\}$, i.e., $x_{i_n} \to x$ as $n \to \infty$. We prove that $\lim\limits_{n \to \infty} x_n = x$. Let $\epsilon > 0$ be arbitrary. There is $N_1 \in \mathbb{N}$ such that

$$|x_{i_n} - x| < \frac{\epsilon}{2}, \text{ whenever } n > N_1.$$

Since $\{x_n\}$ is a Cauchy sequence, there is $N_2 \in \mathbb{N}$ such that

$$|x_n - x_m| < \frac{\epsilon}{2}, \text{ whenever } n, m > N_2.$$

Let $N = \max\{N_1, N_2\}$. If $n > N$, then

$$|x_n - x| \le |x_n - x_{i_n}| + |x_{i_n} - x| < \frac{\epsilon}{2} + \frac{\epsilon}{2} = \epsilon.$$

Therefore, $x_n \to x$ as $n \to \infty$.

455. *Let $\{x_n\}$ be a sequence of real numbers and $0 \le r < 1$ be a real number such that*

$$|x_{n+2} - x_{n+1}| \le r|x_{n+1} - x_n|, \quad \text{for all } n \in \mathbb{N}.$$

Show that $\{x_n\}$ is convergent.

Solution. Let $|x_2 - x_1| = c$. Then, we have

$$|x_3 - x_2| \le r|x_2 - x_1| = rc,$$
$$|x_4 - x_3| \le r|x_3 - x_2| = r^2 c,$$
$$\vdots$$
$$|x_{n+1} - x_n| \le r^{n-1} c.$$

So, we get

$$
\begin{aligned}
|x_{n+k} - x_n| &\le |x_{n+k} - x_{n+k-1}| + \cdots + |x_{n+2} - x_{n+1}| + |x_{n+1} - x_n| \\
&\le r^{n+k-2} c + \cdots + r^n c + r^{n-1} c \\
&= cr^{n-1} \sum_{i=1}^{k} r^{i-1} \\
&\le cr^{n-1} \left(\frac{1}{1-r} \right).
\end{aligned}
$$

Since $\lim\limits_{n \to \infty} \dfrac{c}{1-r} r^{n-1} = 0$, it follows that $\{x_n\}$ is a Cauchy sequence, and hence by the third part of Problem (454), the proof completes.

456. *(1) Let $\{a_n\}$ be a sequence such that*

$$|a_{n+1} - a_n| < 2^{-n}, \quad \text{for all } n \in \mathbb{N}.$$

Prove that $\{a_n\}$ is a Cauchy sequence and hence is a convergent sequence.

(2) Is the result in (1) true if we only assume that $|a_{n+1} - a_n| < \dfrac{1}{n}$, for all $n \in \mathbb{N}$.

Solution. (1) If $m > n$, then

$$|a_m - a_n| \le \sum_{i=n}^{m-1} |a_{i+1} - a_i| < \sum_{i=n}^{m-1} \frac{1}{2^i} = \frac{1}{2^{n-1}} - \frac{1}{2^{m-1}} < \frac{1}{2^{n-1}} \to 0,$$

as $n \to \infty$. Therefore, $\{a_n\}$ is a Cauchy sequence.

(2) No, we can consider the following counter example. Let $a_n = \sum_{i=1}^{n} \frac{1}{i}$. We know that the series $\sum_{i=1}^{\infty} \frac{1}{i}$ is divergent and so $\{a_n\}$ is a divergent sequence. But we have

$$|a_{n+1} - a_n| = \frac{1}{n+1} < \frac{1}{n}, \quad \text{for all } n \in \mathbb{N}.$$

457. *Show that the sequence* $\{a_n\}$ *defined by*

$$a_n = \int_1^n \frac{\cos x}{x^2} dx$$

is a Cauchy sequence and so it is convergent.

Solution. If $n, m \in \mathbb{N}$ and $n > m$, then

$$\left| \int_m^n \frac{\cos x}{x^2} dx \right| \leq \int_m^n \left| \frac{\cos x}{x^2} \right| dx$$
$$\leq \int_m^n \frac{1}{x^2} dx$$
$$= \left(-\frac{1}{x} \right) \Big|_m^n$$
$$= \frac{1}{m} - \frac{1}{n}.$$

Since $\lim_{n \to \infty} \frac{1}{n} = 0$, it follows that for any $\epsilon > 0$, there is $N \in \mathbb{N}$ such that $\left(\frac{1}{m} - \frac{1}{n} \right) < \epsilon$, whenever $n, m > N$. Therefore, for $n > m > N$, we get

$$|a_n - a_m| = \left| \int_m^n \frac{\cos x}{x^2} dx \right| \leq \frac{1}{m} - \frac{1}{n} < \epsilon,$$

which implies that $\{a_n\}$ is a Cauchy sequence.

458. *Let* $\{a_n\}$ *be a decreasing sequence of positive integers. If* $\sum a_n$ *is convergent, show that* $\lim_{n \to \infty} n a_n = 0$.

Solution. Let $s_n = a_1 + a_2 + \cdots + a_n$ be the nth partial sum of the series. Since $\sum a_n$ is convergent, it follows that $\{s_n\}$ is a convergent sequence. We have

$$s_{2n} - s_n = a_{n+1} + \cdots + a_{2n}$$
$$\geq \underbrace{a_{2n} + \cdots + a_{2n}}_{n \text{ times}}$$
$$= n a_{2n}$$
$$= \frac{1}{2} (2n a_{2n}).$$

Since $\lim_{n\to\infty} (s_{2n} - s_n) = 0$, it follows that $\lim_{n\to\infty} 2na_{2n} = 0$.

On the other hand, we have $a_{2n+1} \le a_{2n}$, which implies that

$$(2n + 1)a_{2n+1} \le (2n + 1)a_{2n}$$
$$\le (2n + 2n)a_{2n}$$
$$= 2(2na_n).$$

Thus, $\lim_{n\to\infty} (2n + 1)a_{2n+1} = 0$.

Now, by Problem (430), we conclude that $\lim_{n\to\infty} na_n = 0$.

459. *Let $\{a_n\}$ be a decreasing sequence of positive numbers. Show that the series $\sum_{n=1}^{\infty} a_n$ is convergent if and only if the associated series $\sum_{k=0}^{\infty} 2^k a_{2^k}$ is convergent. This result is known as Cauchy's condensation test.*

Solution. Suppose that

$$s_n = a_1 + a_2 + \cdots + a_n,$$
$$t_k = a_1 + 2a_2 + \cdots + 2^k a_{2^k}.$$

If $n < 2^k$, then

$$s_n \le a_1 + a_2 + \cdots + a_{2^k}$$
$$\le a_1 + (a_2 + a_3) + (a_4 + a_5 + a_6 + a_7) + \cdots$$
$$\quad + (a_{2^k} + a_{2^k+1} + \cdots + a_{2^{k+1}-1}) \tag{7.5}$$
$$\le a_1 + 2a_2 + \cdots + 2^k a_{2^k}$$
$$= t_k.$$

If $n > 2^k$, then

$$s_n \ge a_1 + a_2 + (a_3 + a_4) + \cdots + (a_{2^{k-1}+1} + \cdots + a_{2^k})$$
$$\ge \frac{1}{2}a_1 + a_2 + 2a_4 + \cdots + 2^{k-1} a_{2^k} \tag{7.6}$$
$$= \frac{1}{2}t_k.$$

By (7.5) and (7.6) we conclude that two sequences $\{s_n\}$ and $\{t_n\}$ are either both bounded or both unbounded. Since a non-negative series is convergent if and only if its partial sums is bounded, the proof is completed.

460. *Show that the series*

$$\sum_{n=1}^{\infty} \frac{1}{(n + 1)(n + 2)}$$

is convergent, and find its sum.

Solution. Expanding the general term in partial fractions, we find that

$$\frac{1}{(n+1)(n+2)} = \frac{1}{n+1} - \frac{1}{n+2}.$$

Hence, the nth partial sum of the series is

$$
\begin{aligned}
s_n &= \left(\frac{1}{2} - \frac{1}{3}\right) + \left(\frac{1}{3} - \frac{1}{4}\right) + \left(\frac{1}{4} - \frac{1}{5}\right) + \cdots + \left(\frac{1}{n+1} - \frac{1}{n+2}\right) \\
&= \frac{1}{2} + \left(-\frac{1}{3} + \frac{1}{3}\right) + \left(-\frac{1}{4} + \frac{1}{4}\right) + \cdots + \left(-\frac{1}{n} + \frac{1}{n+1}\right) - \frac{1}{n+2} \\
&= \frac{1}{2} - \frac{1}{n+2}.
\end{aligned}
$$

Therefore,

$$\lim_{n\to\infty} s_n = \lim_{n\to\infty} \left(\frac{1}{2} - \frac{1}{n+2}\right),$$

from which it follows that the given series is convergent with sum $\frac{1}{2}$.

461. *Prove that*

$$\sum_{n=1}^{\infty} \frac{1}{n^2} \leq 2.$$

Solution. Since $(n-1)n < n^2$, it follows that $\dfrac{1}{n^2} < \dfrac{1}{(n-1)n}$, for all $n \geq 2$. So, we have

$$
\begin{aligned}
\sum_{n=2}^{m} \frac{1}{n^2} &< \sum_{n=2}^{m} \frac{1}{(n-1)n} = \sum_{n=2}^{m} \left(\frac{1}{n-1} - \frac{1}{n}\right) \\
&= \left(1 - \frac{1}{2}\right) + \left(\frac{1}{2} - \frac{1}{3}\right) + \cdots + \left(\frac{1}{m-1} - \frac{1}{m}\right) \\
&= 1 - \frac{1}{m}.
\end{aligned}
$$

Consequently, we get

$$\sum_{n=1}^{m} \frac{1}{n^2} < 2 - \frac{1}{m}.$$

This implies that $\displaystyle\sum_{n=1}^{\infty} \frac{1}{n^2} \leq 2$.

462. *Show that the series*

$$\sum_{n=2}^{\infty} \frac{1}{n^2 - 1}$$

is convergent, and find its sum.

Solution. Expanding the general term in partial fractions, we find that

$$\frac{1}{n^2 - 1} = \frac{1}{2}\left(\frac{1}{n-1} - \frac{1}{n+1}\right),$$

Hence, similar to Problem (460), the nth partial sum of the series is

$$s_n = \frac{1}{2}\left(1 + \frac{1}{2} - \frac{1}{n+1} - \frac{1}{n+2}\right).$$

Consequently,

$$\lim_{n\to\infty} s_n = \frac{1}{2} \lim_{n\to\infty} \left(1 + \frac{1}{2} - \frac{1}{n+1} - \frac{1}{n+2}\right),$$

from which it follows that the given series is convergent with sum $\dfrac{3}{4}$.

463. *Show that the series*

$$\sum_{n=1}^{\infty} \frac{n}{n^4 + n^2 + 1}$$

is convergent, and find its sum.

Solution. Since $n^4 < n^4 + n^2 + 1$, it follows that $\dfrac{n}{n^4 + n^2 + 1} < \dfrac{1}{n^3}$. Since $\sum_{n=1}^{\infty} \dfrac{1}{n^3}$ is a convergent p-series, by the comparison test we conclude that the series $\sum_{n=1}^{\infty} \dfrac{n}{n^4 + n^2 + 1}$ is convergent.

In addition, by expanding the general term in partial fractions we obtain

$$\frac{n}{n^4 + n^2 + 1} = \frac{n}{(n^2 + 1)^2 - n^2}$$

$$= \frac{\frac{1}{2}}{n^2 - n + 1} - \frac{\frac{1}{2}}{n^2 + n + 1}.$$

So, the partial sum of the series is

$$s_n = \frac{1}{2} - \frac{\frac{1}{2}}{n^2 + n + 1}.$$

Thus, the sum of the series is equal to $\dfrac{1}{2}$.

464. *Show that*

$$\sum_{n=1}^{\infty} \frac{1}{n(n+1)(n+2)} = \frac{1}{4}.$$

Solution. We can write

$$\frac{1}{n(n+1)(n+2)} = \frac{\frac{1}{2}}{n} - \frac{1}{n+1} + \frac{\frac{1}{2}}{n+2}.$$

Now, we write some terms of the series as follows:

$$a_1 = \frac{1}{2}\left(1 - 1 + \frac{1}{3}\right),$$
$$a_2 = \frac{1}{2}\left(\frac{1}{2} - \frac{2}{3} + \frac{1}{4}\right),$$
$$a_3 = \frac{1}{2}\left(\frac{1}{3} - \frac{2}{4} + \frac{1}{5}\right),$$
$$a_4 = \frac{1}{2}\left(\frac{1}{4} - \frac{2}{5} + \frac{1}{6}\right),$$
$$\vdots$$
$$a_{n-1} = \frac{1}{2}\left(\frac{1}{n-1} - \frac{2}{n} + \frac{1}{n+1}\right),$$
$$a_n = \frac{1}{2}\left(\frac{1}{n} - \frac{2}{n+1} + \frac{1}{n+2}\right).$$

Now, it is easy to see that

$$s_n = a_1 + a_2 + \cdots + a_n = \frac{1}{2}\left(\frac{1}{2} - \frac{1}{n+1} + \frac{1}{n+2}\right) = \frac{1}{4} + \frac{1}{2}\left(\frac{1}{n+2} - \frac{1}{n+1}\right).$$

Therefore, $s_n \to \dfrac{1}{4}$ as $n \to \infty$.

465. *Verify that*

$$\sum_{n=2}^{\infty} \frac{\ln\left(\left(1 + \frac{1}{n}\right)^n (1+n)\right)}{(\ln n^n)\left(\ln(n+1)^{n+1}\right)} = \log_2 \sqrt{e}.$$

Solution. We have

$$\frac{\ln\left(\left(1 + \frac{1}{n}\right)^n (1+n)\right)}{(\ln n^n)\left(\ln(n+1)^{n+1}\right)} = \frac{n\ln(n+1) - n\ln n + \ln(1+n)}{n(n+1)\ln n \ln(n+1)}$$

$$= \frac{1}{n\ln n} - \frac{1}{(n+1)\ln(n+1)}.$$

Hence, the series is a telescoping series and the partial sum of the series is equal to

$$s_n = \frac{1}{2\ln 2} - \frac{1}{(n+1)\ln(n+1)}.$$

Thus, the sum of the series is equal to $\dfrac{1}{2\ln 2}$. Since $\log_a x = \dfrac{\ln x}{\ln a}$, it follows that

$$\frac{1}{2\ln 2} = \frac{\ln \sqrt{e}}{\ln 2} = \log_2 \sqrt{e}.$$

This completes the proof.

466. *Let* $a_n = \dfrac{(-1)^n}{n}$ *and c be any real number. Prove that there is a permutation (one to one and onto function) $\sigma : \mathbb{N} \to \mathbb{N}$ such that*

$$\sum_{n=1}^{\infty} a_{\sigma(n)} = c.$$

Solution. Clearly, $a_{2n} > 0$ and $a_{2n-1} < 0$, for all $n \in \mathbb{N}$. For given c, we choose positive terms in sequential order until their sum exceed c. At this point, we pick negative terms until their sum is less than c. We repeat this process, and this process never stop. Note that all terms of the sequence $\{a_n\}$ will eventually added to the sum in different steps. Let us denote the sum of the m terms by s_m. Indeed, we have

$$s_m = \sum_{n=1}^{m} a_{\sigma(n)}.$$

Now, we have $|c - s_m| \to 0$ as $m \to \infty$.

467. *Let*

$$S_n = \sum_{k=1}^{n} \left(\sqrt{1 + \frac{k}{n^2}} - 1 \right).$$

Prove that $\lim\limits_{n \to \infty} S_n = \dfrac{1}{4}$.

Solution. For every $x > -1$, it is easy to see that $\dfrac{x}{2+x} \le \sqrt{1+x} - 1 \le \dfrac{x}{2}$. So, if we consider $x = \dfrac{k}{n^2}$, then

$$\frac{k}{2n^2 + k} \le \sqrt{1 + \frac{k}{n^2}} - 1 \le \frac{k}{2n^2},$$

and so we obtain

$$\sum_{k=1}^{n} \frac{k}{2n^2 + k} \le \sum_{k=1}^{n} \left(\sqrt{1 + \frac{k}{n^2}} - 1 \right) \le \sum_{k=1}^{n} \frac{k}{2n^2},$$

or equivalently,

$$\sum_{k=1}^{n} \frac{k}{2n^2 + k} \leq S_n \leq \sum_{k=1}^{n} \frac{k}{2n^2}. \tag{7.7}$$

Since $\sum_{k=1}^{n} k = \dfrac{n(n+1)}{2}$, it follows that

$$\lim_{n\to\infty} \sum_{k=1}^{n} \frac{k}{2n^2} = \lim_{n\to\infty} \frac{n(n+1)}{4n^2} = \frac{1}{4}. \tag{7.8}$$

Moreover, by Problem (6), since $\sum_{k=1}^{n} k^2 = \dfrac{n(n+1)(2n+1)}{6}$, it follows that

$$\sum_{k=1}^{n} \frac{k}{2n^2} - \sum_{k=1}^{n} \frac{k}{2n^2 + k} = \sum_{k=1}^{n} \frac{k^2}{2n^2(2n^2 + k)} \leq \sum_{k=1}^{n} \frac{k^2}{4n^4} = \frac{n(n+1)(2n+1)}{24n^4}.$$

Consequently, we have

$$\lim_{n\to\infty} \left(\sum_{k=1}^{n} \frac{k}{2n^2} - \sum_{k=1}^{n} \frac{k}{2n^2 + k} \right) \leq \lim_{n\to\infty} \frac{n(n+1)(2n+1)}{24n^4} = 0.$$

This implies that

$$\lim_{n\to\infty} \left(\sum_{k=1}^{n} \frac{k}{2n^2} - \sum_{k=1}^{n} \frac{k}{2n^2 + k} \right) = 0.$$

Therefore, we conclude that

$$\lim_{n\to\infty} \sum_{k=1}^{n} \frac{k}{2n^2 + k} = \frac{1}{4}. \tag{7.9}$$

Now, by (7.7), (7.8) and (7.9), our proof completes.

468. *Prove that the convergence of $\sum a_n$ implies the convergence of $\sum \dfrac{\sqrt{a_n}}{n}$, if $a_n \geq 0$, for all integer n.*

Solution. By Problem (215), we can write

$$\sum_{k=1}^{n} \frac{\sqrt{a_k}}{k} \leq \left(\sum_{k=1}^{n} \frac{1}{k^2} \right)^{\frac{1}{2}} \cdot \left(\sum_{k=1}^{n} a_k \right)^{\frac{1}{2}}$$

$$\leq \left(\sum_{k=1}^{\infty} \frac{1}{k^2} \right)^{\frac{1}{2}} \cdot \left(\sum_{k=1}^{\infty} a_k \right)^{\frac{1}{2}} < \infty,$$

for all n. Therefore, the partial sum sequence of the non-negative series $\sum \frac{\sqrt{a_k}}{k}$ is bounded. This completes the proof of our statement.

469. *Express the repeating decimal* $0.272727\ldots$ *as a rational number.*

Solution. We can write

$$
0.272727\ldots = \frac{27}{10^2} + \frac{27}{10^4} + \frac{27}{10^6} + \cdots
$$
$$
= \frac{27}{10^2}\left(1 + \frac{1}{10^2} + \frac{1}{10^4} + \cdots\right).
$$

By using the formula for the sum of a convergent geometric series, we get

$$
0.272727\ldots = \frac{27}{100} \cdot \frac{1}{1 - \dfrac{1}{100}} = \frac{27}{99}.
$$

470. *The second term of a geometric series is* 4 *and the sum of the series is* 18. *Determine the first term and ratio of the series.*

Solution. Suppose that

$$
S = r + ra + ra^2 + \cdots = \sum_{n=0}^{\infty} ra^n = \frac{r}{1 - a}.
$$

Then, we have the following system of equations to find the first term r and ratio a:

$$
18 = \frac{r}{1 - a} \quad \text{and} \quad 4 = ra.
$$

Hence, $r = 18(1 - a)$. This implies that $4 = 18(1 - a)a$, or equivalently $9a^2 - 9a + 2 = 0$. So, we obtain $a = \frac{1}{3}$ or $a = \frac{2}{3}$. If $a = \frac{1}{3}$, then $r = 12$. If $a = \frac{2}{3}$, then $r = 6$.

471. *(1) Show that the series* $\sum_{n=1}^{\infty} \frac{1}{e^n}$ *is convergent and find its sum.*

(2) Determine the series $\sum_{n=1}^{\infty} \frac{1}{\cosh n}$ *is convergent or divergent.*

Solution. (1) It is a geometric series, where begins with the term $\frac{1}{e}$, and each successive term is obtained by multiplying the preceding term by the number $\frac{1}{e}$, that is

nth partial sum is $s_n = \dfrac{1}{e} + \dfrac{1}{e^2} + \cdots + \dfrac{1}{e^n}$. Since $\dfrac{1}{e} < 1$, it follows that the series is convergent and we have

$$\lim_{n\to\infty} s_n = \frac{\frac{1}{e}}{1 - \frac{1}{e}} = \frac{1}{e-1},$$

which implies that $\displaystyle\sum_{n=1}^{\infty} \frac{1}{e^n} = \frac{1}{e-1}$.

(2) We have

$$0 < \frac{1}{\cosh n} = \frac{2}{e^n + e^{-n}} < \frac{2}{e^n}.$$

Since $\displaystyle\sum \frac{2}{e^n}$ is convergent, by the comparison test we conclude that $\displaystyle\sum \frac{1}{\cosh n}$ is convergent.

472. *Does the series*

$$\sum_{n=1}^{\infty} \frac{5 + 3\sin 2n}{e^{2n}}$$

converges or diverges?

Solution. For every $n \in \mathbb{N}$, we have $|5 + 3\sin 2n| \le 8$. So,

$$\left| \frac{5 + 3\sin 2n}{e^{2n}} \right| \le \frac{8}{e^{2n}} = 8\left(\frac{1}{e^2}\right)^n.$$

Since $\dfrac{1}{e^2} < 1$, it follows that the series $\displaystyle\sum_{n=1}^{\infty} 8\left(\frac{1}{e^2}\right)^n$ is convergent. Hence, by the comparison test we deduce that $\displaystyle\sum_{n=1}^{\infty} \left| \frac{5 + 3\sin 2n}{e^{2n}} \right|$ is convergent too. Therefore, the desired series is absolutely convergent.

473. *Show that the series*

$$\sum_{n=3}^{\infty} \frac{1}{\Big(\ln(\ln n) \Big)^{\ln n}} \tag{7.10}$$

is convergent.

Solution. We have

$$\Big(\ln(\ln n) \Big)^{\ln n} = e^{\ln\big(\ln(\ln n) \big)^{\ln n}} = e^{\ln n \, \ln\big(\ln(\ln n) \big)} = n^{\ln\big(\ln(\ln n) \big)}.$$

We have $\ln \left(\ln(\ln n) \right) > 3$ when n is large enough. So

$$n^{\ln \left(\ln(\ln n) \right)} > n^3.$$

This implies that

$$\frac{1}{\left(\ln(\ln n) \right)^{\ln n}} < \frac{1}{n^3}.$$

Since $\sum \dfrac{1}{n^3}$ is convergent, it follows that the series (7.10) is convergent.

474. *Suppose that $a_n > 0$ for all $n \in \mathbb{N}$ and*

$$\lim_{n \to \infty} \frac{\ln \left(\dfrac{1}{a_n} \right)}{\ln n} = \ell$$

exists. Prove that

(1) If $\ell > 1$, then $\sum a_n$ is convergent.
(2) If $\ell < 1$, then $\sum a_n$ is divergent.

Solution. (1) Suppose that $1 < p < \ell$ and choose $N \in \mathbb{N}$ such that for each $n > N$, we have

$$\frac{\ln \left(\dfrac{1}{a_n} \right)}{\ln n} > p \text{ or } \ln \left(\frac{1}{a_n} \right) > p \ln n.$$

This implies that

$$\ln \left(\frac{1}{a_n} \right) > \ln n^p,$$

and so $\dfrac{1}{a_n} > n^p$ or

$$a_n < \frac{1}{n^p}.$$

Since $\sum \dfrac{1}{n^p}$ is convergent, by the comparison test we conclude that $\sum a_n$ is convergent.

(2) Suppose that $\ell < 1$ and choose $N \in \mathbb{N}$ such that for each $n > N$, we have

$$\frac{\ln \left(\dfrac{1}{a_n} \right)}{\ln n} < 1 \text{ or } \ln \left(\frac{1}{a_n} \right) < \ln n.$$

This implies that $\dfrac{1}{a_n} < n$ or

$$\frac{1}{n} < a_n.$$

Since $\sum \dfrac{1}{n}$ is divergent, by the comparison test we conclude that $\sum a_n$ is divergent.

475. *Let $\sum a_n$ be a convergent series with non-zero terms. Determine whether $\sum \sqrt{a_n a_{n+1}}$ is convergent.*

Solution. Since $(a_n - a_{n+1})^2 \geq 0$, it follows that

$$4 a_n a_{n+1} \leq a_n^2 + a_{n+1}^2 + 2 a_n a_{n+1}.$$

This implies that

$$\sqrt{a_n a_{n+1}} \leq \frac{a_n + a_{n+1}}{2}.$$

Since

$$\sum \frac{a_n + a_{n+1}}{2} = \frac{1}{2} \left(\sum a_n + \sum a_{n+1} \right)$$

is convergent, by the comparison test, we conclude that $\sum \sqrt{a_n a_{n+1}}$ is convergent.

476. *Determine all $p > 0$ such that*

$$\sum_{n=1}^{\infty} \left(\sqrt{3n^p + 1} - \sqrt{3n^p - 3} \right)$$

is convergent.

Solution. Suppose that $a_n = \sqrt{3n^p + 1} - \sqrt{3n^p - 3}$. Then, we have

$$a_n = \frac{(3n^p + 1) - (3n^p - 3)}{\sqrt{3n^p + 1} + \sqrt{3n^p - 3}} = \frac{4}{\sqrt{3n^p + 1} + \sqrt{3n^p - 3}}$$

$$= \frac{4}{n^{\frac{p}{2}}} \left(\sqrt{3 + \frac{1}{n^p}} + \sqrt{3 - \frac{3}{n^p}} \right).$$

We consider $\sum_{n=1}^{\infty} b_n$, where $b_n = \dfrac{1}{n^{\frac{p}{2}}}$. Then,

$$\lim_{n \to \infty} \frac{a_n}{b_n} = \frac{2}{\sqrt{3}}.$$

So, by the limit comparison test, either both convergent or both divergent. We know that $\sum_{n=1}^{\infty} \frac{1}{n^{\frac{p}{2}}}$ is divergent if $0 < p \leq 2$ and is convergent if $p > 2$. Therefore, $\sum_{n=1}^{\infty} a_n$ is divergent if $0 < p \leq 2$ and is convergent if $p > 2$.

477. *Prove that*

$$\int_1^\infty \left| \frac{\sin x}{x} \right| dx$$

is divergent.

Solution. We have

$$\int_\pi^{(n+1)\pi} \frac{|\sin x|}{x} dx = \sum_{k=1}^{n} \int_{k\pi}^{(k+1)\pi} \frac{|\sin x|}{x} dx$$

$$\geq \sum_{k=1}^{n} \frac{1}{(k+1)\pi} \int_{k\pi}^{(k+1)\pi} |\sin x| dx$$

$$= \sum_{k=1}^{n} \frac{2}{(k+1)\pi}.$$

This implies that the desired integral is divergent.

478. *Show that the series*

$$\sum_{n=1}^{\infty} \frac{n+1}{2^n}$$

is convergent.

Solution. Let $f(x) = \frac{x+1}{2^x}$. Obviously, f is a continuous positive function. Since $f'(x) < 0$, it follows that f is decreasing. Moreover, we have

$$\int_1^\infty \frac{x+1}{2^x} dx = \lim_{u \to \infty} \int_1^u \frac{x+1}{2^x} dx$$

$$= \lim_{u \to \infty} \int_1^u 2^{-x}(x+1) dx$$

$$= \lim_{u \to \infty} \left(-\frac{x \cdot 2^{-x}}{\ln 2} - \frac{2^{-x}}{\ln 2} - \frac{2^{-x}}{(\ln 2)^2} \right)\Big|_1^u$$

$$= \frac{1}{\ln 2} + \frac{1}{2(\ln 2)^2}.$$

Hence, the desired series is convergent, by the integral test, since the associated improper integral is convergent.

479. *Determine the values of $c \in \mathbb{R}$ for which*

$$\sum_{n=1}^{\infty} \left(\frac{cn}{n+1}\right)^n$$

is convergent.

Solution. We observe that

$$\lim_{n\to\infty} \left|\frac{cn}{n+1}\right| = c.$$

Now, by the root test, the series converges for $|c| < 1$ and diverges for $|c| > 1$. If $|c| = 1$, then

$$\lim_{n\to\infty} \left|\frac{cn}{n+1}\right| = \lim_{n\to\infty} \left(\frac{n}{n+1}\right)^n = \frac{1}{e} \neq 0$$

and so the series is divergent.

480. *Determine whether the following series is convergent or divergent:*

$$\sum_{n=0}^{\infty} \left(\frac{n+1}{n+3}\right)^{n^2}.$$

Solution. Let $a_n = \left(\frac{n+1}{n+3}\right)^{n^2}$. Then, we have

$$\sqrt[n]{a_n} = \left(\frac{n+1}{n+3}\right)^n = \left(1 - \frac{2}{n+3}\right)^n.$$

Now, similar to Problem (198), we get

$$\lim_{n\to\infty} \sqrt[n]{a_n} = \lim_{n\to\infty} \left(1 - \frac{2}{n+3}\right)^n = e^{-2} < 1.$$

Thus, the series is convergent by the root test.

481. *Show that if $\{a_n\}_{n=0}^{\infty}$ is a sequence of real numbers such that*

$$\lim_{n\to\infty} \left|\frac{a_{n+1}}{a_n}\right| = L,$$

then

$$\lim_{n\to\infty} |a_n|^{\frac{1}{n}} = L.$$

Solution. By the definition of limit, for every $\epsilon > 0$ there exists N such that

$$\left| \left|\frac{a_{n+1}}{a_n}\right| - L \right| < \epsilon,$$

whenever $n > N$. Thus, we have

$$|a_n| = \frac{|a_n|}{|a_{n-1}|} \ldots \frac{|a_{N+1}|}{|a_N|}|a_N| < (L+\epsilon)^{n-N}|a_N|. \tag{7.11}$$

Take the nth root of both sides of the inequality (7.11). Then, we obtain

$$\sqrt[n]{|a_n|} < (L+\epsilon)^{1-\frac{N}{n}} \sqrt[n]{|a_N|}.$$

Taking $n \to \infty$, then

$$\lim_{n\to\infty} \sqrt[n]{|a_n|} \le L + \epsilon.$$

Since ϵ is arbitrary, we obtain $\lim_{n\to\infty} \sqrt[n]{|a_n|} \le L$. Likewise, we can get $\lim_{n\to\infty} \sqrt[n]{|a_n|} \ge L$.

482. *Verify that* $\lim_{n\to\infty} \dfrac{n}{\sqrt[n]{n!}} = e$.

Solution. Let $a_n = \dfrac{n^n}{n!}$. Then, we have

$$\lim_{n\to\infty} \frac{a_{n+1}}{a_n} = \lim_{n\to\infty} \frac{(n+1)^n}{n^n} = \lim_{n\to\infty} (1+\frac{1}{n})^n = e.$$

Now, we use Problem 481.

483. *Verify that*

$$\lim_{n\to\infty} \frac{\sqrt[n]{(n+1)(n+2)\ldots(2n)}}{n} = \frac{4}{e}.$$

Solution. If $a_n = \dfrac{(n+1)(n+2)\ldots(2n)}{n^n}$, then

$$a_{n+1} = \frac{(n+2)(n+3)\ldots(2n)(2n+1)(2n+2)}{(n+1)^{n+1}}.$$

Hence,

$$\begin{aligned}
\lim_{n\to\infty} \frac{a_{n+1}}{a_n} &= \lim_{n\to\infty} \frac{(2n+1)(2n+2)}{(n+1)} \cdot \frac{n^n}{(n+1)^{n+1}} \\
&= \lim_{n\to\infty} \frac{(2n+1)(2n+2)}{(n+1)(n+1)} \cdot \left(\frac{n}{n+1}\right)^n \\
&= \frac{4}{e}
\end{aligned}$$

Now, it is enough to use Problem 481.

484. *Does the series*

$$\sum_{n=1}^{\infty} \frac{2^n}{n^3 + 3n + 5}$$

converge?

Solution. Since

$$\left(\frac{2^n}{n^3 + 3n + 5}\right)^{\frac{1}{n}} = \frac{2}{(n^3 + 3n + 5)^{\frac{1}{n}}},$$

and

$$\lim_{n \to \infty} (n^3 + 3n + 5)^{\frac{1}{n}} = \lim_{n \to \infty} e^{\frac{\ln(n^3 + 3n + 5)}{n}}$$
$$= e^{\lim_{n \to \infty} \frac{\ln(n^3 + 3n + 5)}{n}}$$
$$= e^L,$$

where

$$L = \lim_{n \to \infty} \frac{\ln(n^3 + 3n + 5)}{n} = \lim_{n \to \infty} \frac{3n^2 + 3}{n^3 + 3n + 5} = 0,$$

we obtain that

$$\lim_{n \to \infty} \left(\frac{2^n}{n^3 + 3n + 5}\right)^{\frac{1}{n}} = 2 > 1.$$

So, by the ratio test we conclude that the series is divergent.

485. *Determine whether the following series is absolutely convergent, conditionally convergent or divergent?*

$$\sum_{n=1}^{\infty} \frac{(-1)^n}{\ln(e^n + e^{-n})}$$

Solution. This is an alternating series. Suppose that $a_n = \dfrac{1}{\ln(e^n + e^{-n})}$. Obviously, we have $\lim_{n \to \infty} a_n = 0$ and $a_{n+1} \le a_n$. So, by the alternating test, the series is convergent. On the other hand, we have

$$\lim_{n\to\infty} \frac{\dfrac{1}{\ln\left(e^n + e^{-n}\right)}}{\dfrac{1}{n}} = \lim_{n\to\infty} \frac{n}{\ln\left(e^n + e^{-n}\right)}$$

$$= \lim_{n\to\infty} \frac{n}{\ln\left(e^n(1 + +e^{-2n})\right)}$$

$$= \lim_{n\to\infty} \frac{n}{n + \ln\left(1 + e^{-2n}\right)}$$

$$= \lim_{n\to\infty} \frac{1}{1 - \dfrac{2e^{-2n}}{1 + e^{-2n}}}$$

$$= 1.$$

Since $\sum \dfrac{1}{n}$ is divergent, it follows that $\sum \dfrac{1}{\ln\left(e^n + e^{-n}\right)}$ is divergent. Therefore, the desired series is conditionally convergent.

486. *Determine whether the following series is absolutely convergent, conditionally convergent, or divergent?*

$$\sum_{n=1}^{\infty}(-1)^n\left(\frac{\pi}{2} - \tan^{-1} n\right).$$

Solution. If $f(x) = \dfrac{\pi}{2} - \tan^{-1} x$ and $a_n = \dfrac{\pi}{2} - \tan^{-1} n$, then

$$f'(x) = \frac{-1}{1 + x^2} < 0.$$

Since f is decreasing, it follows that $a_{n+1} \le a_n$. Moreover, $\lim\limits_{n\to\infty} a_n = 0$. So, by the alternating test, we conclude that the series is convergent.

Now, f is a continuous positive function which is decreasing on the interval $[1, \infty)$. We have

$$\int_1^\infty f(x)dx = \int_1^\infty \left(\frac{\pi}{2} - \tan^{-1} x\right)dx = \lim_{u\to\infty} \int_1^u \left(\frac{\pi}{2} - \tan^{-1} x\right)dx$$

$$= \lim_{u\to\infty} \left(\frac{\pi}{2}x - x\tan^{-1} x + \frac{1}{2}\ln(1 + x^2)\right)\Big|_0^u$$

$$= \lim_{u\to\infty} \left(\frac{\pi}{2}u - u\tan^{-1} u + \frac{1}{2}\ln(1 + u^2)\right)$$

$$= \lim_{u\to\infty} \left(u\left(\frac{\pi}{2} - \tan^{-1} u\right) + \frac{1}{2}\ln(1 + u^2)\right)$$

Since $\int_1^\infty \left(\dfrac{\pi}{2} - \tan^{-1} x\right)dx$ is divergent, by the integral test we deduce that

$$\sum_{n=1}^{\infty} \left(\frac{\pi}{2} - \tan^{-1} n\right).$$

is divergent. Therefore, the desired series is conditionally convergent.

487. *Let $\sum a_n$ be an absolutely convergent series and $\{b_n\}$ be a convergent sequence of numbers such that $b_n \to L$ as $n \to \infty$. If $L \neq 0$, show that*

(1) There is a natural number N such that $\frac{1}{2}|L| < |b_n| < \frac{3}{2}|L|$, whenever $n > N$;

(2) The series $\sum \dfrac{a_n}{b_n}$ is absolutely convergent.

Solution. (1) Let $\epsilon = \frac{1}{2}|L|$. Then, there exists $N \in \mathbb{N}$ such that $|b_n - L| < \epsilon$, whenever $n > N$. Hence, we have

$$\left| |b_n| - |L| \right| \leq |b_n - L| < \frac{1}{2}|L|,$$

and so $-\frac{1}{2}|L| < |b_n| - |L| < \frac{1}{2}|L|$. This implies that $\frac{1}{2}|L| < |b_n| < \frac{3}{2}|L|$.

(2) By the first part, there exists $N \in \mathbb{N}$ such that $\frac{1}{2}|L| < |b_n|$, whenever $n > N$. This implies that $\dfrac{1}{|b_n|} < \dfrac{2}{|L|}$. Consequently, we get

$$\left|\frac{a_n}{b_n}\right| < \frac{2}{|L|}|a_n|, \quad \text{whenever } n > N.$$

Since $\sum \dfrac{2}{|L|}|a_n|$ is convergent, by the comparison test, $\sum \left|\dfrac{a_n}{b_n}\right|$ is convergent. Therefore, $\sum \dfrac{a_n}{b_n}$ is absolutely convergent.

488. *Find the interval of convergence of the power series*

$$\sum_{n=0}^{\infty} \frac{(-1)^n (3x + 2)^n}{3^n (2n + 1)^2}.$$

Solution. Assume that $a_n = \dfrac{(-1)^n (3x + 2)^n}{3^n (2n + 1)^2}$. Then, we have

$$\lim_{n\to\infty}\left|\frac{a_{n+1}}{a_n}\right| = \lim_{n\to\infty}\left|\frac{(-1)^{n+1}(3x+2)^{n+1}}{3^{n+1}(2n+3)^2} \cdot \frac{3^n(2n+1)^2}{(-1)^n(3x+2)^n}\right|$$

$$= \frac{1}{3}\lim_{n\to\infty}\frac{(2n+1)^2}{(2n+3)^2}|3x+2|$$

$$= \frac{1}{3}|3x+2|.$$

So the series is absolutely convergent if $|3x+2| < 3$ or equivalently $-\frac{5}{3} < x < \frac{1}{3}$.

When $x = \frac{1}{3}$, the series is $\sum_{n=0}^{\infty} \frac{(-1)^n}{(2n+1)^2}$, which is a convergent series.

When $x = -\frac{5}{3}$, the series is $\sum_{n=0}^{\infty} \frac{1}{(2n+1)^2}$, which is a convergent series.

Therefore, the interval of convergence is $[-\frac{5}{3}, \frac{1}{3}]$.

489. *Find the radius of convergence and interval of convergence of the power series*

$$\sum_{n=2}^{\infty} \frac{n^3 x^{2n}}{(\ln n)^n}$$

Solution. We have

$$\lim_{n\to\infty} \sqrt[n]{\left|\frac{n^3 x^{2n}}{(\ln n)^n}\right|} = \lim_{n\to\infty} \frac{(\sqrt[n]{n})^3}{\ln n}|x|^2 = 0$$

for every value of x. Note that $\sqrt[n]{n} \to 1$ as $n \to \infty$. Therefore, the series is absolutely convergent for all x, that is, it has radius of convergence ∞ and the interval of convergence is $(-\infty, \infty)$.

490. *If a, b and c are positive, find the radius of convergence of hypergeometric series*

$$1 + \frac{ab}{c}x + \frac{a(a+1)b(b+1)}{2!c(c+1)}x^2 + \frac{a(a+1)(a+2)b(b+1)(b+2)}{3!c(c+1)(c+2)}x^3 + \dots.$$

Solution. We apply ratio test. Then, we have

$$\lim_{n\to\infty} \frac{\frac{a(a+1)\dots(a+n)b(b+1)\dots(b+n)}{(n+1)!c(c+1)\dots(c+n)}|x|^{n+1}}{\frac{a(a+1)\dots(a+n-1)b(b+1)\dots(b+n-1)}{n!c(c+1)\dots(c+n-1)}|x|^n}$$

$$= \lim_{n\to\infty} \frac{(a+n)(b+n)}{(n+1)(c+n)}|x| = |x|.$$

Therefore, the radius of convergence is equal to 1.

491. *Let*

$$f(x) = \sum_{n=0}^{\infty} \frac{x^n}{3^n (n^2 + 1)}.$$

(1) Show that $\dfrac{4}{3} < f(2) < 2.$

(2) Show that $\dfrac{2}{3} < f(-2) < \dfrac{4}{5}.$

Solution. (1) We have

$$f(2) = \sum_{n=0}^{\infty} \frac{2^n}{3^n (n^2 + 1)} > 1 + \frac{2}{3.2} + \frac{2^2}{3^2.5} = \frac{64}{45} > \frac{4}{3}$$

and

$$f(2) = 1 + \sum_{n=1}^{\infty} \frac{2^n}{3^n (n^2 + 1)} < 1 + \frac{1}{2} \sum_{n=1}^{\infty} \left(\frac{2}{3}\right)^n = 1 + \frac{1}{2} \cdot \frac{\frac{2}{3}}{1 - \frac{2}{3}} = 2.$$

(2) We have

$$f(-2) = \sum_{n=0}^{\infty} (-1)^n \frac{2^n}{3^n (n^2 + 1)} < 1 - \frac{1}{3} + \frac{4}{45} = \frac{34}{45} < \frac{4}{5}.$$

and

$$f(-2) = \sum_{n=0}^{\infty} (-1)^n \frac{2^n}{3^n (n^2 + 1)} > 1 - \frac{1}{3} + \frac{4}{45} - \frac{8}{270} = \frac{98}{145} > \frac{2}{3}.$$

492. *(1) Find a series solution of the equation*

$$y'' + y = 0, \quad -\infty < x < \infty. \tag{7.12}$$

(2) Show that we can write the solution of Eq. (7.12) as follows:

$$y = c \sin(x + \theta). \tag{7.13}$$

(3) Find any function f *such that*

$$f(a - x) = f'(x), \quad -\infty < x < \infty \text{ and } a \text{ is fixed.} \tag{7.14}$$

Solution. (1) We look for a solution of the form

$$y = \sum_{n=0}^{\infty} a_n x^n. \tag{7.15}$$

Differentiating term by term yields

$$y' = \sum_{n=1}^{\infty} n a_n x^{n-1},$$

$$y'' = \sum_{n=2}^{\infty} n(n-1) a_n x^{n-2}. \tag{7.16}$$

Substituting the series (7.15) and (7.16) in Eq. (7.12) gives

$$\sum_{n=2}^{\infty} n(n-1) a_n x^{n-2} + \sum_{n=0}^{\infty} a_n x^n = 0.$$

This implies that

$$\sum_{n=0}^{\infty} \left((n+2)(n+1) a_{n+2} + a_n \right) x^n = 0.$$

Thus, we conclude that $(n+2)(n+1) a_{n+2} + a_n = 0$, for $n = 0, 1, 2, \ldots$. Consequently, if $n = 2k$, then

$$a_n = a_{2k} = \frac{(-1)^k}{(2k)!} a_0, \ k = 1, 2, 3, \ldots$$

and if $n = 2k + 1$, then

$$a_n = a_{2k+1} = \frac{(-1)^k}{(2k+1)!} a_1, \ k = 1, 2, 3, \ldots$$

Thus, we get

$$y = a_0 + a_1 x - \frac{a_0}{2!} x^2 - \frac{a_1}{3!} x^3 + \cdots + \frac{(-1)^n a_0}{(2n)!} x^{2n} + \frac{(-1)^n a_1}{(2n+1)!} x^{2n+1} + \cdots$$

$$= a_0 \sum_{n=0}^{\infty} \frac{(-1)^n}{(2n)!} x^{2n} + a_1 \sum_{n=0}^{\infty} \frac{(-1)^n}{(2n+1)!} x^{2n+1}.$$

Note that the first series is the Taylor series of $\cos x$ and the second series is the Taylor series of $\sin x$. Therefore, we obtain the solution

$$y = a_0 \cos x + a_1 \sin x, \tag{7.17}$$

where a_0 and a_1 are arbitrary constants.

(2) We consider the points $O(0,0)$ and $P(a_1, a_0)$ in $\mathbb{R}^2$. Then, we have

$$a_1 = a \cos\theta, \quad a_0 = a \sin\theta, \tag{7.18}$$

where θ is the angle between x-axis and the segment OP, and $a = \sqrt{a_0^2 + a_1^2}$. Now, by combining formulas (7.17) and (7.18) we obtain $y = a \cos x \sin\theta + a \sin x \cos\theta$, which simplifies to $y = a \sin(x + \theta)$.

(3) If f has second derivative, taking derivative from (7.14) gives

$$-f'(a - x) = f''(x). \tag{7.19}$$

Changing x to $a - x$ in (7.14) gives

$$f(x) = f'(a - x). \tag{7.20}$$

Now, from (7.19) and (7.20) we get

$$f''(x) + f(x) = 0. \tag{7.21}$$

By part (2), the solution of (7.21) is $f(x) = a \sin(x + \theta)$. Consequently, we have $a \sin(a - x + \theta) = a \cos(x + \theta)$. If $a \neq 0$, then $\sin(a - x + \theta) = \cos(x + \theta)$, and so $\theta = \dfrac{\pi}{4} - \dfrac{a}{2} + k\pi$. Therefore, we conclude that

$$f(x) = (-1)^k a \sin\left(x + \frac{\pi}{4} - \frac{a}{2}\right).$$

493. *Prove that*

$$\sum_{n=1}^{\infty} n^k x^n = \frac{P_k(x)}{(1-x)^{k+1}}, \tag{7.22}$$

where $P_k(x)$ is a polynomial of degree k, the term of lowest degree is x and that of highest degree is x^k, for all positive integer k.

Solution. We prove (7.22) by mathematical induction. If $k = 1$, then

$$\sum_{n=1}^{\infty} nx^n = x \sum_{n=1}^{\infty} nx^{n-1} = x \sum_{n=1}^{\infty} \frac{d}{dx}(x^n)$$

$$= x \frac{d}{dx}\left(\sum_{n=1}^{\infty} x^n\right) = x \frac{d}{dx}\left(\frac{x}{1-x}\right) = \frac{x}{(1-x)^2}.$$

Suppose that (7.22) is true for k, then we prove it for $k + 1$. Since $x^{k+1}x^n = x(nn^k x^{n-1}) = x \frac{d}{dx}(n^k x^n)$, we obtain

$$\sum_{n=1}^{\infty} n^{k+1} x^n = x \frac{d}{dx} \sum_{n=1}^{\infty} n^k x^n$$

$$= x \frac{d}{dx}\left(\frac{P_k(x)}{(1-x)^{k+1}}\right)$$

$$= x\left(\frac{P_k'(x)(1-x)^{k+1} + (k+1)(1-x)^k P_k(x)}{(1-x)^{2k+2}}\right)$$

$$= x\left(\frac{P_k'(x)(1-x) + (k+1) P_k(x)}{(1-x)^{k+2}}\right)$$

$$= \frac{(k+1)x P_k(x) + x(1-x) P_k'(x)}{(1-x)^{k+2}}$$

The numerator has x as the lowest degree term from $x P_k'(x)$ and $(k+1)x^{k+1} - kx^{k+1} = x^{k+1}$ as the highest degree term from $(k+1) P_k(x) - x P_k'(x)$. This completes the proof.

494. *(1) Prove that $e^x > 1 + x$, for all $x \in \mathbb{R}$.*
(2) Show that the series

$$\sum_{n=1}^{\infty} \ln\left(1 + \frac{1}{n^2}\right)$$

is convergent.

Solution. (1) Let $f(x) = e^x$. By using Taylor's theorem, we can write

$$e^x = 1 + x + \frac{e^\xi}{2!}x^2,$$

where ξ is a number between 0 and x. Since $\frac{e^\xi}{2!}x^2 \geq 0$, it follows that $e^x > 1 + x$.

(2) From (1), we get $x \geq \ln(1 + x)$, for all $x > -1$. So, we have

$$\ln\left(1 + \frac{1}{n^2}\right) \leq \frac{1}{n^2}.$$

Since $\sum \dfrac{1}{n^2}$ is convergent, by the comparison test, we conclude that the desired series is convergent.

495. *Let f be a real valued function such that the second derivative of f exists and bounded on* [0, 1].

(1) Prove that if the series $\displaystyle\sum_{k=1}^{\infty} f(\tfrac{1}{k})$ *is convergent, then* $f(0) = f'(0) = 0$.

(2) Conversely, show that if $f(0) = f'(0) = 0$, *then the series* $\displaystyle\sum_{k=1}^{\infty} f(\tfrac{1}{k})$ *is convergent.*

Solution. (1) Since $\displaystyle\sum_{k=1}^{\infty} f(\tfrac{1}{k})$ is convergent, it follows that $\displaystyle\lim_{k\to\infty} f(\tfrac{1}{k}) = 0$. Since f is continuous, we obtain $f\left(\displaystyle\lim_{k\to\infty} \tfrac{1}{k}\right) = 0$. This implies that $f(0) = 0$. Now, by using Taylor's theorem, there is a point t_k between 0 and $\dfrac{1}{k}$ such that

$$f(\tfrac{1}{k}) = f(0) + f'(0)\dfrac{1}{k} + f''(t_k)\dfrac{1}{k^2}. \tag{7.23}$$

or

$$f'(0)\dfrac{1}{k} = f(\tfrac{1}{k}) - f''(t_k)\dfrac{1}{k^2}. \tag{7.24}$$

By assumption, we can suppose that $|f''(x)| \le M$, for all $x \in [0, 1]$. Thus,

$$|f''(t_k)| \cdot \dfrac{1}{k^2} \le M \cdot \dfrac{1}{k^2}, \tag{7.25}$$

and comparison test implies that $\displaystyle\sum_{k=1}^{\infty} f''(t_k)\dfrac{1}{k^2}$ is convergent. On the other hand, $\displaystyle\sum_{k=1}^{\infty} f(\tfrac{1}{k})$ is convergent by assumption. Therefore, by Eq. (7.24), $f'(0)\displaystyle\sum_{k=1}^{\infty} \dfrac{1}{k}$ must be convergent and it is impossible, unless $f'(0) = 0$.

(2) By Eq. (7.23) and $f(0) = f'(0) = 0$, we obtain $|f(\tfrac{1}{k})| = |f''(t_k)|\dfrac{1}{k^2} \le M \cdot \dfrac{1}{k^2}$, for all k. Now, comparison test implies that $\displaystyle\sum_{k=1}^{\infty} f(\tfrac{1}{k})$ is convergent.

496. *Show that for all positive integer a,*

$$\sum_{n=0}^{\infty} \frac{(-1)^n}{na+1} = \int_0^1 \frac{1}{x^a+1}\,dx, \quad (a > 0). \qquad (7.26)$$

Then, use Eq. (7.26) to find the sum of the following series:

$$1 - \frac{1}{3} + \frac{1}{5} - \frac{1}{7} + \cdots.$$

Solution. We know that if $-1 < x < 1$, then

$$\frac{1}{x+1} = 1 - x + x^2 - \cdots + (-1)^n x^n + \cdots = \sum_{n=0}^{\infty} (-1)^n x^n.$$

So, we get

$$\frac{1}{x^a+1} = 1 - x^a + x^{2a} - \cdots + (-1)^n x^{na} + \cdots = \sum_{n=0}^{\infty} (-1)^n x^{na}.$$

Then, integrating this series term by term from 0 to 1, we find that

$$\int_0^1 \frac{1}{x^a+1}\,dx = \lim_{u \to 1^-} \int_0^u \frac{1}{x^2+1}\,dx = \sum_{n=0}^{\infty} \int_0^1 (-1)^n x^{na}\,dx = \sum_{n=0}^{\infty} \frac{(-1)^n}{na+1}.$$

Now, suppose that $n = 2$ in (7.26). Then, we observe that

$$\int_0^1 \frac{1}{x^2+1}\,dx = \sum_{n=0}^{\infty} \frac{(-1)^n}{2n+1} = 1 - \frac{1}{3} + \frac{1}{5} - \frac{1}{7} + \cdots.$$

But $\int_0^1 \frac{1}{x^2+1}\,dx = \tan^{-1} x \big|_0^1 = \frac{\pi}{4}.$

497. *Prove that*

(1) for every $n \in \mathbb{N}$,

$$0 < e - \sum_{k=0}^{n} \frac{1}{k!} < \frac{1}{n \cdot n!},$$

(2) the number e is irrational.

Solution. (1) Since $e = \sum_{k=0}^{\infty} \frac{1}{k!}$, it follows that

$$e - \sum_{k=0}^{n} \frac{1}{k!} = \sum_{k=n+1}^{\infty} \frac{1}{k!}$$

$$= \frac{1}{n!} \left(\frac{1}{n+1} + \frac{1}{(n+1)(n+2)} + \cdots + \frac{1}{(n+1)(n+2)\ldots(n+k)} + \cdots \right)$$

$$< \frac{1}{n!} \left(\frac{1}{n+1} + \frac{1}{(n+1)^2} + \cdots + \frac{1}{(n+1)^k} + \cdots \right)$$

$$= \frac{1}{n!} \cdot \frac{\frac{1}{n+1}}{1 - \frac{1}{n+1}}$$

$$= \frac{1}{n!} \cdot \frac{1}{n}.$$

(2) Let e be rational and $e = \dfrac{a}{b}$ for some $a, b \in \mathbb{N}$. By part (1), we have

$$0 < \frac{a}{b} - \sum_{k=0}^{n} \frac{1}{k!} < \frac{1}{n \cdot n!},$$

or equivalently

$$0 < \frac{a \cdot n!}{b} - \sum_{k=0}^{n} \frac{n!}{k!} < \frac{1}{n} \leq 1. \tag{7.27}$$

If $n \geq b$, then the middle term of (7.27) is an integer. This is a contradiction, since the middle term lies between 0 and 1.

498. *Let $f(x) = \sin \left(k \sin^{-1} x \right)$ for $|x| < 1$.*

(1) Show that $(1 - x^2) f''(x) - x f'(x) + k^2 f(x) = 0$.
(2) Show that

$$(1 - x^2) f^{(n+2)}(x) - (2n + 1) x f^{(n+1)}(x) + (k^2 - n^2) f^n(x) = 0, \tag{7.28}$$

for $n = 0, 1, 2, \ldots$.
(3) Find the Maclaurin series for $f(x)$.
(4) If k is an odd number, prove that $f(x)$ is a polynomial of degree k.
(5) By using part (3), determine the following limit

$$\lim_{x \to 0} \frac{\sin \left(k \sin^{-1} x \right)}{x}.$$

Solution. (1) First, we determine $f'(x)$ and $f''(x)$. We have

$$f'(x) = \frac{k}{\sqrt{1 - x^2}} \cos \left(k \sin^{-1} x \right),$$

$$f''(x) = \frac{kx}{(1 - x^2)\sqrt{1 - x^2}} \cos \left(k \sin^{-1} x \right) - \frac{k^2}{1 - x^2} \sin \left(k \sin^{-1} x \right),$$

and hence

$$
(1 - x^2)f''(x) - xf'(x) + k^2 f(x)
$$

$$
= (1 - x^2)\left(\frac{kx}{(1 - x^2)\sqrt{1 - x^2}} \cos\left(k \sin^{-1} x\right) - \frac{k^2}{1 - x^2} \sin\left(k \sin^{-1} x\right)\right)
$$

$$
- x\left(\frac{k}{\sqrt{1 - x^2}} \cos\left(k \sin^{-1} x\right)\right) + k^2 \sin\left(k \sin^{-1} x\right)
$$

$$
= \frac{kx}{\sqrt{1 - x^2}} \cos\left(k \sin^{-1} x\right) - k^2 \sin\left(k \sin^{-1} x\right) - \frac{kx}{\sqrt{1 - x^2}} \cos\left(k \sin^{-1} x\right)
$$

$$
+ k^2 \sin\left(k \sin^{-1} x\right) = 0.
$$

(2) We use mathematical induction. According to part (1), the equality is true for $n = 0$. Suppose that (7.28) is true for n. We prove (7.28) for $n + 1$. We take derivative from (7.28). Then, we obtain

$$
(1 - x^2)f^{(n+3)}(x) - 2xf^{(n+2)}(x) - (2n + 1)f^{(n+1)}(x)
$$

$$
+ (2n + 1)xf^{(n+2)}(x) + (k^2 - n^2)f^{(n+1)}(x) = 0.
$$

This implies that

$$
(1 - x^2)f^{(n+3)}(x) + (2n + 3)xf^{(n+2)}(x) + \left(k^2 - (n + 1)^2\right)f^{(n+1)}(x) = 0,
$$

and this completes the proof.

(3) We have

$$
f(0) = \sin\left(k \sin^{-1} 0\right) = 0,
$$

$$
f'(0) = k \cos\left(k \sin^{-1} 0\right) = n,
$$

$$
f''(0) = 0,
$$

$$
\vdots
$$

$$
f^{n+2}(0) = (n^2 - k^2)f^{(n)}(0).
$$

Since $f''(0) = 0$, it follows that all even derivatives vanish. The odd derivatives are

$$
f'''(0) = (1 - k^2)f'(0) = (1 - k^2)k
$$

$$
f^{(5)}(0) = (3^2 - k^2)f'''(0) = (3^2 - k^2)(1 - k^2)k,
$$

and so on. Therefore, we get

$$
\sin\left(k \sin^{-1} x\right) = kx + \frac{(1 - k^2)}{3!}x^3 + \frac{(3^2 - k^2)(1 - k^2)}{5!}x^5 + \dots.
$$

(4) If k is odd, then the series will stop and consequently the function is a polynomial of degree k.

(5) We have

$$\lim_{x \to 0} \frac{\sin\left(k \sin^{-1} x\right)}{x} = \lim_{x \to 0} \frac{kx + \dfrac{(1-k^2)}{3!}x^3 + \dots}{x} = k.$$

499. *Estimate*

$$\int_0^{\frac{1}{2}} \cos x^2 dx$$

to three decimal place accuracy.

Solution. We know that the Maclaurin series of $\cos x$ is $\sum\limits_{n=0}^{\infty} \dfrac{(-1)^n}{(2n)!} x^{2n}$. Then, we deduce that

$$\cos x^2 = \sum_{n=0}^{\infty} \frac{(-1)^n}{(2n)!} x^{4n}$$

$$= 1 - \frac{1}{2!}x^4 + \frac{1}{4!}x^8 - \frac{1}{6!}x^{12} + \dots.$$

Integrating term by term gives

$$\int_0^{\frac{1}{2}} \cos x^2 dx = \left(x - \frac{1}{2!} \cdot \frac{1}{5}x^5 + \frac{1}{4!} \cdot \frac{1}{9}x^9 - \frac{1}{6!} \cdot \frac{1}{13}x^{13} + \dots\right)\Bigg|_0^{\frac{1}{2}}$$

$$= \frac{1}{2} - \frac{1}{2!} \cdot \frac{1}{5}\left(\frac{1}{2}\right)^5 + \frac{1}{4!} \cdot \frac{1}{9}\left(\frac{1}{2}\right)^9 - \frac{1}{6!} \cdot \frac{1}{13}\left(\frac{1}{2}\right)^{13} + \dots.$$

Calculation shows that the term

$$\frac{1}{9} \cdot \frac{1}{4!} \cdot \left(\frac{1}{2}\right)^9$$

is the first term that is less than 0.0005. Therefore, the estimate of required integral is

$$\frac{1}{2} - \frac{1}{320} = \frac{159}{320} \approx 0.497.$$

500. *Use the power series for* $\tan^{-1} x$ *to prove the following expression for* π *as the sum of a series:*

$$\pi = 2\sqrt{3} \sum_{n=0}^{\infty} \frac{(-1)^n}{(2n+1)3^n}.$$

Solution. We know that

$$\frac{1}{1+x} = 1 - x + x^2 - x^3 + \dots, \quad (|x| < 1). \tag{7.29}$$

Changing x to x^2 in (7.29), we find that

$$\frac{1}{1+x^2} = 1 - x^2 + x^4 - x^6 + \ldots, \quad (|x| < 1). \tag{7.30}$$

Term by term integration of (7.30) gives

$$\tan^{-1} x = \int_0^x \frac{1}{1+t^2} dt = x - \frac{x^3}{3} + \frac{x^5}{5} - \frac{x^7}{7} + \ldots. \tag{7.31}$$

Now, if we put $x = \dfrac{1}{\sqrt{3}}$ in (7.31), then we obtain

$$\tan^{-1} \frac{1}{\sqrt{3}} = \frac{1}{\sqrt{3}} - \left(\frac{1}{\sqrt{3}}\right)^3 \cdot \frac{1}{3} + \left(\frac{1}{\sqrt{3}}\right)^5 \cdot \frac{1}{5} - \left(\frac{1}{\sqrt{3}}\right)^7 \cdot \frac{1}{7} + \ldots$$

Hence, we deduce that

$$\frac{\pi}{6} = \frac{1}{\sqrt{3}}\left(1 - \left(\frac{1}{\sqrt{3}}\right)^2 \cdot \frac{1}{3} + \left(\frac{1}{\sqrt{3}}\right)^4 \cdot \frac{1}{5} - \left(\frac{1}{\sqrt{3}}\right)^6 \cdot \frac{1}{7} + \ldots\right).$$

Thus, we have

$$\pi = 2\sqrt{3}\left(1 - \frac{1}{3} \cdot \frac{1}{3} + \left(\frac{1}{3}\right)^2 \cdot \frac{1}{5} - \left(\frac{1}{3}\right)^3 \cdot \frac{1}{7} + \ldots\right),$$

and the proof completed.

501. If $f(x) = \tan^{-1}$, find $f^{(65)}(0)$.

Solution. In Problem (500) we observed that

$$\tan^{-1} x = \sum_{n=0}^{\infty} \frac{(-1)^n}{2n+1} x^{2n+1}, \quad (|x| < 1).$$

In general we have

$$a_n = \frac{f^{(n)}(0)}{n!},$$

or equivalently $f^{(n)}(0) = a_n \cdot n!$. So, we obtain $f^{(63)}(0) = a_{63} \cdot (63)!$. Since $a_{63} = \dfrac{(-1)^{31}}{63} = -\dfrac{1}{63}$, it follows that $f^{(63)}(0) = -(62)!$.

7.3 Exercises

Easier Exercises

By the definition of limit, prove that:

1. $\lim\limits_{n\to\infty} \dfrac{(-1)^n}{n} = 0,$

2. $\lim\limits_{n\to\infty} \dfrac{1}{\sqrt[3]{n}} = 0,$

3. $\lim\limits_{n\to\infty} \dfrac{3n-1}{n+4} = 3,$

4. $\lim\limits_{n\to\infty} \dfrac{n+2}{n^2-2} = 0.$

Evaluate the given limit:

5. $\lim\limits_{n\to\infty} \cos\left(\dfrac{\ln n}{n}\right),$

6. $\lim\limits_{n\to\infty} \sin^{-1}\left(\dfrac{2n-1}{4n}\right),$

7. $\lim\limits_{n\to\infty} \left(\sqrt{n^2+n} - \sqrt{n^2-3n}\right),$

8. $\lim\limits_{n\to\infty} \dfrac{3^n n!}{n^n}.$

9. Given the sequence $\{a_n\}$ such that

$$a_n = \dfrac{1 - \left(1 - \dfrac{1}{n}\right)^a}{1 - \left(1 - \dfrac{1}{n}\right)^b},$$

where a and b are constant and $b \neq 0$. Determine if the sequence is convergent or divergent. If the sequence is convergent, find its limit.

10. After verifying that the sequence $\left\{\left(1 + \dfrac{1}{n}\right)^n\right\}$ is strictly increasing, while $\left\{\left(1 + \dfrac{1}{n}\right)^{n+1}\right\}$ is strictly decreasing, show that

$$\left(1 + \dfrac{1}{n}\right)^n < e < \left(1 + \dfrac{1}{n}\right)^{n+1}.$$

What is the disadvantage of using this double inequality to estimate the number e?

11. Suppose that the sequence $\{\cos nx\}$ is convergent for some $x \in \mathbb{R}$. Show that $x = 2k\pi$ for some integer k.

12. Let $\{a_n\}$ be a sequence defined by $a_n = n^\alpha(1 + \beta)^{-n} \sin n$, for all $n \in \mathbb{N}$, where α and β are fixed positive real numbers. Show that $\{a_n\}$ converges.

13. (a) Suppose that f is a differentiable function on $[0, 1]$ and $f(0) = 0$. Define the sequence $\{a_n\}$ by the rule $a_n = nf\left(\dfrac{1}{n}\right)$. Show that $\lim\limits_{n\to\infty} a_n = f'(0)$.

    (b) Find the limit of $a_n = n \tan^{-1}\left(\dfrac{1}{n}\right)$.

14. Let $a_1 = 1$ and
$$a_{n+1} = \frac{1}{2 + a_n}, \quad \text{for } n \in \mathbb{N}.$$

(a) Show that $\{a_n\}$ satisfies Cauchy criterion.
(b) Find the limit of $\{a_n\}$.

15. Let $\{a_n\}$ be a sequence defined by $a_1 = 1$, $a_2 = 2$ and
$$a_{n+2} = \frac{3}{4}a_n + \frac{1}{4}a_{n+1}, \quad \text{for } n \in \mathbb{N}.$$

(a) Show that $\{a_n\}$ converges.
(b) Find the limit of $\{a_n\}$.

16. Let $\{a_n\}$ be a sequence defined by $a_1 = 0$, $a_2 = 3$ and
$$a_n = \frac{2a_{n-1} + a_{n-2}}{3},$$

for all $n \geq 3$. Determine whether the sequence $\{a_n\}$ is convergent or divergent.

17. Let b_k denote the number of prime numbers less than or equal to k. Let $a_1 = 2$, $a_2 = 3$ and
$$a_n = \sum_{k=3}^{n} \frac{1}{b_k},$$

for all $n \geq 3$. Determine whether the sequence $\{a_n\}$ is convergent or divergent.

18. Consider the sequence $0.2, \ 0.22, \ 0.222, \ \ldots$, then by writing this sequence as a sequence of partial sums of a series, find the limit of this sequence.

19. Let $\{a_n\}$ be a sequence of positive numbers such that $\lim\limits_{n \to \infty} a_n = a$. Prove that
$$\lim_{n \to \infty} \sqrt[n]{a_1 a_2 \ldots a_n} = a.$$

20. Let $|a_n - a| < t_n$ for large n and $\lim\limits_{n \to \infty} t_n = 0$. Show that $\lim\limits_{n \to \infty} a_n = a$.

21. Let $\lim\limits_{n \to \infty} a_n = a$ (finite) and for $\epsilon > 0$, $|a_n - t_n| < \epsilon$ for large n. Show that $\lim\limits_{n \to \infty} t_n = a$.

22. If $\{a_n\}$ is unbounded and monotonic, prove that either $\lim\limits_{n \to \infty} a_n = \infty$ or $\lim\limits_{n \to \infty} a_n = -\infty$.

23. (a) If $a_n = b_n$ except for finitely many values of n, prove that $\sum a_n$ and $\sum b_n$ diverge or converge together.
    (b) Let $b_{n_k} = a_k$ for some increasing sequence $\{n_k\}$ of positive integers, and $b_n = 0$ if n is any other positive integer. Show that $\sum a_n$ and $\sum b_n$ diverge or converge together, and that in the latter case they have the same sum.

Determine whether the given series is convergent or divergent:

24. $\displaystyle\sum_{n=2}^{\infty} \frac{(\ln n)(\sin n)}{\left(n + \sin^2 n\right)^{\frac{5}{4}}}$,

25. $\displaystyle\sum_{n=1}^{\infty} \sqrt{n}\left(1 - n \sin(\frac{1}{n})\right)$,

26. $\displaystyle\sum_{n=1}^{\infty} \frac{n \sin n}{e^n}$,

27. $\displaystyle\sum_{n=2}^{\infty} \frac{\ln 2 + \cdots + \ln n}{n}$,

28. $\displaystyle\sum_{n=0}^{\infty} \frac{(3n)!}{3^{3n} n!(n+1)!(n+3)!}$,

29. $\displaystyle\sum_{n=1}^{\infty} \frac{n!}{2^{n!}}$,

30. $\displaystyle\sum_{n=1}^{\infty} \left(\sqrt{n+2} - 2\sqrt{n+1} + \sqrt{n}\right)$,

31. $\displaystyle\sum_{n=1}^{\infty} (n+2)\left(1 - \cos(\frac{1}{n})\right)$,

32. $\displaystyle\sum_{n=4}^{\infty} \frac{1}{n(\ln n)(\ln(\ln n))}$,

33. $\displaystyle\sum_{n=1}^{\infty} \left(\tan^{-1}(n+1) - \tan^{-1} n\right)$,

34. $\displaystyle\sum_{n=1}^{\infty} \frac{1 \cdot 3 \cdot 5 \ldots (2n-1)}{2 \cdot 4 \cdot 6 \ldots (2n)}$,

35. $\displaystyle\sum_{n=1}^{\infty} \frac{1 \cdot 3 \cdot 5 \ldots (2n-1)}{3 \cdot 6 \cdot 9 \ldots (3n)}$,

36. $\displaystyle\sum_{n=1}^{\infty} \left(\frac{\pi}{2} - \tan^{-1} n\right)$,

37. $\displaystyle\sum_{n=1}^{\infty} e^{-2n} \sinh n$,

38. $\displaystyle\sum_{n=2}^{\infty} \frac{\cos n\theta}{\sqrt{n^3 - 1}}$,

39. $\displaystyle\sum_{n=2}^{\infty} \frac{1}{n\sqrt{\ln n}}$,

40. $\displaystyle\sum_{n=2}^{\infty} \frac{1}{(\ln n)^{\ln n}}$,

41. $\displaystyle\sum_{n=3}^{\infty} \frac{1}{(\ln n)^{\ln(\ln n)}}$,

42. $\displaystyle\sum_{n=2}^{\infty} \left(1 - \frac{1}{n}\right)^{n^2}$,

43. $\displaystyle\sum_{n=1}^{\infty} (n+1)\left(\frac{1 + \sin\left(\frac{n\pi}{6}\right)}{3}\right)^n$,

44. $\displaystyle\sum_{n=1}^{\infty} \left(\frac{n!}{n^n}\right)^n$,

45. $\displaystyle\sum_{n=1}^{\infty} \left(1 + \frac{1}{2} + \cdots + \frac{1}{n}\right)^n$,

46. $\displaystyle\sum_{n=1}^{\infty} \frac{1+2}{1-2} + \frac{1+2+4}{1-2+4} + \frac{1+2+4+8}{1-2+4-8} + \ldots$

47. Let $\{a_n\}$ be a decreasing sequence such that $a_n \to 0$. Show that

$$a_1 - a_2 \leq \sum_{n=1}^{\infty} (-1)^{n+1} a_n \leq a_1.$$

48. Suppose that $a_n > 0$ for all $n \in \mathbb{N}$ and $\sum a_n$ converges. Show that the series

$$\sum_{n=1}^{\infty} \left(1 - \frac{\sin a_n}{a_n}\right)$$

converges.

Determine whether the given series is absolutely convergent, conditionally convergent or divergent:

49. $\displaystyle\sum_{n=1}^{\infty}(-1)^n\frac{\cos n}{n^2}$,

54. $1-\dfrac{1}{101}+\dfrac{1}{201}-\dfrac{1}{301}+\cdots,$

50. $\displaystyle\sum_{n=1}^{\infty}(-1)^{n-1}\frac{6^n}{5^{n+1}}$,

55. $\displaystyle\sum_{n=1}^{\infty}\frac{n\sin n\theta}{2^n}$,

51. $\displaystyle\sum_{n=1}^{\infty}(-1)^n\frac{n}{e^{2n}}$,

56. $\displaystyle\sum_{n=1}^{\infty}(-1)^n\frac{1}{n(\ln n)^2}$,

52. $\displaystyle\sum_{n=1}^{\infty}(-1)^{n+1}\ln\left(\frac{n+1}{n}\right),$

57. $\displaystyle\sum_{n=2}^{\infty}\frac{n\sin\theta}{n^2+(-1)^n}$,

53. $\displaystyle\sum_{n=1}^{\infty}\frac{2\sin\left(\frac{2n\pi}{3}\right)}{\sqrt{3n}}$,

58. $\displaystyle\sum_{n=1}^{\infty}\frac{1\cdot 3\cdot 5\ldots(2n+1)}{4\cdot 6\cdot 8\ldots(2n+4)}\sin n\theta,$

59. $\dfrac{1}{\sqrt{1\cdot 2}}-\dfrac{1}{\sqrt{2\cdot 3}}+\dfrac{1}{\sqrt{3\cdot 4}}-\dfrac{1}{\sqrt{4\cdot 5}}+\cdots,$

60. $1+\dfrac{1}{4}-\dfrac{1}{9}-\dfrac{1}{16}+\dfrac{1}{25}+\dfrac{1}{36}-\dfrac{1}{49}-\dfrac{1}{64}+\cdots.$

Find the radius of convergence and interval convergence of the given power series:

61. $\displaystyle\sum_{n=0}^{\infty}\frac{n+3}{(n+1)(n+4)}(x+2)^n$,

67. $\displaystyle\sum_{n=0}^{\infty}10^n\left(\frac{x-1}{3}\right)^n$,

62. $\displaystyle\sum_{n=0}^{\infty}\frac{(n!)^3}{(3n)!}x^{3n}$,

68. $\displaystyle\sum_{n=0}^{\infty}(\sinh 2n)x^n$,

63. $\displaystyle\sum_{n=0}^{\infty}\frac{n^2}{3^n}(x-1)^n$,

69. $\displaystyle\sum_{n=1}^{\infty}\left(\frac{5^n}{n}+\frac{3^n}{n^2}\right)x^n$,

64. $\displaystyle\sum_{n=0}^{\infty}\frac{4^{n+1}}{n+3}x^{2n}$,

70. $\displaystyle\sum_{n=2}^{\infty}\frac{1}{n\ln n}(x+1)^n$,

65. $\displaystyle\sum_{n=1}^{\infty}n^{\sqrt{n}}x^n$,

71. $\displaystyle\sum_{n=2}^{\infty}\frac{(-1)^{n+1}}{n(\ln n)^2}x^n$,

66. $\displaystyle\sum_{n=2}^{\infty}n^{\ln n}x^n$,

72. $\displaystyle\sum_{n=0}^{\infty}\left(2+\sin\left(\frac{n\pi}{6}\right)\right)^n(x+2)^n.$

73. Suppose that k is a positive integer and $\sum a_n x^n$ has radius of convergence R. Show that the $f(x^k)=\sum a_n x^{kn}$ has radius of convergence $R^{\frac{1}{k}}$.
74. If g is a rational function defined for all non-negative integers, then $\sum a_n x^n$ and $\sum a_n g(n)x^n$ have the same radius of convergence.
75. Given that power series $f(x)=\sum a_n x^n$ satisfies $f'(x)=-2xf(x)$ and $f(0)=1$. Find the sequence $\{a_n\}$. Do you recognize f?

76. If $f(x) = \sum\limits_{n=0}^{\infty} a_n x^n$, $(|x| < 1)$ and $a_n \geq 0$, prove that

$$\sum_{n=0}^{\infty} a_n = \lim_{x \to 1^-} f(x).$$

Harder Exercises

77. Let $f : (0, 2) \to \mathbb{R}$ be twice differentiable and $f\left(1 - \dfrac{1}{n}\right) = 1$, for all $n \in \mathbb{N}$.
    Evaluate $f''(1)$.

78. Let $f : [0, 1] \to \mathbb{R}$ be a function and

$$a_n = f\left(\frac{1}{n}\right) - f\left(\frac{1}{n+1}\right),$$

for all $n \in \mathbb{N}$. Show that

(a) If f is continuous, then $\sum a_n$ converges.

(b) If f is differentiable and $|f'(x)| < \dfrac{1}{2}$, for all $x \in [0, 1]$, then $\sum a_n \sqrt{n} \cos n$ converges.

79. Let $f : [0, 1] \to \mathbb{R}$ be a function and

$$a_n = f\left(\sin \frac{\pi}{n}\right) - f\left(\sin \frac{\pi}{n+1}\right),$$

for all $n \in \mathbb{N}$.

(a) If f is continuous, show that $\sum a_n$ converges and find its sum.

(b) If f is differentiable and $|f'(x)| < 1$, for all $x \in [0, 1]$, discuss the convergence/divergence of $\sum |a_n|$.

80. Let $r > -1$ be a real number and $f(x) = (1 + x)^r$. Then,

$$f^{(n)}(x) = r(r-1) \ldots (r-n+1)(1+x)^{r-n}.$$

If we define

$$\binom{r}{0} = 1 \text{ and } \binom{r}{n} = \frac{r(r-1) \ldots (r-n+1)}{n!}, \quad n \geq 1,$$

then

$$\frac{f^{(n)}(0)}{n!} = \binom{r}{n}.$$

We call $\binom{r}{n}$ the *generalized binomial coefficient*.

(a) Show that

$$\lim_{n\to\infty} \left(1 - \frac{\alpha}{1}\right)\left(1 - \frac{\alpha}{2}\right) \cdots \left(1 - \frac{\alpha}{n}\right) = 0 \text{ if } \alpha > 0.$$

Hint: Look at the logarithm of the absolute value of product.

(b) From (a), conclude that

$$\lim_{n\to\infty} \binom{r}{n} = 0 \text{ if } r > -1.$$

81. Fix a positive number c. Choose $a_1 > \sqrt{c}$, and define $a_2, a_3, \ldots$, by the recursion formula

$$a_{n+1} = \frac{1}{2}\left(a_n + \frac{c}{a_n}\right).$$

(a) Prove that $\{a_n\}$ decreases monotonically and that $\lim_{n\to\infty} a_n = \sqrt{c}$.

(b) Put $e_n = a_n - \sqrt{c}$ and show that

$$e_{n+1} = \frac{e_n^2}{2a_n} < \frac{e_n^2}{2\sqrt{c}},$$

so that setting $b = 2\sqrt{c}$,

$$e_{n+1} < b\left(\frac{e_1}{b}\right)^{2^n}, \quad (n \in \mathbb{N}).$$

(c) This is a good algorithm for computing square roots, since the recursion formula is simple and convergence is extremely rapid. For example, if $c = 3$ and $a_1 = 2$, show that $\dfrac{e_1}{b} < \dfrac{1}{10}$ and that therefore

$$e_5 < 4 \times 10^{-16} \text{ and } e_6 < 4 \cdot 10^{-32}.$$

82. Suppose that f is twice differentiable on $[a, b]$, $f(a) < 0$, $f(b) > 0$, $f'(x) > \delta > 0$, and $0 \le f''(x) \le M$ for all $x \in [a, b]$. Let c be the unique point in (a, b) at which $f(c) = 0$. Complete the details in the following outline of *Newton's method* for computing c.

(a) Choose $a_1 \in (c, b)$ and define $\{a_n\}$ by

$$a_{n+1} = a_n - \frac{f(a_n)}{f'(a_n)}.$$

Interpret this geometrically, in terms of a tangent to the graph of f.
(b) Prove that $a_{n+1} < a_n$ and that $\lim_{n\to\infty} a_n = c$.
(c) Use Taylor's theorem to show that

$$a_{n+1} - c = \frac{f''(t_n)}{2 f'(a_n)} (a_n - c)^2,$$

for some $t_n \in (c, a_n)$.
(d) If $A = \dfrac{M}{2\delta}$, deduce that

$$0 \le a_{n+1} - c \le \frac{1}{A} \left(A(a_1 - c) \right)^{2^n}.$$

(e) Show that Newton's method amounts to finding a fixed point of the function g defined by

$$g(x) = x - \frac{f(x)}{f'(x)}.$$

How does $g'(x)$ behave for x near c?
(f) Put $f(x) = x^{\frac{1}{3}}$ on $(-\infty, \infty)$ and try Newton's method. What happen?

83. Show that

$$\sum_{n=1}^{\infty} \frac{(-1)^{n+1}}{n} = \frac{1}{2} \sum_{n=1}^{\infty} \frac{1}{n(2n-1)}.$$

84. Prove that

$$\sum_{n=1}^{\infty} \frac{1}{n^p}$$

is convergent when $p > 1$ and divergent when $p \le 1$.
85. Let

$$S(p) = \sum_{n=1}^{\infty} \frac{1}{n^p},$$

where $p > 1$. Show that

$$\frac{1}{(p-1)(N+1)^{p-1}} < S(p) - \sum_{n=1}^{N} \frac{1}{n^p} < \frac{1}{(p-1)N^{p-1}}.$$

86. (a) If $0 < 2\epsilon < \theta < \pi - 2\epsilon$, prove that

$$\lim_{n\to\infty} \frac{|\sin\theta| + |\sin 2\theta| + \cdots + |\sin n\theta|}{n} \ge \frac{\sin\epsilon}{2}.$$

(b) Show that

$$\sum \frac{\sin n\theta}{n^p}$$

converges conditionally if $0 < p \le 1$ and $\theta \ne k\pi$, where $k \in \mathbb{Z}$.

87. Suppose that f is positive, decreasing and integrable on $[1, \infty)$, and let

$$a_n = \sum_{k=1}^{n} f(k) - \int_1^n f(x)dx.$$

(a) Show that $\{a_n\}$ is non-increasing and non-negative and

$$0 < \lim_{n \to \infty} a_n < f(1).$$

(b) From (a), deduce that

$$\gamma = \lim_{n \to \infty} \left(1 + \frac{1}{2} + \frac{1}{3} + \cdots + \frac{1}{n} - \ln n\right)$$

exists, and $0 < \gamma < 1$ (γ is *Euler's constant*, $\gamma \approx 0.577$).

88. Prove that if the series

$$\sum_{n=1}^{\infty} a_n$$

is convergent to S, then the series

$$(a_1 + \cdots + a_{n_1}) + (a_{n_1+1} + \cdots + a_{n_2}) + (a_{n_2+1} + \cdots + a_{n_3}) + \cdots$$

obtained by intersecting parenthesis in the first series, is also convergent to S.

89. Show that the series

$$\left(\frac{3}{2} - \frac{4}{3}\right) + \left(\frac{5}{4} - \frac{6}{5}\right) + \cdots + \left(\frac{2n+1}{2n} - \frac{2n+2}{2n+1}\right) + \cdots$$

is convergent whereas the series

$$\frac{3}{2} - \frac{4}{3} + \frac{5}{4} - \frac{6}{5} + \cdots$$

obtained from the first series by omitting parenthesis is divergent.

90. If $\sum a_n$ is an absolutely convergent series and $\sum b_n$ is a rearrangement of $\sum a_n$, prove that $\sum b_n$ is convergent and the sum of the two series are the same.

91. If $\sum a_n$ is conditionally convergent, prove that the series of its positive terms and the series of its negative terms are both divergent.

92. *Dirichlet's test*: If $S_n = \sum\limits_{i=1}^{n} a_i$ is bounded and $\{b_n\}$ decreases to 0, prove that $\sum a_n b_n$ is convergent.

93. *Abel's test*: If $\sum a_n$ is convergent and the sequence $\{b_n\}$ is monotone and bounded, prove that $\sum a_n b_n$ is convergent.

94. If $\sum a_n$ and If $\sum |b_n - b_{n+1}|$ are convergent, prove that $\sum a_n b_n$ is convergent.

95. If $\{a_n\}$ is decreasing and $\lim\limits_{n \to \infty} b_n = 0$, prove that $\sum a_n \sin n\theta$ is convergent for all real number θ, and $\sum a_n \cos n\theta$ is convergent for all real number θ other than multiples of 2π.

96. If $\sum |a_n|^2$ and $\sum |b_n|^2$ are convergent, prove that $\sum a_n b_n$ is convergent.

97. Suppose that

$$\sum_{n=1}^{\infty} c_n, \quad \text{where } c_n > 0$$

is convergent and

$$r_n = \sum_{i=n+1}^{\infty} c_i.$$

Prove the following

(a) $\sum\limits_{n=1}^{\infty} \dfrac{c_n}{\sqrt{r_{n-1}}}$ is convergent.

(b) $\sum\limits_{n=1}^{\infty} \dfrac{c_n}{r_{n-1}}$ is divergent.

Hint:

$$\frac{c_n}{\sqrt{r_{n-1}}} = \frac{r_{n-1} - r_n}{\sqrt{r_{n-1}}} = \frac{(\sqrt{r_{n-1}} - \sqrt{r_n})(\sqrt{r_{n-1}} + \sqrt{r_n})}{\sqrt{r_{n-1}}}.$$

98. For $x > 0$, let

$$a_n = \left| e^x - \sum_{k=0}^{n} \frac{x^k}{k!} \right|, \quad (n = 0, 1, 2, \ldots).$$

Using Taylor's theorem show that the series $\sum a_n$ converges.

99. Suppose that

$$f(x) = \sum_{n=0}^{\infty} a_n (x - c)^n$$

has radius of convergence R and $0 < r < R_1 < R$. Show that there is an integer k such that

$$\left| f(x) - \sum_{n=0}^{k} a_n (x - c)^n \right| \le \left(\frac{r}{R_1} \right)^{k+1} \frac{R_1}{R_1 - r},$$

whenever $|x - c| \le r$.

100. Suppose that $f(x) = \sum_{n=0}^{\infty} a_n (x - c)^n$, $|x - c| < R$ and $f(t_n) = 0$, where $t_n \ne c$
    and $\lim_{n \to \infty} t_n = c$. Show that $f(x) = 0$ ($|x - c| < R$).

101. Express

$$\int_1^x \frac{\ln t}{t - 1} dt$$

    as a power series in $x - 1$ and find the radius of convergence of the series.

102. Obtain the Maclaurin series for $\sin^{-1} x$ and then deduce that

$$\sum_{n=0}^{\infty} \binom{2n}{n} \frac{1}{2^{2n}(2n + 1)} = \frac{\pi}{2}.$$

103. (a) Use the formula for the tangent of the difference of two angles to show
    that

$$\tan \left(\tan^{-1}(n + 1) - \tan^{-1}(n - 1) \right) = \frac{2}{n^2}.$$

    (b) Show that

$$\sum_{n=1}^{N} \tan^{-1} \left(\frac{2}{n^2} \right) = \tan^{-1}(N + 1) + \tan^{-1} N - \frac{\pi}{4}.$$

    (c) Find the value of

$$\sum_{n=1}^{\infty} \tan^{-1} \left(\frac{2}{n^2} \right).$$

104. Let a be any positive number

$$\ln a = 2 \left(b + \frac{b^3}{3} + \frac{b^5}{5} + \frac{b^7}{7} + \dots \right),$$

    where

$$b = \frac{a - 1}{a + 1},$$

    and use this formula to approximate $\ln 3$ to four decimal place.

105. Prove the following generalization of the second derivative test for a local extremum. If

$$f'(a) = f''(a) = \cdots = f^{(n-1)}(a) = 0 \ (n \geq 2),$$

and if the nth derivative $f^{(n)}(a)$ is finite and different from zero, then f has a strict local minimum at a if n is even and $f^{(n)}(a) > 0$ and a strict local maximum at a if n is even and $f^{(n)}(a) < 0$, but no extremum at a if n is odd.

Appendix A
Tables of Integrals

Tables A.1, A.2, A.3, A.4 and A.5 are the list of basic integral formulas.

Table A.1 Some elementary forms

1. $\int x^r dx = \dfrac{x^r}{r+1} + C \ (r \neq -1)$	2. $\int \dfrac{1}{x} dx = \ln|x| + C$
3. $\int \sin dx = -\cos x + C$	4. $\int \cos x dx = \sin x + C$
5. $\int \sec^2 dx = \tan x + C$	6. $\int \csc^2 x dx = -\cot x + C$
7. $\int \sec x \tan x dx = \sec x + C$	8. $\int \csc x \cot x dx = -\csc x + C$
9. $\int \tan x dx = -\ln|\cos x| + C$	10. $\int \cot x dx = \ln|\sin x| + C$
11. $\int \sec x dx = \ln|\sec x + \tan x| + C$	12. $\int \csc x dx = \ln|\csc x - \cot x| + C$
13. $\int e^x dx = e^x + C$	14. $\int a^x dx = \dfrac{a^x}{\ln a} + C$
15. $\int \sinh x dx = \cosh x + C$	16. $\int \cosh x dx = \sinh x + C$
17. $\int \operatorname{sech}^2 x dx = \tanh x + C$	18. $\int \operatorname{csch}^2 x dx = -\coth x + C$
19. $\int \operatorname{sech} x \tanh x dx = -\operatorname{sech} x + C$	20. $\int \operatorname{csch} x \coth x dx = -\operatorname{csch} x + C$
21. $\int \sin^2 x dx = \dfrac{1}{2}x - \dfrac{1}{4}\sin 2x + C$	22. $\int \cos^2 x dx = \dfrac{1}{2}x + \dfrac{1}{4}\sin 2x + C$
23. $\int \tan^2 x dx = \tan x - x + C$	24. $\int \cot^2 x dx = -\cot x - x + C$

© The Editor(s) (if applicable) and The Author(s), under exclusive license
to Springer Nature Singapore Pte Ltd. 2020
B. Davvaz, *Examples and Problems in Advanced Calculus: Real-Valued Functions*,
https://doi.org/10.1007/978-981-15-9569-1

Table A.2 Inverse trigonometric functions

25. $\displaystyle\int \sin^{-1} x\,dx = x\sin^{-1} x + \sqrt{1-x^2} + C$

26. $\displaystyle\int \cos^{-1} x\,dx = x\cos^{-1} x - \sqrt{1-x^2} + C$

27. $\displaystyle\int \tan^{-1} x\,dx = x\tan^{-1} x - \dfrac{1}{2}\ln(x^2+1) + C$

28. $\displaystyle\int \cot^{-1} x\,dx = x\cot^{-1} x + \dfrac{1}{2}\ln(x^2+1) + C$

29. $\displaystyle\int \sec^{-1} x\,dx = x\sec^{-1} x - \ln|x + \sqrt{x^2-1}| + C$

30. $\displaystyle\int \csc^{-1} x\,dx = x\csc^{-1} x + \ln|x + \sqrt{x^2-1}| + C$

Table A.3 Forms containing logarithms and exponentials

31. $\displaystyle\int \ln x\,dx = x\ln x - x + C$

32. $\displaystyle\int x^n \ln x\,dx = \dfrac{x^{n+1}}{n+1}\left(\ln x - \dfrac{1}{n+1}\right) + C$

33. $\displaystyle\int \dfrac{1}{x\ln x}\,dx = \ln|\ln x| + C$

34. $\displaystyle\int (\ln x)^n\,dx = x(\ln x)^n - n\int (\ln x)^{n-1}\,dx$

35. $\displaystyle\int xe^x\,dx = xe^x - x + C$

36. $\displaystyle\int x^n e^x\,dx = x^n e^x - n\int x^{n-1} e^x\,dx$

37. $\displaystyle\int e^{ax}\cos bx\,dx = e^{ax}\dfrac{a\cos bx + b\sin bx}{a^2+b^2} + C$

38. $\displaystyle\int e^{ax}\sin bx\,dx = e^{ax}\dfrac{a\sin bx - b\cos bx}{a^2+b^2} + C$

Table A.4 Forms containing hyperbolic functions

39. $\displaystyle\int \tanh x\,dx = \ln(\cosh x) + C$

40. $\displaystyle\int \coth x\,dx = \ln|\sinh x| + C$

41. $\displaystyle\int \operatorname{sech} x\,dx = 2\tan^{-1}(e^x) + C$

42. $\displaystyle\int \operatorname{csch} x\,dx = \ln\left|\tanh \dfrac{x}{2}\right| + C$

43. $\displaystyle\int \sinh^2 x\,dx = \dfrac{1}{4}\sinh 2x - \dfrac{x}{2} + C$

44. $\displaystyle\int \cosh^2 x\,dx = \dfrac{1}{4}\sinh 2x + \dfrac{x}{2} + C$

45. $\displaystyle\int \tanh^2 x\,dx = x - \tanh x + C$

46. $\displaystyle\int \coth^2 x\,dx = x - \coth x + C$

47. $\displaystyle\int x\sinh x\,dx = x\cosh x - \sinh x + C$

48. $\displaystyle\int x\cosh x\,dx = x\sinh x - \cosh x + C$

49. $\displaystyle\int e^{ax}\sinh bx\,dx = e^{ax}\dfrac{a\sinh bx - b\cosh bx}{a^2-b^2} + C$

50. $\displaystyle\int e^{ax}\cosh bx\,dx = e^{ax}\dfrac{a\cosh bx - b\sinh bx}{a^2-b^2} + C$

Table A.5 Forms containing trigonometric functions

51. $\displaystyle\int \sin^n x\, dx = -\frac{1}{n}\sin^{n-1} x \cos x + \frac{n-1}{n}\int \sin^{n-2} x\, dx$

52. $\displaystyle\int \cos^n x\, dx = \frac{1}{n}\cos^{n-1} x \sin x + \frac{n-1}{n}\int \cos^{n-2} x\, dx$

53. $\displaystyle\int \tan^n x\, dx = \frac{1}{n-1}\tan^{n-1} x - \int \tan^{n-2} x\, dx$

54. $\displaystyle\int \cot^n x\, dx = -\frac{1}{n-1}\cot^{n-1} x - \int \cot^{n-2} x\, dx$

55. $\displaystyle\int \sec^n x\, dx = \frac{1}{n-1}\sec^{n-2} x \tan x + \frac{n-2}{n-1}\int \sec^{n-2} x\, dx$

56. $\displaystyle\int \csc^n x\, dx = -\frac{1}{n-1}\csc^{n-2} x \cot x + \frac{n-2}{n-1}\int \csc^{n-2} x\, dx$

57. $\displaystyle\int \sin ax \cdot \sin bx\, dx = -\frac{\sin(a+b)x}{2(a+b)} + \frac{\sin(a-b)x}{2(a-b)} + C$

58. $\displaystyle\int \cos ax \cdot \cos bx\, dx = \frac{\sin(a+b)x}{2(a+b)} + \frac{\sin(a-b)x}{2(a-b)} + C$

59. $\displaystyle\int \sin ax \cdot \cos bx\, dx = -\frac{\cos(a+b)x}{2(a+b)} - \frac{\cos(a-b)x}{2(a-b)} + C$

60. $\displaystyle\int x \sin x\, dx = \sin x - x \cos x + C$

61. $\displaystyle\int x \cos x\, dx = \cos x + x \sin x + C$

62. $\displaystyle\int x^2 \sin x\, dx = 2x \sin x + (2 - x^2) \cos x + C$

63. $\displaystyle\int x^2 \cos x\, dx = 2x \cos x + (x^2 - 2) \sin x + C$

64. $\displaystyle\int x^n \sin x\, dx = -x^n \cos x + n \int x^{n-1} \cos x\, dx$

65. $\displaystyle\int x^n \cos x\, dx = x^n \sin x - n \int x^{n-1} \sin x\, dx$

66. $\displaystyle\int \sin^m x \cos^n x\, dx$

$\displaystyle = -\frac{\sin^{m-1} x \cos^{n+1} x}{m+n} + \frac{m-1}{m+n}\int \sin^{m-2} x \cos^n x\, dx$

$\displaystyle = \frac{\sin^{m+1} x \cos^{n-1} x}{m+n} + \frac{n-1}{m+n}\int \sin^m x \cos^{n-2} x\, dx$

Table A.6 Forms containing $\sqrt{x^2 \pm a^2}$

67. $\displaystyle\int \frac{1}{\sqrt{x^2 \pm a^2}}\,dx = \ln|x + \sqrt{x^2 \pm a^2}| + C$

68. $\displaystyle\int \sqrt{x^2 \pm a^2}\,dx = \frac{x}{2}\sqrt{x^2 \pm a^2} \pm \frac{a^2}{2}\ln|x + \sqrt{x^2 \pm a^2}| + C$

69. $\displaystyle\int x^2\sqrt{x^2 \pm a^2}\,dx = \frac{x}{8}(2x^2 \pm a^2)\sqrt{x^2 \pm a^2} - \frac{a^4}{8}\ln|x + \sqrt{x^2 \pm a^2}| + C$

70. $\displaystyle\int \frac{\sqrt{x^2 + a^2}}{x}\,dx = \sqrt{x^2 + a^2} - a\ln\left|\frac{a + \sqrt{x^2 + a^2}}{x}\right| + C$

71. $\displaystyle\int \frac{\sqrt{x^2 - a^2}}{x}\,dx = \sqrt{x^2 - a^2} - a\sec^{-1}\left|\frac{x}{a}\right| + C$

72. $\displaystyle\int \frac{\sqrt{x^2 \pm a^2}}{x^2}\,dx = -\frac{\sqrt{x^2 \pm a^2}}{x} + \ln|x + \sqrt{x^2 \pm a^2}| + C$

73. $\displaystyle\int \frac{x^2}{\sqrt{x^2 \pm a^2}}\,dx = \frac{x}{2}\sqrt{x^2 \pm a^2} - \frac{\pm a^2}{2}\ln|x + \sqrt{x^2 \pm a^2}| + C$

74. $\displaystyle\int \frac{1}{x\sqrt{x^2 + a^2}}\,dx = -\frac{1}{a}\ln\left|\frac{a + \sqrt{x^2 + a^2}}{x}\right| + C$

75. $\displaystyle\int \frac{1}{x\sqrt{x^2 - a^2}}\,dx = \frac{1}{a}\sec^{-1}\left|\frac{x}{a}\right| + C$

76. $\displaystyle\int \frac{1}{x^2\sqrt{x^2 \pm a^2}}\,dx = -\frac{\sqrt{x^2 \pm a^2}}{\pm a^2 x} + C$

77. $\displaystyle\int \frac{1}{(x^2 \pm a^2)^{\frac{3}{2}}}\,dx = \frac{x}{\pm a^2\sqrt{x^2 \pm a^2}} + C$

78. $\displaystyle\int (x^2 \pm a^2)^{\frac{3}{2}}\,dx$

$\displaystyle = \frac{x}{8}(2x^2 \pm 5a^2)\sqrt{x^2 \pm a^2} + \frac{3a^4}{8}\ln|x + \sqrt{x^2 \pm a^2}| + C$

In Table A.6 we may replace

$$\ln\left(x + \sqrt{x^2 + a^2}\right) \text{ by } \sinh^{-1}\frac{x}{a}$$

$$\ln\left|x + \sqrt{x^2 - a^2}\right| \text{ by } \cosh^{-1}\frac{x}{a}$$

$$\ln\left|\frac{x + \sqrt{x^2 + a^2}}{x}\right| \text{ by } \sinh^{-1}\frac{a}{x}$$

(See Tables A.7 and A.8).

Table A.7 Forms containing $\sqrt{a^2 - x^2}$

79. $\displaystyle\int \frac{1}{\sqrt{a^2 - x^2}} dx = \sin^{-1} \frac{x}{a} + C$

80. $\displaystyle\int \sqrt{a^2 - x^2} dx = \frac{x}{2}\sqrt{a^2 - x^2} + \frac{a^2}{2} \sin^{-1} \frac{x}{a} + C$

81. $\displaystyle\int x^2 \sqrt{a^2 - x^2} dx = \frac{x}{8}(2x^2 - a^2)\sqrt{a^2 - x^2} + \frac{a^4}{8} \sin^{-1} \frac{x}{a} + C$

82. $\displaystyle\int \frac{\sqrt{a^2 - x^2}}{x} dx = \sqrt{a^2 - x^2} - a \ln \left| \frac{a + \sqrt{a^2 - x^2}}{x} \right| + C$

$\displaystyle = \sqrt{a^2 - x^2} - a \cosh^{-1} \frac{a}{x} + C$

83. $\displaystyle\int \frac{\sqrt{a^2 - x^2}}{x^2} dx = -\frac{\sqrt{a^2 - x^2}}{x} - \sin^{-1} \frac{x}{a} + C$

Table A.8 Forms containing $a^2 \pm x^2$

84. $\displaystyle\int \frac{1}{a^2 + x^2} dx = \frac{1}{a} \tan^{-1} \frac{x}{a} + C$

85. $\displaystyle\int \frac{1}{a^2 - x^2} dx = \frac{1}{2a} \ln \left| \frac{x + a}{x - a} \right| + C$

86. $\displaystyle\int \frac{1}{x^2 - a^2} dx = \frac{1}{2a} \ln \left| \frac{x - a}{x + a} \right| + C$

References

1. Apostol, T.M.: Calculus. Vol. I: One-Variable Calculus, with an Introduction to Linear Algebra, 2nd edn. Blaisdell Publishing Co. Ginn and Co., Waltham, Mass.-Toronto, Ont.-London (1967)
2. Apostol, T.M.: Mathematical Analysis. Addison-Wesley Publishing Company (1975)
3. Bartle, R.G.: The Elements of Real Analysis, 2nd edn. Wiley (1976)
4. Buck, R.C.: Advanced Calculus, 3rd edn. Waveland Pr Inc. (2003)
5. Duren, W.L.: Calculus and Analytic Geometry. Xerox College Publishing (1972)
6. Ellis, R., Gulick, D.: Calculus with Analytic Geometry, 5th edn. Holt Rinehart & Winston (1998)
7. Kuhfittig, P.: Technical Calculus with Analytic Geometry. Brooks/Cole, Cengage Learning, Boston, MA (2013)
8. Leithold, L.: The Calculus with Analytic Geometry. Harper & Row, Publishers (1981)
9. Morris, C.C., Stark, R.M.: Fundamentals of Calculus. Wiley (2015)
10. Ross, K.A.: Elementary Analysis: The Theory of Calculus. Undergraduate Texts in Mathematics. Springer, New York-Heidelberg (1980)
11. Rudin, W.: Principles of Mathematical Analysis, 3rd edn. International Series in Pure and Applied Mathematics. McGraw-Hill Book Co., New York-Auckland-Düsseldorf (1976)
12. Silverman, R.A.: Calculus with Analytic Geometry. Prentice-Hall, Inc. (1985)
13. Simmons, G.F.: Calculus with Analytic Geometry, 2nd edn. McGraw-Hill Education (1996)
14. Stewart, J.: Calculus, 7th edn. Cengage Learning (2012)
15. Thomas, G.B., Finney, R.L.: Calculus and Analytic Geometry. Addison Wesley Publishing Company (1995)
16. Trench, W.F.: Advanced Calculus. Harper & Row, New York (1978)

B. Davvaz, *Examples and Problems in Advanced Calculus: Real-Valued Functions*,
https://doi.org/10.1007/978-981-15-9569-1

Index

Printed in the United States
by Baker & Taylor Publisher Services